WHERE FUTURES CONVERGE

WHERE FUTURES CONVERGE
KENDALL SQUARE AND THE MAKING OF A
GLOBAL INNOVATION HUB

ROBERT BUDERI

The MIT Press
Cambridge, Massachusetts
London, England

© 2022 Robert Buderi

All rights reserved. No part of this book may be reproduced in any form by any electronic or mechanical means (including photocopying, recording, or information storage and retrieval) without permission in writing from the publisher.

The MIT Press would like to thank the anonymous peer reviewers who provided comments on drafts of this book. The generous work of academic experts is essential for establishing the authority and quality of our publications. We acknowledge with gratitude the contributions of these otherwise uncredited readers.

This book was set in Minion Pro and Neue Haas Grotesk by New Best-set Typesetters Ltd. Printed and bound in the United States of America.

The author acknowledges with gratitude the support of the Alfred P. Sloan Foundation in the research and writing of this book.

Library of Congress Cataloging-in-Publication Data

Names: Buderi, Robert, author.
Title: Where futures converge : Kendall Square and the making of a global innovation hub / Robert Buderi.
Description: Cambridge, Massachusetts : The MIT Press, [2022] | Includes bibliographical references and index.
Identifiers: LCCN 2021019874 | ISBN 9780262046510 (hardcover)
Subjects: LCSH: Economic development—Massachusetts—Cambridge. | Enterprise zones—Massachusetts—Cambridge. | Research parks—Massachusetts—Cambridge. | Biotechnology industries—Massachusetts—Cambridge. | Technological innovations—Massachusetts—Cambridge.
Classification: LCC HC108.C19 B83 2022 | DDC 330.9744/4—dc23
LC record available at https://lccn.loc.gov/2021019874

10 9 8 7 6 5 4 3 2 1

To the friends, colleagues, and supporters who made Xconomy possible. A heartfelt thank you to all!

CONTENTS

INTRODUCTION 1.0, OR PREFACE: THE KENDALL THEORY OF BIOGEOGRAPHY ix
INTRODUCTION 2.0: THE BEST PLACE TO HAVE THESE PROBLEMS xiii

1   THE MOST INNOVATIVE SQUARE ~~MILE~~ KILOMETER ON EARTH 1

2   MODEL OF INNOVATION 9

3   THE FIRST ECONOMIC VISION FOR KENDALL SQUARE IS A BUST 21

4   CHARLES DAVENPORT AND THE SQUARE'S TRANSFORMATION 25

5   KENDALL BECOMES KENDALL 33

6   DAVENPORT'S FAILED DREAM OPENS THE DOOR FOR "NEW TECHNOLOGY" 37

7   "A CANOPY OF INDUSTRIAL HAZE" 47

8   RAD LAB: KENDALL SQUARE'S TIPPING POINT 55

   SPOTLIGHT: THE F&T—PLACE-MAKING'S FIRST PLACE 65

9   URBAN MARSHLAND TO URBAN RENEWAL 69

10 KENDALL, WE HAVE A PROBLEM 77

11 TECH SURGE: LOTUS TO AI ALLEY 87

12 THE ORDINANCE AND BIOGEN 99

13 BEGINNINGS OF GENE TOWN 111

14 BUBBLE DAYS: MEDIA LAB TO AKAMAI 119

   SPOTLIGHT: LITA NELSEN ON TECHNOLOGY LICENSING AND HOW "CLUSTERS FEED THEMSELVES" 133

15 CAMBRIDGE INNOVATION CENTER: KENDALL SQUARE'S STARTUP HEART 137

   SPOTLIGHT: INNOVATION SPACE ZONING—HOW KENDALL SQUARE HOPES TO KEEP ITS STARTUP COMMUNITY VIBRANT 147

16 "NIBBER": BIG PHARMA UPS THE ANTE 151

    SPOTLIGHT: BOB LANGER—PERSONIFICATION OF KENDALL SQUARE'S SECRET SAUCE 159

17 HOMEGROWN BIOTECHS MAKE THEIR MARK 163

18 ROAD TO THE BROAD 179

19 THE CORPORATIZATION OF KENDALL SQUARE 193

20 VENTURE MIGRATION AND THE TECH STARTUP SQUEEZE 205

    SPOTLIGHT: FLAGSHIP PIONEERING—KENDALL SQUARE COMPANY CREATOR 217

21 FORTY MISSING COMPANIES 223

22 700 MAIN: THE STORY OF KENDALL SQUARE—IN ONE BUILDING 237

23 NEXUS OF COLLABORATION 249

    SPOTLIGHT: MAPPING THE MODERNA NETWORK 257

24 CHALLENGES AND REGIONAL ADVANTAGE 261

25 VOICES OF THE SQUARE 273

26 ELEVEN DECISIONS THAT SHAPED KENDALL SQUARE 283

27 LESSONS AND OBSERVATIONS 287

28 CONVERGENCE AND CONSILIENCE 295

    ACKNOWLEDGMENTS 303
    LIST OF INTERVIEWS 305
    NOTES 309
    BIBLIOGRAPHY 337
    INDEX 359

INTRODUCTION 1.0, OR PREFACE
THE KENDALL THEORY OF BIOGEOGRAPHY

Kendall Square is arguably the world's greatest and densest innovation hub, an enormously complex ecosystem of people, ideas, companies, offices, and laboratories. So what better way to dive into the story of Kendall Square than to talk to one of the world's greatest experts on biological ecosystems?

That was my thinking almost as soon as Victor (Vic) McElheny, a compendium of knowledge who used to run the Knight Science Journalism program at the Massachusetts Institute of Technology (MIT), mentioned that he regularly saw Ed Wilson at the assisted living community they both called home in nearby Lexington, Massachusetts. I say "almost as soon" because it took me a few seconds to work out that "Ed" was the scientist far more widely known as E. O. Wilson.

Wilson, a professor emeritus at Harvard University and two-time winner of the Pulitzer Prize, was one of my heroes. I got to know him briefly when I was a fellow at MIT in the mid-1980s, and even have a signed copy of one of his first twenty-three books, *Biophilia*, on my shelf. He was probably the foremost expert on ants ever, and he and Princeton scientist Robert MacArthur developed the theory of island biogeography to explain how ecosystems develop and are sustained on islands—proposing that the number of species an island holds is a function of the rate at which established species grow extinct and new species colonize it.

This sounds a lot like the innovation ecosystem of Kendall Square, where startups come and go, some growing to giants, others finding their niche or providing fodder for the next business species. You can see why I thought it might be interesting to sit down with Wilson and perhaps get some fresh insights into ways to approach my topic. Vic set up a date for lunch in their facility's community dining room.

Wilson, then approaching his ninetieth birthday (he passed away in December 2021, at 92), was putting the finishing touches on his twenty-fourth book. He also had been getting an increasing number of requests to speak to business audiences about his work. "They want to know if they can use biological success stories in principles that allow them to improve the success of their own enterprises," he explained.[1]

In fact, he said, there are "biological analogs" to innovation ecosystems like Kendall Square. Consider the Matabele ant, named after a fierce warrior tribe of southern Africa. This ant has evolved specifically to kill and eat termites. Matabele ants, as Wilson described them, almost poetically, "have these heavy, helmet-like heads, very heavily armored. They have sharp needle-like mandibles. They're good fighters." The ants swarm onto termite mounds, "and pretty soon the surface around the termite nest, and into it, is littered with the battle blood of termite soldiers." The ants then carry the bodies of the termite soldiers to their home. "They specialize on eating the dead from the battlefield," Wilson related.

I could easily see parallels to today's business world, and I wondered who the Matabele ants—and termites—of Kendall Square might be.

Wilson's more important point, though, was that the key property of ecosystems is sustainability. "They've been able to exist long periods of time without going extinct," he said. Species are able to come together in ways that have "a mutually positive result. For example, that could include predators that keep down the prey species in sustainable numbers so the prey species don't wipe out their food."

But sustainability doesn't necessarily mean stability. On the contrary, Wilson pointed out, ecosystems are almost constantly changing. Species go extinct and new species arrive and evolve. The outside world brings cataclysmic events like earthquakes, pandemics, the crash of continental plates, or the waxing and waning of ice ages.

To me, that picture of ecosystems in the wild was fascinatingly similar to the story of Kendall Square. As you will read, the square was booming in the early 1900s, leading the country and even the world in soap-making, confectionary, printing and publishing, road-building technology, high-tech rubber, and much more. But a mere few decades later, virtually all of those once-dominant leaders were gone. Similarly, in the 1980s, Kendall Square was hailed for its *AI Alley*, where a slew of startups out of MIT were supposedly blazing the way to a very bright and smart future in artificial intelligence, only to go almost entirely extinct within a decade.

How and why did this innovation ecosystem change so rapidly? The many threads to this story include companies displaced by upstart innovators with better technology or business models and companies forced to move because land or labor were more affordable or available elsewhere. Others may have been ahead of their times or doomed by strategic and management mistakes.

Today, biotechnology dominates Kendall Square. But that too could change. Firms could move out of town to hold the line on costs now that Kendall Square real estate is among the most expensive in the world. Another global downturn or other disaster could drive many out of business in a kind of mass extinction event (I wrote this sentence well before COVID-19 swept the world and seemed to increase the value of biotechnology companies—but the pandemic nevertheless illustrates the point). It's also possible to imagine someone inventing a radically new way to discover drugs that makes today's approach obsolete, that emerging fields will somehow transcend today's firms, or that a backlash against drug prices will upend the business models of companies in Kendall Square. In some later chapters, I've included insights from an array of leaders in different fields who try to look ahead to what might drive the Kendall Square ecosystem twenty-five to fifty years from now. Not all of the possibilities involve biotech.

Perhaps most importantly in my effort to chronicle the origins and evolution of this dynamic ecosystem, I've tried to take Wilson's advice to heart. As he pointed out, "We won't make any progress until we know what the species are, any more than you wouldn't make any progress with a business community if you didn't know really much about the product or the people making it." So at its heart, this book is the collected stories of the people and ideas that turned less than one square kilometer of land across the Charles River from Boston into one of the most powerful innovation engines the world has ever seen.

INTRODUCTION 2.0
THE BEST PLACE TO HAVE THESE PROBLEMS

Katie Rae is a longtime force of nature in Boston's high-tech startup ecosystem, and she now heads The Engine, a venture capital firm and accelerator formed by MIT. She calls Kendall Square "the most connected place in the world." When students or entrepreneurs come to her with issues trying to launch or grow or save a company, Rae tells them to stop whining and get to work. "You are in the best place to have these problems. Ask for help. It's sitting right there."[1]

Which points to one of several big incongruities associated with Kendall Square. While many feel that it's an almost unparalleled hub of innovation, on the surface it appears one of the least welcoming and dullest. As MIT Media Lab professor and urban expert Kent Larson says: "People from all over the world think of Kendall Square as a model innovation district. MIT is surrounded by hundreds of startups, and the area is an international center for biotech, robotics, and highly creative commercial activity. But, in fact, as a community it's quite dysfunctional. Except for MIT resources, the district lacks a pharmacy, healthcare facility, and other amenities that people need in daily life. The area has extremely low residential density, and the available housing targets mostly affluent people–not the young entrepreneurs who are the lifeblood of the district. Because of its residential shortfall, most people live elsewhere, which limits opportunities for after-hours serendipitous exchange of ideas that is so important for innovation. Not surprisingly, Kendall Square feels dead at night and on weekends."[2]

And one of my favorite descriptions comes from Joost Bonsen, an MIT lecturer who has long been a fixture of the local innovation scene—among other things, instituting Venture Nights, a networking event for entrepreneurs held at the Muddy Charles Pub on the MIT campus. Says Bonsen of Kendall Square: "It's one of the most interesting places, but stunningly unlivable and ugly. It's like gentrification gone rogue."[3]

I have lived in Kendall Square for eight-plus years. I have worked here for more than twenty. I have written two other books here, I ran a magazine here, started my own online news media company here. I sold it, then the new owner moved us to Boston—all the way to Boston! I was walking home one day through Kendall Square when a scientist-turned-professor-and-entrepreneur (there are quite a few of those out in the Kendall Square wild, but this one was named David Edwards) saw me and hurried out the door of his restaurant (paid for by one of his startup successes). He waved me inside and recruited me to run a nonprofit he and two other scientist-entrepreneurs had created, in essence bringing me back home. As you'll read, a big part of what planners, developers, and all sorts of business and city leaders are trying to do in Kendall Square these days—pre- and post-COVID—is to increase the rate of "human collisions." Now that's a Kendall Square human collision story.

So as I thought about how to approach this book, I decided to give in to that personal connection, at least in part. This is a book, then, about a place, written by someone who lived and worked in that place for a long time. Someone who has known many of the people involved for years, some for decades. It's a look at what they're doing that's filled with hopefully insightful and occasionally funny stories about them while also providing a window on the future, in the context of what I believe is an unprecedented view of how things got the way they are.

But it's not just a feel-good book. Kendall Square has some major challenges ahead of it—and there are some very smart people who think it's in danger of losing its soul. "Boy, anybody who thinks that what is flourishing today is going to be flourishing tomorrow has to be out of their mind," says Ovadia Robert "Bob" Simha, director of planning for MIT for forty years and a vocal—some might say virtuoso—critic of some plans now unfolding. "We have a dumbbell city here—we have rich and poor. We've eviscerated the middle class."[4]

A similar warning comes from Lee Farris, vice chair of the Cambridge Residents Alliance, which has been closely involved in the Kendall Square development plans and in particular the effort to include more affordable housing. Despite helping win some adjustments that have improved the situation, she told a radio interviewer from NPR affiliate WBUR, "There's lots of market-rate, luxury-priced apartments being built, but there's not enough affordable housing being constructed."[5]

C. A. Webb, the super-thoughtful president of the Kendall Square Association (KSA), takes such concerns to heart—and notes that the shortcomings go beyond housing to the makeup of the workforce itself. "There is some really interesting and important tension around equity, you know, for whom?" she says. "For Kendall Square to be the definitive epicenter for innovation in the world, to truly steward this place as an innovation epicenter, all of our organizations, all of our people, are going to have to put diversity, equity, and inclusion at the center. And, really, no one's gotten it right."[6]

I come back to these concerns, and look at others, in a more comprehensive way at several places in this book. First up is a look at Kendall Square as it exists today—and where it's headed. As I really dived into the project in 2019, before the pandemic struck, there was what seemed to me a transformative amount of activity going on in the square—cranes and bulldozers churning away, skyscrapers going up, designs and plans unfolding for an array of new buildings and open spaces. Much of this was being driven by MIT, owner of a huge amount of real estate in and around the campus. "We may actually be able to create a square in Kendall Square," quipped Steve Marsh, managing director of real estate in the Massachusetts Institute of Technology Investment Management Company (MITIMCo), to some three hundred attendees of a public hearing.[7]

One of my first interviews was with Susan Hockfield, former president of MIT, at her office at the Koch Institute for Integrative Cancer Research at MIT, about a ten-minute stroll from my condo. I walked in through the lobby, filled with art and exhibits, and elevatored up to the fourth floor, where Hockfield shares an office suite with Phil Sharp—a Nobel Laureate who cofounded Biogen and Alnylam, both headquartered in Kendall Square.

After finding my way to Hockfield's office, I was greeted by Bingley, a rambunctious golden retriever. Hockfield explained she normally left her pet at home but decided to bring him in that day. Once he quieted down, she told me the story of her awakening of sorts to the Kendall Square situation—*plight* might be a better word—soon after she came to MIT from Yale University in 2004 to take the reins as president. "So I arrive here, and I remember standing out on Main Street and saying to someone, 'Why does this look so awful?' And I said, 'Who owns these dirt parking lots?' 'MIT,' the response came back. I said, 'You're kidding me. We can do something. I know that we can actually do something, so what's the plan? We need a plan.' And so we put together a plan. That's kind of the short version."[8]

That plan, called MIT 2030, is manifested in part by the unbelievable building boom in Kendall Square that's taking place as I write this and that is slated to continue through at least 2025 (and those are just the current plans, in case you're wondering why it doesn't go all the way to 2030). The more complete, but still greatly abbreviated, version of Hockfield's story is that there had been years of efforts already to revitalize Kendall Square when she arrived. As she continued: "Something new was ready to happen, and perhaps my enthusiasm for the Kendall Square dream in terms of what it would do, certainly for MIT, but for the region, for the Commonwealth, for the nation, for the world—to me was as clear as day. And so it was easy for me to say, 'Let's do it.' But truth be told, my role could be considered to be just pouring some gasoline on the fire."

This sliver of Cambridge continues to change fast and furiously. Some people even think it's a little out of control. Welcome to Kendall Square.

1 THE MOST INNOVATIVE SQUARE ~~MILE~~ KILOMETER ON EARTH

What is Kendall Square? Geographically, it's a tiny plot of land that when outlined on a map looks kind of like a narrow-headed robot with a straight nose and pompadour haircut tipped on its side. It's located on what was once largely marsh and mud flats along the banks of the Charles River across from Boston, on the eastern edge of Cambridge. The Boston Consulting Group once famously called it "the most innovative square mile on earth."[1] And that's the phrase the Kendall Square Association proudly wields as a slogan, substituting *the planet* for *earth*. But I've gone over the map many times, and even given some dispute over the exact boundaries, the total area is really just about a square kilometer.

Another way of looking at it comes from Mark Levin, former CEO of Cambridge-based Millennium Pharmaceuticals (an early genomics biotech now part of Takeda) and cofounder of Third Rock Ventures (a leading life sciences venture firm that has backed several companies in the square). "Do you know the book *The Soul of a New Machine*, about building a new minicomputer?" he asks, referring to Tracy Kidder's 1981 Pulitzer Prize winner. "It was this concept of you put these people together in a small space and they're interacting all day long with different disciplines and they've got an open mind about things. Kendall Square is really *The Soul of a New Machine*—on a square size—where everybody's packed in and you meet people, you're energized by things, you collaborate with everybody, and everybody's changing jobs. It's just a very tight, small group of people that are really energized around doing great science and, most importantly, making a difference for patients."[2]

Those people are packed into a rectangle roughly bounded by Charles Street on the north, Bristol Street to the west, the river on the southeast, and the Massachusetts Institute of Technology to the southwest—everything within about a ten-minute walk of the Kendall/MIT subway station.

Let's take a tour, starting at my own condo building at Binney Street and First Street on the map's far right side. Directly opposite, overlooking the Charles River, are the Esplanade condominiums designed by Moshe Safdie. The Mediterranean-style, cascading terraces facing the river are derived from Safdie's *Habitat*, built for the 1967 World's Fair in Montreal. Safdie is also the architect who designed many of Kendall Square's landmark features, including the red brick Marriott Hotel complex opposite MIT, with its piazza and rooftop urban park. Left of the Esplanade, as we face the river,

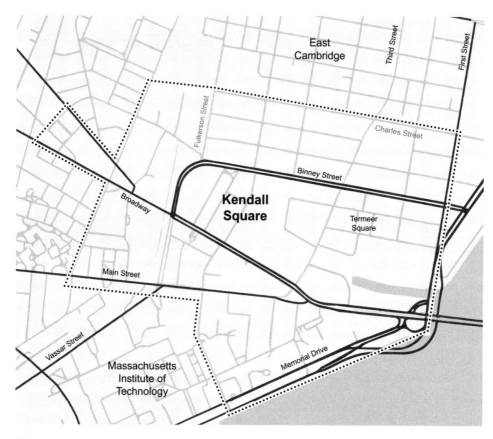

Figure 1 Kendall Square does not have clearly defined or agreed-upon boundaries. This map shows the core of the square's innovation ecosystem.

is a drab office building once occupied by software maker Lotus Development. Today, its tenants include Flagship Pioneering, whose five-hundred-odd employees conceive, nurture, and hatch five to seven life science startups every year. The most famous of its offspring, as I write this, is Moderna, which in December 2020 became just the second company to have a COVID-19 vaccine approved by the FDA.

There's much more on Flagship and many of the people and companies we'll meet on this quick tour throughout this book. For now, let's head up Binney Street, away from the Esplanade, where virtually everything has been built new or refurbished in the past few years, apart from the Church of Jesus Christ of Latter-day Saints. On my left is the home of Sanofi Genzyme and gene therapy company Bluebird Bio. Genzyme, a trailblazing local biotech, was founded in 1981 and was one of the world's largest biotechnology companies when purchased by Sanofi in 2011 for $20 billion. Bluebird won its first approval in 2019, in Europe, for a treatment for a rare genetic disorder called *beta thalassemia*. The treatment replaces the regular blood transfusions normally needed to manage the disease with a one-time therapy.[3]

On my right, at 75 Binney Street, is the headquarters of IBM Watson Health, a leader in the fusion of healthcare and artificial intelligence. The same building, which opened in 2018, houses a branch of IBM Research and Big Blue's cybersecurity command, IBM Resilient.

I cross the street and continue up Binney with big pharma powerhouse Bristol Myers Squibb on my port side; Facebook occupies part of this building as well. Just beyond it, I cut left through a neatly landscaped passageway and head across Linskey Way onto West Kendall Street. I pass two office and lab buildings occupied by Alnylam Pharmaceuticals, one of the earliest RNA interference (RNAi) companies. *RNA interference* refers to the ability of RNA molecules to interfere with the expression of harmful genes, blocking their disease-causing effects. In 2018, after some sixteen years and its "dark ages," Alnylam became the first company to have an RNAi drug approved by the FDA—a treatment for a deadly rare disease called *hereditary transthyretin amyloidosis*. As of early 2021, three other drugs had been approved, with more in the pipeline, including preclinical studies of an antiviral intended to target and kill COVID-19.

Just opposite the second Alnylam building is one of the few outdoor gathering areas in Kendall Square, a small park with a winter ice skating rink that transforms into a place for concerts, with tables for lunch and conversation during warmer months. This is Henri A. Termeer Square, honoring the longtime Genzyme head who passed away suddenly in 2017. Termeer was also a mentor to young entrepreneurs—a key figure in forging the strong networks key to the area's life science success (see chapter 23).

A sculpture of Termeer, seated, legs crossed, one open hand inviting conversation, was dedicated via livestream in December 2020. He directly faces the award-winning building that long served as Genzyme's headquarters before Sanofi moved it around the corner. This glass-walled structure, with at least one indoor garden on each of its twelve floors, has been taken over by Japanese drugmaker Takeda, which through several high-profile acquisitions has become the largest life science employer in Massachusetts. It moved its global R&D headquarters to Cambridge in 2015, following that up four years later by shifting its US headquarters and commercial operations from Chicago to Massachusetts. When Takeda's board of directors met in Cambridge in 2019, it was the first directors' meeting outside Japan in the company's then 238-year history.

Next, after zagging past the Kendall Square Association offices, I turn left onto Third Street and walk by the Cambridge Innovation Center (CIC), an early coworking space founded by two MIT graduates. The CIC has expanded to three buildings in the square (as well as to Boston and other cities worldwide). But this is the main one. All told, the Cambridge branches currently hold some seven hundred companies, and since the company's 1999 founding, the organization as a whole has nurtured five-thousand-plus startups.

Third Street dead-ends at Main Street. Straight in front of me is MIT's campus. The institute isn't technically part of Kendall Square, even though its spirit enervates the streets. So we'll turn right on Main, away from the Charles River. Main Street is, well, the main street of Kendall Square. It isn't a grand avenue by any means—just two lanes, with bike paths on each side. On my right, opposite the MIT campus, is the Marriott Hotel and a swath of office buildings called Kendall Center. Google is the dominant tenant, with plans to expand its footprint from about 1,800 employees to 3,000 in coming years, even after COVID. Google entered the square in 2005, with its then-stealthy acquisition of Android, which had a small Cambridge presence. Focus areas of Google's local operation have included Google Search, Google Chrome, and YouTube, as well as smaller efforts in image search, networking, Google Play, and the Go programming language.

On the left, or MIT, side of the street, upheaval and construction reign. Three new multistory structures were in various stages of completion heading into 2021, two of them incorporating the early 1900s factory buildings that had been there, as part of either their base or facade. Eventually, two other buildings, not yet under construction, will go up on their flanks. As a group, this quintet, plus one other building on the other side of Main Street, will offer lab and startup space, more than 750 graduate student and general public apartment units, corporate R&D facilities (including quarters for Boeing, Apple, and IBM's Cambridge research arm), MIT's admissions office, and the new home of the MIT Museum. It will also become a focal point for many of MIT's entrepreneurship and innovation activities. I go into more depth about this construction, including some illustrations, in the next chapter. MIT officials like to call it a "new front door."

Two blocks farther down Main Street, we come to where Main Street meets Vassar Street in what venture capitalist Juan Enriquez, writing in *WIRED*, dubbed "the world's most innovative intersection."[4]

On my right is the Whitehead Institute, where biomedical researchers are uncovering basic mechanisms of diseases like autism, Parkinson's disease, and diabetes. Beside it stands the Broad Institute, a genomics and biological powerhouse that is helping spearhead the Human Cell Atlas project, an international effort to map every cell in the human body. Over a weekend in March 2020, the institute retooled its vast genomic sequencing operation to COVID testing—on its way to becoming one of country's largest testing centers, able to process up to one hundred thousand tests a day. A year later, in March 2021, the institute announced a major initiative to employ machine learning and artificial intelligence to combat disease through its newly formed Eric and Wendy Schmidt Center, named after the former Google CEO and his wife, who donated $150 million—a figure matched in a gift to the institute's general endowment by namesake donors Eli and Edythe Broad.[5]

On my left, just across Main Street from the Whitehead and Broad Institutes, is the Koch Institute mentioned in Introduction 2.0, the mission of which centers on uniting biology and engineering to fight cancer. Besides Phil Sharp and Susan Hockfield, its denizens include chemical engineer Robert "Bob" Langer, a prolific inventor who has founded or cofounded forty-one companies, including Moderna and Teal Bio, formed during the COVID pandemic to develop advanced mask technology; and Sangeeta Bhatia, the twenty-fifth person—and second woman, after Caltech Nobel Laureate Frances Arnold—elected to all three National Academies: of Sciences, Engineering, and Medicine. Bhatia herself has helped found five companies and has teamed with

Hockfield and fellow MIT professor Nancy Hopkins to launch an important effort to help more women faculty join the entrepreneurship ranks (see chapter 21).

Also at or abutting this intersection: MIT's Stata Center, which holds the school's Computer Science and Artificial Intelligence Laboratory (CSAIL); the World Wide Web Consortium (W3C), where web inventor Tim Berners-Lee works; MIT's Department of Brain and Cognitive Sciences; the McGovern Institute for Brain Research; a Novartis research complex; and Moderna's headquarters, among other entities or groups. Just one block back is the nineteen-story headquarters of Akamai. The company's servers handle a large percentage of global internet content delivery, and its Network Operations Command Center (NOCC; pronounced *knock*) provides real-time, 24-7 monitoring of internet traffic worldwide.

The Broad Institute's founder, mathematician Eric Lander, was a New York whiz kid who garnered the top prize in the Westinghouse science competition in 1974. The second-place finisher was Virginia teenager Tom Leighton, Akamai's cofounder and CEO. From his eighteenth-floor office adorned with *Star Trek* memorabilia, Leighton can literally look down on the office of his former rival.[6]

But's let's keep going just a bit farther. Two short blocks later, we come to 700 Main Street, a three-story structure with a red brick facade. The oldest industrial building in Cambridge, its journey reflects the essence or spirit of Kendall Square. Over the years, it has played host to a leading railroad car innovator, the invention of the pipe wrench, the first long-distance telephone call, Polaroid founder Edwin Land's private lab, and a range of biotechnology innovations.[7]

I've devoted a whole chapter (chapter 22) to the story of 700 Main. Today, the address is home to startup space LabCentral. Walking through the bright green door labeled just "700," we're in the midst of some seventy fledgling companies tackling multiple forms of cancer; neurological, genetic, and autoimmune diseases; cardiovascular ailments; new forms of imaging; and a lot more.

With that quick tour, we've walked basically the entire width of Kendall Square, not quite in a straight line—covering just over one mile. I hope this brisk stroll captures the flavor of this unique place, though it just scratches the surface of what's going on. But there's another way to experience the exceptional nature of Kendall Square—by looking at its aggregate numbers. Some sixty-six thousand people lived, worked, and learned in Kendall Square pre-pandemic, according to the Kendall Square Association. Most of those fell into the "working" category, as only a few thousand, at most, actually live in the square; no one seems to have a good count. As for the total number of businesses, MIT officials estimated in late 2018 that some three hundred large tech and biotech companies operated in Kendall Square. But those figures don't include smaller startups or the scores of venture capital firms, law offices, accountants, and other service providers, so the total number likely well exceeds one thousand.

Billions of dollars flow into Kendall Square to fuel this innovation engine. The seven hundred–plus startups housed inside the Cambridge Innovation Center pre-pandemic, together with their precursors, have alone raised nearly $8 billion in venture capital and strategic investment since 2001, the CIC says. I've never seen an accurate figure of the total deal flow in the square—and it depends on whether you are counting venture money, public stock offerings, acquisition prices, and so on. But the figure for the past decade is likely in the hundreds of billions, if not more than one trillion.

These big numbers of companies and dollars, packed into barely a square half-mile, are accompanied by something that's far more challenging to measure—a culture that's hard to find anywhere else, argues Fiona Murray, professor of entrepreneurship and associate dean for innovation and inclusion at the MIT Sloan School of Management. Indeed, she notes, people come from around the world to learn about that culture in hopes of fostering it in their regions: pre-COVID, MIT Sloan hosted visiting delegations almost every week. "It's one of the most special innovation ecosystems in the world, but it's a little easier to get your head around because it's more geographically concentrated than Silicon Valley," Murray says. "You can experience the culture just by walking through Kendall Square."[8]

The square's ethos is special, she says, in part because it often encompasses *tough tech*. These are technology problems that can't be addressed quickly through apps or social media and often take a decade or longer to bring to market—things like developing a new drug, creating new robotics or other hardware systems, or tackling big energy and environmental problems. "I think Kendall Square—and greater Boston—is a very impact-oriented ecosystem, so people have a shared desire to do things that have impact in the world," Murray adds.

> An estimated 97 companies headquartered or with a large presence in Massachusetts were working on some aspect of COVID-19 in early 2021, according to MassBio. Thirty-two of them were in Cambridge, and twenty-three of those were in Kendall Square.

> "Silicon Valley gives us social media apps.
>   Massachusetts gives us a 94.5% effective Covid vaccine."—Tweet by Mike Volpe, CEO of Lola.com[9]

It's this quality that drew inventor and computer scientist Danny Hillis back to Kendall Square after nearly twenty years in California. Hillis was an MIT doctoral student when he cofounded parallel computing company Thinking Machines in 1982, which for most of its life was headquartered in Kendall Square. In 1996, Hillis migrated to the Golden State, but returned in 2015, tired of the West Coast tech scene. When he first got to California, the internet was just taking off. "One of the things that happened was that because so many people made money so quickly, the West Coast began to attract people kind of like the gold rush people who were attracted to the idea of making money," he told me shortly after his return. "That slants the whole conversation of even the technical people. If you sit in a restaurant in Palo Alto, at every table you hear people talking about their mezzanine financing, their strategy, their seed investment—it's all about the financial side of things. Whereas you sit in a restaurant in

Kendall Square you hear people talking about CRISPR [the latest gene-editing technology], and gene drive and deep learning methods. It's about the idea."[10]

So how did this special place come to be? The story of Kendall Square is one of relentless change and evolution, like many of the natural ecosystems that E. O. Wilson has spent his life studying. "When you go all the way back two-and-a-half centuries, you realize there's this really fascinating, dramatic, common theme in Kendall Square that perpetuates itself economic cycle after economic cycle, technology after technology," says Travis McCready, the first executive director of the Kendall Square Association and, later, CEO of the Massachusetts Life Sciences Center. "It's just this really crazy mashup of innovation, entrepreneurship, technology, leaders and leadership, politics, and real estate development—it's all there."[11]

McCready loves history and had his own favorite walking tour that he'd use to pitch Kendall Square to major corporations, research organizations, and various officials. In rapid-fire fashion, he'd go to 700 Main and talk about Alexander Graham Bell and the telephone, and Edwin Land and instant photography. He'd stop by the home of Boston Woven Hose and Rubber, which revolutionized firefighting and, later, bicycle tires. He would talk about the first recombinant DNA ordinance and how that encouraged Biogen's move to the square, kicking off its rise as an epicenter of biotechnology. He would show visitors the haunts of once high-flying tech companies like Lotus and Thinking Machines.

"Boom, boom, boom, boom," as McCready sums up his highlight reel of a tour. "It ended with a punchline, which was: 'You're looking at other places, right? Do you want to go someplace where innovation is new within the last fifteen to twenty years, or do you want to go someplace where innovation is so hardwired that we've been doing it for 250 years? Where do you want to be when there's a recession? Do you want to be in a place that has to ask itself whether it's going to invest in innovation, or do you want to be in a place that has gone through dozens of recessions—and every time reinvests in innovation?"

"Kendall Square," says McCready, "is almost literally—not just figuratively—America's first innovation district."

2 MODEL OF INNOVATION

On a freezing but clear winter day, I bundled and booted up and made a journey to what can be seen as the zeitgeist of Kendall Square—a dynamic model of the area called CityScope Volpe. The model graphically illustrates the density and fervent activity of today's square. But it also offers a glimpse of its future, showing the ambitious redevelopment of a crucial 14.5-acre parcel in the heart of the ecosystem. The plan represents a sort of seminal vision for what the square could be: Silicon Valley meets Paris in a confined area. That future is what I came to contemplate.

The model's home is in the City Science zone of the MIT Media Lab. To get there, I meandered through the sprawling complex, past offices and labs, side-stepping a ping-pong table and wandering along hallways with well-worn couches, gym-type lockers, and video game consoles bearing names like Astro City and Karate Chimp. City Science is nestled just about as close to the center of the legendary lab as something can get—almost smack-dab at the intersection of the original 1985, I. M. Pei–designed Wiesner Building and the expansion edifice by Japanese architect Fumihiko Maki that opened some twenty-four years later. The City Science zone is about half a basketball court in size with glass walls that face outward to the corridor, allowing passersby to view much of what's inside. Peering in, you can see projects in various stages of completion—some from the present day, others presenting a window into the cutting edge of the past.

In fact, City Science seems almost like a mashup of a museum of technology and a crystal ball. Among the artifacts on view are the GreenWheel, an "electronic assist" device dating to 2008 that nested inside a standard bicycle wheel to provide extra power when needed—and the CityCar, a project that began in 2003 and was initially sponsored mainly by General Motors. The CityCar was an ultra-small, all-electric vehicle that folded to enable three of the cars to fit straight-on into a single parallel parking space. It had robot wheels and drive-by-wire, and the entire front opened so that the driver and a passenger could step directly ahead onto the curb. It spawned a commercial prototype called the Hiriko Fold. Alas, the Spanish consortium that had licensed the technology itself folded due to lack of financing.

I also lingered by some of City Science's contemporary projects. One is a mock up that led to a spinoff company called Ori Living. Instead of smart or autonomous cars, it envisions smart interiors—such as an entire studio apartment that transforms from a living room to a bedroom, walk-in closet, or work area complete with desk with the push of a button—perfect for the ultramillennial or Gen Z crowd.

But what really catches the eye here are the interactive models of the CityScope project—kind of 3D, in-person SimCity-ish models—on display at the front of the big room. The models offer views of real cities, neighborhoods, or even slivers of neighborhoods like a commuter train station. One CityScope model shows a large segment of Andorra la Vella, capital city of the principality of Andorra, in the Pyrenees between France and Spain. Making use of mobile phone location data, tiny lights flash to depict telecom activity from busy times like the Tour de France, so you can see where people most likely congregated. It isn't actual tracking, but you know roughly where callers were at different times. That enables researchers to simulate what real people might have done—and from that explore ways to improve such things as traffic flow.

Another CityScope project zeroes in on an envisioned bus rapid transit (BRT) station in Boston, and still another showcases several square blocks of Barcelona.

The biggest model by far, though, is the one I came to see—CityScope Volpe. It takes up almost all of a glass tabletop six feet square. A large monitor on a separate table by the side of the model shows two images. On the right is a 3D computer graphics version of the model itself. On the left is a radar plot showing data related to various types of land uses and their density, diversity, and proximity. It's peppered with labels: Residential, Coworking, Educational, Parks, Healthy Food, Security, and Public Transportation. Under the radar plot are boxes and bar charts simulating the social and environmental performance of proposed designs, such as how they might impact shared-use autonomous mobility versus people opting for privately owned vehicles.

CityScope Volpe is a reconstruction of the heart of Kendall Square. It's named after the John A. Volpe National Transportation Systems Center, a research facility built in the mid-1960s for NASA before being taken over by the Department of Transportation that has occupied some of the most desired real estate in Massachusetts for decades. The complex harbors a monstrosity of a main government building surrounded by five smaller, equally drab structures and a lot of parking tarmac—all virtually inaccessible to the general public. What might have looked like puddles around it after a rain could just as easily have been the drool from developers. Now, at long last, it's about to be torn down in the latest chapter of the evolution of Kendall Square.[1]

In January 2017, the federal government accepted a $750 million bid by MIT to purchase and develop most of the 14.5-acre parcel. The plan is to first erect a new Volpe Center headquarters facility that consolidates all of the work previously sprawled across the six buildings into one structure occupying just four acres, complete with rooftop solar panels and LEED gold certification. Parking will be largely underground, freeing up spaces for public walkways to connect with the surrounding neighborhood. Maya Lin, the architect-artist who designed the Vietnam Veterans Memorial in Washington, DC, will create a public landscape that will represent the Doppler effect by featuring grassy mounds that undulate in a way evocative of sound waves.

Next, once the reimagined Volpe Center is complete sometime in 2022 or thereabout, as the CityScope model somewhat awkwardly shows, MIT will start redeveloping the site's remaining ten acres—transforming it into a mixed-use area with

Figure 2 Rendering of early 2021 plans for developing the Volpe complex (center of drawing). Plans include four apartment buildings, among them the large tower with spire; four office and lab structures; a community center; and the new Volpe Transportation Center.
Source: MITIMCo, Elkus Manfredi Architects.

residential units, lab and commercial space, ground-floor retail, a community center, and a job connector that helps train and prepare people in the community for work businesses in the area have to offer. A quartet of apartment buildings will include some 1,400 residences—among them 280 permanently subsidized affordable housing units and another twenty apartments devoted to middle-income families.

Dreams for Volpe, though, form just a small part of what's ahead for Kendall Square. One side of the Volpe property borders Broadway and faces the Marriott Hotel. The other side of the hotel looks out over the plaza we passed on our tour in the last chapter, the MBTA Kendall/MIT subway station, or T stop, and then across Main Street to the MIT campus. Until the fall of 2018, a row of buildings lined the side of Main Street directly opposite the plaza and the Marriott, forming a kind of buffer or perimeter between the city and the MIT campus. These buildings harbored a mix of retail, business, and academic operations—a Fidelity branch, a post office, a florist, a coffee and sandwich shop, the MIT Press Bookstore, and the school's Security Studies department, among other things. Behind the buffer buildings, accessible through some small side streets, lay a series of gray, uncovered parking lots reserved for MIT staffers.

Before the Volpe Center effort even got underway, that row of structures along Main Street was partly torn down, and the parking lots dug up, to create underground parking and lay foundations for a series of new buildings that are part of another massive Kendall Square facelift. I gave a high-level picture of that upheaval during our walking tour—but here is a more detailed account. This sister project of sorts is called the Kendall Square Initiative, a plan formally approved by Cambridge back in May

Figure 3 Rendering of the six new or expanded buildings under the Kendall Square Initiative. *Source:* MITIMCo.

2016 for six buildings.[2] They will contain up to 888,000 square feet of office space; a residence tower with 454 graduate student apartments (building 4 in figure 3); 300 public residential housing units (building 1); three commercial research facilities; a significant expansion of startup space provider LabCentral; the new home of the MIT Museum; the school's admissions office; and MIT InnovationHQ, where MIT is consolidating an array of entrepreneurship and innovation programs and initiatives.

The ground floor of several of these edifices will host a variety of restaurants and retail stores, including a long-sought-after pharmacy.[3] The development also creates nearly 2 acres of open space, including walkways and public gardens behind and between buildings—much of it covering the area once taken up by university parking lots. As noted in chapter 1, the intent is to transform what was previously a random, uninspiring, and uninviting eastern entrance to an iconic university—about a half-mile walk from the school's longtime main entrance at 77 Massachusetts Avenue—into what MIT calls a "new front door."

"When somebody comes out of the T, they will no longer ask where MIT is," quipped former MIT executive vice president and treasurer Israel Ruiz at a public hearing about the plans.

The first of these structures, the new graduate student residence, opened in late 2020 amid the coronavirus pandemic. On its heels was a centerpiece of the Kendall Square Initiative—a seventeen-story, indented, cube-like office tower opposite the Marriott at 314 Main Street (building 5). This is a special address for MIT geeks, 3.14

being the first three digits of pi. Dubbed Pi Tower, it's expected to be partially open by mid-2021—but MIT didn't have to build it for folks to come. The MIT Museum, relocating its headquarters from the relative backwaters near Central Square, will take up the second and third floors and much of the ground level—where it will be flanked by a café and the MIT Press Bookstore on one side and a restaurant on the other. The remaining fourteen floors were put up for lease. Well before above-ground construction began, Boeing became the first big tenant, announcing it would rent four floors as an R&D center for its subsidiary Aurora Flight Sciences, an MIT spinout developing unmanned aerial vehicles and autonomous flight systems that Boeing acquired in 2017. Capital One then grabbed three floors, and Apple snatched up one—later adding three more. Subsequently, IBM rented space for the MIT-IBM Watson AI Lab—part of a ten-year, $250 million commitment in which Big Blue will fund research addressing a variety of issues around artificial intelligence. Several other companies signed leases in Pi Tower in 2020. Although COVID delayed construction several months, the building was on track to be fully booked by its unveiling around June 2021. The MIT Museum was eyeing early 2022 for its public opening.

As part of the Kendall Square Initiative, an existing historic building at 292 Main Street, in the same block as the graduate student residence tower but not shown in the illustration, has been renovated to serve as a gateway between MIT and Kendall Square.[4] The ground floor will house the new MIT Welcome Center, which anticipates handling over fifty thousand visitors per year. MIT's admissions office will occupy the second floor. The five floors above it form MIT InnovationHQ, which will offer coworking and maker space for students, faculty, alumni, and the wider community and will also house a variety of campus programs that support innovation and entrepreneurship. "It's the sort of new space that will become the heart of our innovation community, welcoming stakeholders to join MIT as it moves ideas to impact," says Fiona Murray.[5]

The iconic clock tower building (building 3) is being renovated and expanded mostly as lab space: LabCentral and Bayer will be key tenants. It is set to open in mid-2022, roughly in concert with the building 1 apartments. The other two structures, a lab complex (building 2) and small retail building (building 6), are not scheduled for completion until at least 2023.

That's just a snapshot of the buildings making up the Kendall Square Initiative. Meanwhile, Boston Properties is building an eighteen-story glass tower across Main Street from the core MIT development that's reminiscent of a squat ocean liner turned on one end. Google is signed as the main tenant. Down the road a bit is a third upstart tower—this one for internet powerhouse Akamai, which was born and raised in Kendall Square. Dedicated in November 2019, it stands nineteen stories tall and is almost sci-fi in design, with odd angles and what look like observation ports jutting out of it. And a couple of long blocks from that, MIT plans to construct a home for its new Stephen A. Schwarzman College of Computing—dubbed almost instantly the *AI college* when it was announced in fall 2018. The building, part of MIT's $1 billion commitment to addressing the future of AI and computing, is slated for completion sometime in 2023.[6]

It all adds up to 1,700 new housing units, tens of thousands of square feet of new office space, dozens of new research facilities and retail shops, and several acres of parkland, walkways, and other public spaces. Even more to the joy of those who have long lived in Kendall Square, the square's first supermarket opened in November 2019. Until it opened, those in the square might have been blessed with huge brainpower and the opportunity to work on some of the world's toughest problems, but they had no place to buy milk.[7]

*   *   *

All of the new projects I described, with the exception of the AI college building, are incorporated into the CityScope Volpe model, which is run by Media Lab professor Kent Larson, who specializes in urban planning technology. While Larson was not involved in any of the projects ongoing in Kendall Square, the model—employing a variety of planning, analysis, and prediction techniques—can be used to research and evaluate the likely effects this major increase in density will have on the already superdense square. Both Larson and Joost Bonsen, the MIT lecturer mentioned in the second introduction who teaches a course with him, showed me around at different times.

"This is a tool for editing the city," explains Bonsen. "The whole point here is to look at how things are and how they could be."[8]

Figure 4 CityScope Volpe model viewed from the north, looking toward the Charles River. The Volpe site is the open area in the middle. City Science researchers Luis Alonso (front) and Arnaud Grignard manipulate land uses, density, and mobility systems using optically tagged LEGO bricks. A continuously updated digital display models various performance metrics. Photo: Kent Larson.

In the model, each LEGO block represents an area of twenty-two meters by twenty-two meters (roughly seventy-five feet square) and one story tall. A camera beneath the table reads barcodes on the underside of each piece or segment so that when a section is moved, the CAD file shown on the screen above the table is updated.

This setup allows students and researchers to both see at a glance how things are currently, and evaluate new configurations—including actual plans as in the case of various Kendall Square projects. Not enough housing in a certain area? Just swap it for some office space, and so on. Another feature of CityScope allows researchers to simulate the effects of such moves on energy usage, traffic, density of people, flows of investment capital, and more through a series of color-coded lights that are beamed from an overhead projector. For instance, you can ask questions about various mobility options, such as: What happens to parking spaces and therefore the amount of people who drive versus take public transit, bicycle, or use ride-sharing options as density increases? What is the end result on commute times or public transit capacity? How does all this impact the neighborhood and wider community? Larson calls the simulation feature "a fingerprint of the health of the community."[9] He points out that other, more ambitious designs the researchers have modeled showcase how much closer the square could come to "civic homeostasis"—with housing better matching local jobs and thereby minimizing commuting, and retail areas offering all the basic amenities of a city neighborhood. "This would create a vibrant 24/7 innovation community, not just an innovation district," he says.

Using the model, students explored different scenarios associated with Kendall Square's latest development plans. Armed with police incident reports, one student found that the larger intersections, with higher foot traffic, actually had more incidents, such as mugging, than quieter areas. Another ongoing study tracks tweets from Kendall Square denizens. "About 10 percent of tweets have the GPS coordinates embedded in the tweet, so then you can locate that tweet," Larson explains. On his laptop, he showed me some results. "So here's Third Street, which is active," he says. "The Media Lab is bright, CSAIL is bright, the Cambridge Innovation Center glows. But the government's Volpe Center for Transportation is dark, showing very little Twitter use." The tweet-tracking provides researchers with a rough proxy for following the flow of human activity, making it possible to see what areas are busiest, compare daytime to nighttime patterns, and ask questions about whether things need to be changed.

* * *

I was fascinated by the model. Staring down like some Goliath looking at a Lilliputian city, you get a much more vivid sense of the sheer density of the ecosystem. The Media Lab mockup, along with a more complete, architectural model in the MITIMCo ninth-floor offices at One Broadway, also depict the intensive efforts to make Kendall Square a more attractive and vibrant place. When Susan Hockfield became president of MIT in 2004, she relates, "I don't think it was dangerous to walk into Kendall Square, just unpleasant." And there were very few services. One day, she recalls, "my husband needed a haircut, and he walked over on a Saturday and he said, 'Arrgh.' No place to get a haircut—or anything else."[10]

She was determined to change that. "You secure a neighborhood by people living there," Hockfield explains. "The idea of having a barbershop, having a drugstore, having a little grocery . . . is all about what living there means."

MITIMCo's managing director, Steve Marsh, describes the goal as "trying to activate the streets." MITIMCo's architectural model depicts the efforts on that front far more vividly than the Media Lab's LEGO pieces. It shows a pedestrian zone lined on both sides with restaurants and shops running into the Volpe site along a street called Broad Canal Way. (This small roadway currently allows cars and stops when it hits Volpe at Third Street.) The plaza formed where Broad Canal Way will meet the new Volpe Center is what Marsh had in mind when he spoke about finally creating a square in Kendall Square.[11]

The end goal of all this, says erstwhile MIT executive vice president Israel Ruiz, is "creating the right framework for which the most amazing discoveries occur. I like to say, well, we are now connecting this industry with that industry. This person with that person. They met at a coffee shop. They met at a flower shop. They met at a park, at an open forum. And we're now trying to think about our spaces—and by our spaces I mean the academic spaces as well as the labs, as well as the now-commercial spaces that MIT owns—as ways to activate these conversations. We've used the term *human collisions*. How do we increase those human collisions? How do we increase the conversations? Because we know that conversations lead to new discoveries and new paradigm shifts."[12]

*   *   *

Looking over the CityScope and MITIMCo models, though, it's easy to glimpse the problems, challenges, and, well, stress that may lie ahead on that path to a more vibrant Kendall Square. Take something as simple as sunlight. When Larson's students did a lighting and shading analysis, they found that the two new massive towers on Main Street will almost completely occlude the beautiful public garden on the rooftop of the garage abutting the Marriott Hotel. Similarly, Safdie's public piazza on the Main Street/MIT side of the Marriott would lose its afternoon sun—and find itself in a sort of wind tunnel. The plaza is a signature part of Kendall Square, a meeting and conversation spot with benches, space for occasional outdoor concerts, tiled plaques honoring famous entrepreneurs from an aborted effort called the Entrepreneur Walk of Fame, and an entrance to the MBTA station. It will be interesting to see if it's used less when ensconced in shadow and swept more by wind.

Staring at the models, it's also hard not to think of the almost soulless nature of Kendall Square. "Kendall Square is really a terrible place," observed noted architect Charles Correa. "You've got these huge monsters, each autonomous, with its own internal logic, paying no attention to the street or to the others around it. It's a case study of what's wrong with whatever we've been doing for the last fifty years or seventy years."[13]

Correa tried to add some life to the scene with his McGovern Institute for Brain Research building, with a distinctive planar design featuring lots of glass and a huge atrium intended to complement the vivid and colorful Frank Gehry–designed Stata Center across the street.

But despite their creative architecture, these structures haven't done much to change the area's overall aesthetic. And there are plenty of detractors of the new buildings being planned. The Kendall Square Initiative is "rife with bad planning, bad design, and illusory benefits" and "an architecture that is offensive to the eye and impractical for generations to come," argues longtime MIT planning head Simha. I sat down with the now-retired firebrand at a high-end bakery and cafe called Tatte on Third Street. He has written blistering criticisms of the new plans, almost building by building. Among his complaints: Pi Tower housing the MIT Museum, Boeing, Capital One, and Apple "will create an ugly canyon."[14] Overall, he says, the plan "will produce buildings that are ugly, out of scale with human beings and are antithetical with environmental and energy principals that a great university should aspire to."[15]

Simha is an outlier, perhaps the most vehement critic of the current Kendall Square plans. But while much of the historic bleakness has been vanquished, the lasting sterility of the square is something those who work and, especially, live there roll their eyes and sigh about; indeed, this soullessness is such a truth that it's almost a badge of honor they all share. A steady influx of restaurants and bars in recent years, along with a smattering of new apartment buildings, has helped a lot, but it still hasn't made much of a dent in the overall situation.

Figure 5 Sign placed on the sidewalk just outside Cambridge Spirits, Kendall Square's only liquor store. *Source:* Charles Marquardt.

Box 1
Charlie's Sign

> The mantra of *live, work, play* has been cited for years around Kendall Square. There has long been a spattering of condos and apartments, and in recent years their numbers have grown, accompanied by a prepandemic rise in restaurants, bakeries, and coffee shops. But the square is still mostly a *work* place. A lotta work. A little bit of live. A little bit of play.
>
> It was big news for Kendall Square denizens on September 16, 2013, then, when Charles Marquardt opened Cambridge Spirits, a small wine, beer, and liquor store along Broad Canal Way. The store filled a void in the neighborhood, but only a small one. Charlie soon put a little folding blackboard outside the store echoing his fellow residents' complaints about what was still needed.
>
> The Kendall Square Initiative launched by MIT aimed to fill the unchecked pharmacy and grocery boxes. Indeed, a Roche Bros. supermarket opened in late 2019 as part of the building 1 development. A pharmacy was set to open at 238 Main Street, but not before 2022.

But there's also a deeper, even more serious concern hanging over the square as it tries to bulldoze, crane, landscape, and Maya Lin its way into a warmer and fuzzier technofuture: Can the innovation engine keep running at the same breakneck pace?

Look hard at the CityScope model of the square, or retrace the steps of my walking tour from chapter 1, and one disturbing fact jumps out. Aside from a few notable exceptions—among them biotechs like Biogen, Alnylam, Moderna, and Bluebird Bio, and a few tech companies such as Akamai—none of the firms occupying significant space in Kendall Square are homegrown. Instead they are giants like Google, Microsoft, Facebook, Bristol Myers Squibb, and Takeda, or, as in the case of IBM Resilient and Sanofi Genzyme, were homegrown independent companies later bought by giants.

The swarming of these behemoths, which also include Apple, Facebook, Amazon, and Dell EMC on the tech side, and just about every big pharma in the world on the life science side, is creating jobs and bolstering tax rolls for the city and state. But the influx is also driving up rents, soaking up talent, and making some things harder for startups and nontechnical people. A more hidden worry is the question of whether all these giants add to the soul of the place or are slowly sucking the ever-loving creative and innovative life from it.

Consider these statistics from Juan Enriquez's 2015 *WIRED* article about the intersection of Vassar and Main: "In the past decade alone, 18 out of every 100 MIT alumni launched companies, eight per cent of them near MIT. Massachusetts now hosts 6,900 startups launched by MIT alumni. Combined, they represent 26 per cent of all state sales."[16] A big worry now is whether these startups will keep coming to Kendall Square. Or does that even matter?

It matters to C. A. Webb. Concerns about rising rents driving out startups go back to at least 2012, she says. "Obviously, as rents were going up, so many of the startups have had to leave. Except for the very, very wealthy ones that their VCs are saying, 'Well, sure, you can spend that much on overhead because you need to be down the street from that board member or advisor.'" Skyrocketing prices have also affected many midsized companies, leaving a less-than-ideal mix of real estate options, she says. "There is an appreciation for the importance of maintaining a cross-section of stages so that the district doesn't just become a big corporate campus for pharmas."[17]

Thanks to a lot of dialog among a lot of parties—including MIT, the city, the KSA, and citizen groups like the Cambridge Residents Alliance and the East Cambridge Planning Team—all of these issues are being addressed at least to some degree in current plans. For instance, Cambridge now requires a percentage of new space in Kendall Square to be set aside for both community programs and entrepreneurs. Still, she says, "I think that time will tell whether we were really shortsighted on not being more intentional in maintaining a home for startups and scale-ups in Kendall."

* * *

Ensuring space for startup and medium-sized companies isn't the only concern facing Kendall Square. In early 2019, Webb's association kicked off a multiyear strategic initiative with three pillars of focus: diversity and inclusion; place-making, encompassing more retail, public parks, and other spaces for interaction and livability; and transportation reform—because getting to and from Kendall Square remains a major problem. An oft-cited Medium post about the KSA's transportation campaign was entitled: "You Can't Find the Cure for Cancer While Sitting in Traffic."[18]

Such concerns are not likely to be resolved quickly, of course: many are covered in more detail in later chapters. But if there is an overarching lesson from the long and often tumultuous history of Kendall Square, it is that the square has constantly, and usually successfully, remade itself. And, in truth, a lot of the progress has happened by accident. That's how innovation works. I was reminded of this when visiting Joost Bonsen's cluttered office in the MIT Media Lab.

"Kendall Square is the notorious prediction failure zone," Bonsen proclaims. "Everything that the square was predicated on, all the developments, didn't come true. Basically, they bet wrong essentially every step of the way. And yet the place—like a phoenix," he adds. "Despite essentially every error possible, this place is successful."

It all started back in the 1790s, long before MIT came to Cambridge.

3   THE FIRST ECONOMIC VISION FOR KENDALL SQUARE IS A BUST

The spot of land that would become Kendall Square had two fundamental advantages from the start. One was the immigrant, revolutionary spirit of Boston that fueled innovative ideas. The other: simple geography. The location of what is now the square, on the water, with cheap land tantalizingly close to Boston's commerce center and workforce pool just across the Charles River and between Boston and the educational mecca of Harvard College, made it almost ideal as a place for industry to grow and thrive.

Kendall Square was a crossroads waiting to happen.

It took a while—the wait lasted more than 150 years—for this promise to be even partially realized, however. At first, all the action on the other side of the Charles from Boston went down in the early colonial village of Cambridge, established in 1630 on a low drumlin upriver from Boston Harbor, where it would be hard to reach for enemy warships and also easy to defend against land assault. Harvard College was founded there in 1636. At the time, the shoreline between Harvard Square and Boston—including the location of today's Kendall Square—was undeveloped marsh and mud flats, where people hunted for clams and oysters or harvested salt hay to feed livestock. In fact, before much of it was filled in, what we now call the Charles River was part of an extensive waterway known as the Back Bay or Oyster Bank Bay, for the oyster grounds lining its shores along the Cambridge side of the estuary.

The marsh contained a smattering of small drumlins, along with some forest and woods. At low tide, the oyster-rich, muddy flats would be exposed. At high tide, the drumlins became almost like a series of small islands, cut off from the mainland. Although a few cart paths had been forged by the shellfish hunters, the entirety of what is now Cambridge east of Harvard Square was virtually uninhabited before the end of the Revolutionary War in 1783. "There were only three farms east of Harvard Yard in all of Cambridgeport and East Cambridge by the end of the revolution," recounts Charles Sullivan, executive director of the Cambridge Historical Commission and about as passionate a scholar and storyteller of Cantabridgian history as exists. "Kendall Square was no place at all."[1]

But shrewd businessmen and land speculators saw potential in that marshland. An early settlement took place on a large drumlin called Graves's Neck, named after its first settler, Thomas Graves, who moved there in 1629. This isolated spit of solid land later became the heart of the village of East Cambridge.

Over the years, as the population grew and land was filled in, Kendall Square began to encroach on East Cambridge—and even though the square is not technically part of East Cambridge, many residents consider it to be. Graves's Neck started about where Second Street is today—a few blocks back from the CambridgeSide mall—and went up to about Sixth Street. A waterway called Miller's River—since almost completely filled in—opened into a wide tidal bay and separated it from Charlestown proper on the east. The drumlin extended south and west to about where Charles Street is today. The Kendall Square area anywhere near the modern-day banks of the Charles River, including what is now MIT, consisted almost entirely of uninhabited wetlands known as the Great Marsh.[2]

* * *

Boston in those early days lasting through the American Revolution was itself almost an island, an oddly shaped peninsula that jutted up into Boston Bay via a razor-thin stretch of land called Boston Neck that connected to the mainland around Roxbury, some five miles south and a smidge to the west. With two prominences on the peninsula—one on the west pointing at Graves's Neck, the other on the east looking at Charlestown—it resembled a geological version of an animal's head, maybe a cross between a cow and a rabbit.

For years, the only way to travel between Boston and Cambridge without a boat was via the Boston Neck. Relates Charlie Sullivan, "Anyone from the north or northwest going to Boston, taking their produce or driving their flocks or whatever, had to go the long way around—through Harvard Square, across the river by the stadium, down through Brookline Village, Dudley Square, and up Washington Street into Boston." That was roughly an eight-mile trip from the center of Cambridge. The other choice for travelers was to take the ferry between Charlestown and Boston and then journey by the northern roads that connected with Harvard Square. The Charlestown ferry route cut the distance in half compared to the journey via Boston Neck, but wasn't as convenient for all purposes.

The obvious solution was a bridge making a direct road connection between Boston and Cambridge—right over the Great Marsh land that would become Kendall Square.

The first bridge across the Charles, from Boston to Charlestown farther down river, was completed in 1786, replacing the ferry. But it was still a roundabout way to Harvard. Sometime in the early 1790s, a group of investors—it's reasonable to think of them as one of Boston's early angel syndicates—formed a corporation with the goal of building a for-profit toll bridge directly linking Boston with Cambridge. One of their lot was Francis Dana, chief justice of the Massachusetts Supreme Court from 1791 to 1806, the first American ambassador to Russia, and a signer of the Articles of Confederation. Dana lived east of Harvard Square, and his family land included much of what is now mid-Cambridge and lower Cambridgeport—land in between Kendall Square and Harvard Square.[3]

The plan for a toll bridge raised some alarms. Massachusetts native and American founding father John Adams, already vice president of the United States and only a few years away from becoming the country's second president, openly fretted about the fervor around the land speculation, predicting reality would be achieved via "a few bankruptcies which may daily be expected."[4] One Boston newspaper, the *Columbian Centinel*, part of a group of local papers that later merged and eventually became the *Boston Herald*, warned that the envisioned bridge would corrupt Harvard students by making Boston too accessible, thus offering "vices that may corrupt their minds, and destroy their health, reputation and usefulness forever."[5]

The plan went ahead anyway, however, and the so-called West Boston Bridge opened in November 1793. The backers offered shares to the public and sold the two-hundred-share allotment within three hours. It served as what Sullivan calls a *desire line* that provided the first direct and easy access between Boston and Cambridge, connecting them through Kendall Square where the Longfellow Bridge is today and reducing the distance between Boston and Harvard Square to just three miles.

"It all starts with the West Boston Bridge," says Gavin Kleespies of the Massachusetts Historical Society. "Everything just comes together in Kendall Square. It's a central place where lots of things converge."

The original bridge consisted of two main parts: a 3,483-foot-long timber bridge connecting Boston to the first patch of dry land on the Cambridge side, then a 3,344-foot causeway across more of the Great Marsh to Pelham's Island, a drumlin at roughly the site of today's Lafayette Square near Central Square. The bridge first spurred residential and business development mostly around Pelham's Island. Marshland was drained, commercial buildings were erected, and house lots were laid out.

Then in 1805, two turnpikes were authorized—the Cambridge and Concord Turnpike that ran along what is now Broadway direct to Concord, and the Middlesex Turnpike that became modern-day Hampshire Street and connected to the mill towns along the Merrimack River and north to New Hampshire.[6] Both ran into the heart of modern Kendall Square, around the intersection of Broadway and Hampshire today. At roughly the same time, mostly in 1805 and 1806, a network of canals was dug. Most came up from the Charles River, but one cut straight across from East Cambridge, providing access to Miller's River and the relatively new Middlesex Canal coming down from the north.[7] For Kendall Square, the most important of these waterways was the Broad Canal. Roughly a hundred feet wide, it ran for almost a mile from just down river of the West Boston Bridge all the way to between Portland and Berkshire Streets, a bit past where One Kendall Square is today. One-quarter of its original length still exists, offering kayak, paddleboard, and canoe rentals for those wishing to explore the Charles River.

The new transportation networks helped transform the area into a shipping crossroads. "The first economic vision for Kendall Square," relates Sullivan, "was that all of the produce and timber and building stone from New Hampshire would be able to come down from the Merrimack River and into the Charles River, via the Middlesex Canal—and a group of investors envisioned that the shoreline near Kendall Square could become a transshipment point."

To help realize that ambition, the Broad Canal's promoters convinced the US Congress in 1805 to designate the town of Cambridge a port of delivery, thereby giving rise to the name of Cambridgeport. Today, Cambridgeport is a neighborhood mostly west of MIT. But at the time, it included much of what is now Kendall Square and other areas, and became one of four early nineteenth-century villages making up the town. Perfectly situated to be the key middleman or staging ground for cargoes coming into the newly christened port, the section around the Broad Canal seemed poised to boom. The area as a whole was known as Lower Port, because it was lower on the Charles River than other wharves. What later became Kendall Square was dubbed Dock Square.[8]

*   *   *

But this first big economic vision for Kendall Square was derailed virtually from the start by international conflict and competition.

Both Britain and France had been seizing American commercial ships during the Napoleonic Wars that began in 1803 and ended with Napoleon's defeat at Waterloo in 1815. Most infuriating to the nascent United States, Britain began engaging in *impressment*—seizing American sailors and forcing them to join the Royal Navy. Wishing to avoid military conflict, the outraged US had responded with the Embargo Act of 1807, signed into law by Thomas Jefferson that December, which prohibited American ships from traveling to foreign ports—hoping the ensuing hardship on all parties would convince France and Britain to allow America to resume trading as a neutral party. The act turned out to be a disaster, forcing ships in New England and the Atlantic seaboard to remain idle in ports. It was not lifted until March of 1809. Then came the War of 1812 and British blockades of American ports, causing additional economic hardship.

As a result, hopes for a major transshipment center around Dock Square and Cambridgeport were dashed. Some commercial trade took place in the Lower Port, thanks to the turnpikes and the West Boston Bridge, but far less than what investors and planners had anticipated. The six brick storehouses they had built in 1806 on Broadway near the corner of Court Street—now called Third Street, almost exactly where the Cambridge Innovation Center, the Marriott Hotel, and the Volpe Transportation Center are today—"stood alone in the marshes for several decades," writes historian Susan Maycock.[9] It didn't help that the new railroads initially bypassed Cambridge and that the vast indoor Quincy Market opened in Boston in 1826, attracting vendors and shoppers from all over. More than twenty years later, in 1847, the publishers of Cambridge's city directory still rued the course of events. The double whammy of Quincy Market and the proliferation of the railroads "have almost annihilated the extensive trade which was formerly carried on between the Port and the country towns, even as far back as the borders of Vermont and New Hampshire," the directory lamented.[10]

As Charlie Sullivan puts it, "Kendall Square just became a place to pass through on the way to Boston until probably the 1850s." Then it all got crazy.

4 CHARLES DAVENPORT AND THE SQUARE'S TRANSFORMATION

In 1812, in the small village of Newton Upper Falls, Massachusetts, a boy was born who would forever change the story of Kendall Square. His name was Charles Davenport. He started modestly enough, finding work in 1828 as an apprentice woodworker at age sixteen with George W. Randall, a Cambridgeport builder of carriages and coaches.

Davenport's talent and keen work ethic attracted the attention of a prominent local businessman named Capt. Ebenezer Kimball, who owned the nearby Pearl Street Hotel.[1] The captain also operated a livery stable, as well as a stage coach line offering service between Cambridge and Boston. His coaches initially ran twice a day, but then Kimball defied conventional wisdom and ran them hourly—providing a significant boost to his business.[2]

Kimball's commercial eye saw potential synergies between Davenport's skills and his existing stage coach business. In 1832, soon after the youth completed his apprenticeship, the two formed a carriage-building partnership, "with Kimball supplying the capital and Davenport supplying the coach-building 'smarts,'" according to an online history by the Mid-Continent Railway Museum. One of their first moves was to buy out Davenport's old boss, Randall, bringing in a core staff of "two journeymen and four apprentices working to complete the numerous orders they received."[3] Those orders quickly came to include all the coaches for the captain's stage line.

The business, initially located on Massachusetts Avenue in the heart of Central Square, thrived. It quickly took over "the blacksmith, the painter, the harnessmaker, and the trimmer all of which were located adjacent to the coach and carriage shop at 579–587 Massachusetts Avenue," according to the railway museum history.[4] Within the first two years, Kimball and Davenport began building the first omnibuses in New England. An *omnibus* was a horse-drawn carriage not unlike a stagecoach, but typically able to accommodate more passengers on shorter journeys. Some had two decks, precursors to today's double-decker buses. (The word *bus* is a shortened form of *omnibus*.)

In 1834, the business also made the leap to building railway cars—offering a groundbreaking new design. For the company's first order for the Boston & Worcester Railroad, "Mr. Davenport designed a car with the passageway running its entire length and permitting the passengers to all face the same way. This was the beginning of the American type of railroad passenger car," noted a profile of the city's early business leaders commissioned by a local bank.[5] Adds the Cambridge Historical Society, "While the physical design of the center-aisle train car was revolutionary for its increased efficiency, it also had major social consequences for travelling. A person no longer needed to pay for an entire compartment to travel, or even one of a few expensive seats. With rows of identical seats, anyone who could afford a single ticket could travel, and people all sat together as equals in the car."[6]

That first design provided a single entranceway in the middle of the car, with a kind of running board spanning the entire length of the car on both sides of the entrance. Within a few years, Davenport changed the design to add doors at each end of a railway car, making for easier and less congested access. By about 1837, Kimball and Davenport's railcars came with toilets and separate ladies' washrooms. These cars also featured reversible seats, which eliminated the need to turn cars around at the end of the line so that travelers could face forward, as well as the "Davenport drawbar," which reduced the shock of starting a train and for which the young inventor had received a patent in 1835.

Thanks to such advances, the enterprise became hugely successful. In 1837, just a few years after entering the railcar business with four-wheeled cars able to carry twenty-four passengers, the business expanded to crafting eight-wheeled cars with sixty seats—and then to sixteen-wheelers accommodating seventy-six passengers.

Kimball died in 1839, at the age of forty-six or forty-seven; the cause is unclear. He had four sons and two daughters, and Davenport almost immediately created a new partnership with Kimball's son-in-law, Albert Bridges. The fledgling business was called Davenport & Bridges, signifying that Davenport was the chief partner.[7]

Despite Kimball's death, the business continued to prosper, "becoming perhaps the preeminent builder of railway cars in their times," according to the railway museum.[8] The new partnership soon outgrew Kimball & Davenport's original digs—and the spot chosen for its new headquarters was 700 Main Street, an existing three-story brick structure at the corner of Osborn and Main in what is now considered part of Kendall Square.

It's hard to overstate the significance of the 700 Main building, which is why I devote a whole chapter to its rich history after the Davenport era. More than any other commercial building, its journey reflects the essence or spirit of Kendall Square. As the Cambridge Historical Commission describes it, 700 Main stands as "the oldest industrial structure in Cambridge, and the complex as a whole illustrates the evolution of Cambridge's industrial economy, from heavy manufacturing to photographic imaging to biotechnology."[9]

In 1842, when Davenport & Bridges moved in, 700 Main (along with six single-story wooden workshops erected on the east side of the property) became the factory for building some of the poshest, and costliest, railroad cars in the country. "The size of the cars forms a pleasant room, handsomely painted, with floor matting, with windows secured from jarring, and with curtains to shield from the blazing sun. We should have said rooms; for in four out of six cars, (the other two being designed only for male passengers,) [sic] there is a ladies' apartment, with luxurious sofas for seats. . . . These cars are so hung on springs, and are of such large size, that they are freed from most of the jar, and especially from the swinging motion so disagreeable to most railroads," wrote the *American Railroad Journal* in 1842.[10] Or as the inaugural issue of *Scientific American*, dated August 28, 1845, reported: "Davenport & Bridges . . . have recently introduced a variety of excellent improvements in the construction of trucks, springs, and connections, which are calculated to avoid atmospheric resistance, secure safety and convenience, and contribute ease and comfort to passengers, while flying at the rate of 30 or 40 miles per hour."[11]

Within about five years, Davenport & Bridges laid plans to expand further, occupying the entire block between Portland Street and Osborn Street—what's now called the Osborn Triangle. Two long, two-story wings were added, along with eight smaller buildings. One wing served as a foundry and blacksmith shop, complete with sixteen forges. Its counterpart became a machine shop. All told, the factory employed more than a hundred workers and was the largest in Cambridge at the time. A drawing of the C. Davenport & Co. Car Works from 1854 shows the sprawling complex with an uninterrupted view due south to the Charles River. Not even a road appears between the factory and the waterway.

Despite the thriving and what seemed secure business, though, Davenport & Bridges hit an unexpected rough patch. The company had accepted payment for many orders largely in the form of railroad stock—and when some key customers ran into financial difficulty, they not only canceled existing orders, but their stock became mostly

Figure 6 Davenport Car Works in 1854, with nothing between it and the Charles River.

worthless. In October 1949, faced with mounting liabilities, Davenport & Bridges was forced to suspend business. The partnership was dissolved the following May.

Davenport himself was hit hard. One account noted, "Mr. Davenport was a heavy holder of railroad stock and when stocks greatly depreciated in 1849 he lost over three hundred thousand dollars."[12] In today's dollars, that was akin to losing about $10 million.

But he rebounded quickly. In 1850, Davenport reorganized the business—still at 700 Main Street. "Within a few months the plant was reopened and was soon running to full capacity, this time under the name of C. Davenport & Company. By the close of 1851 the firm had done a business of well over half a million dollars. Within a few years Mr. Davenport had amassed another comfortable fortune and his creditors lost not one cent from the suspension of the firm."[13]

In 1855, at just forty-three years old, Davenport "inexplicably" sold his business and retired a wealthy man, living for a time at Fountain Hill, a country estate he built in nearby Watertown, and undertaking a series of world travels.[14] Davenport's part in the Kendall Square story was far from over, though—and his travels would prove an inspiration.

* * *

Looking back, the success of Edward Kimball's and Charles Davenport's railroad car business and its move to 700 Main Street in 1842 was the harbinger of Kendall Square's expansion into a powerful manufacturing center. Previously, the bulk of the action had been in Cambridgeport around Central Square and Lafayette Square, where the causeway ended and many of the farmers and others coming from the interior to trade in Boston found taverns and inns to stay in.[15] That's where Kimball ran the Pearl Street Hotel and operated his "hourly" stage coaches between Cambridge and Boston—along Main Street and across the new West Boston Bridge.

But Davenport's railway car factory at 700 Main Street was just the first of many manufacturing operations that sprang up in today's Kendall Square area. Most of the early development occurred on the northern, drier side of Main Street—"a polyglot assortment of storage buildings, workers houses, saloons and wheelwrights."[16] Up in the heart of Dock Square and the Lower Port, the triangle of land between Broadway and Main Street capped by Portland Street on one end and the Charles River on the other had been filled in completely shortly before Davenport's move to Main Street: this includes the area that today is home to the Marriott Hotel and the Cambridge Innovation Center. It wasn't just businesses that stood there, though. At the time, the triangle also contained residential homes: it even had its own elementary school that was erected in 1836, about where Google's big office tower is now at Three Cambridge Center. Meanwhile, Davenport's factory became the iron foundry Allen & Endicott when Davenport retired in 1855.

Two events then accelerated growth. The first was the Civil War, which boosted demand for the output of the area's factories. The second was the arrival of rail service to Kendall Square in 1868, connecting the city's small but growing industrial companies with lines serving the Boston area from the west and north.[17] This was the 8.5-mile long Grand Junction branch of the Boston & Albany Railroad. The tracks, which are still active, loop around from East Boston, near where Logan Airport is today, then cut down through the heart of Kendall Square, alongside Draper Laboratory and under the McGovern Institute for Brain Research, then run almost parallel to the MIT campus before crossing the Charles River in Cambridgeport.

With the railroad's arrival, the stage was well and truly set for the ecosystem's next era. More and more manufacturers came to the area, some starting new, but others moving operations from Boston to the cheaper land available in Kendall Square, as well as Cambridgeport and East Cambridge on its flanks. "Cambridge is offering all of this marsh land that can be filled," notes Sullivan.

Many of these businesses were smokestack operations that set up shop along Broadway and the Broad Canal, taking advantage of cheaper deliveries of coal from ships to fire their factories. Alive with activity, the Lower Port at last began to fulfill the original vision and ambitions of the builders of the West Boston Bridge and the canals to create a center for storage and shipment of bulk materials. Wharves able to handle products such as stone, coal, and lumber lined the river side of the causeway and also the Broad Canal and its sister waterways. "By the middle of the nineteenth century, there's beginning to be a lot of coastal traffic and lumber from Maine and the Maritimes [in eastern Canada]," recounts Sullivan. "Coal was being brought from Nova Scotia first, and then from Virginia, by schooner—and these kinds of activities are picking up on the canals, as was envisioned before the embargoes leading up to the War of 1812."[18]

The Charles River was tidal before the Charles River Dam and its lock system went operational in 1910, at the site of today's Museum of Science.[19] So all during this period, except for in the main channel, the river basically drained out at low tide: canal boats and coastal schooners would just rest on the muddy bottom, waiting for the water to rise again.

The companies that grew up around the canals reflected the demands of a surging America. As Cambridge boasted in a lengthy book produced to celebrate its fiftieth anniversary as a city in 1896, "The advantage of tide-water so near at hand, and the cheapest possible water freights for coal and raw materials and for the delivery of manufactured products in all parts of the world, add to the attractions offered in Cambridge to great manufacturing industries."[20]

In the first part of the 1800s, the small amount of manufacturing was largely centered around soap, leather, and cordage. The middle of the century saw the rise of new businesses, led by glass and brick factories, as well as the railroad car and carriage-making operation of Charles Davenport—and the city's population more than quadrupled from 2,453 residents in 1800 to 12,340. By 1890, the official population had swollen to 70,026. Dominating the industrial landscape were foundries, rubber factories, furniture makers, and clothing companies, joined by soap makers, brick and tile concerns, and confectioners. An important publishing industry had also sprouted up—including bookbinding and book and pamphlet printing. In its vanguard was the Riverside Press, founded in 1852 by longtime Cantabridgian and future city mayor Henry Houghton, who made it a division of his newly formed H. O. Houghton & Co. In 1872, Houghton took on a partner, George Mifflin, and renamed the business Houghton, Mifflin & Co. The Riverside Press, set on the Charles River in Cambridgeport, employed over six hundred people by 1886.[21]

Besides emerging as a manufacturing powerhouse, the city expanded its reputation for innovation and entrepreneurship—thanks to both some of its companies and various individuals who lived or worked there. A few, such as Houghton, even became household names or the makers of world-leading products. Virtually all operated in what would become Kendall Square or neighboring Cambridgeport, typically drawn by apprenticeship opportunities or cheap land and proximity to Boston.

The modern sewing machine, for example, was invented by a man named Elias Howe, who worked in a second-floor workshop along Main Street, roughly one block closer to Central Square than Charles Davenport, and at virtually the same time. A Massachusetts native, Howe had come to Cambridge for an apprenticeship as a teenager around 1837 and hit upon the idea for his sewing machine while serving in the shop of a master mechanic who manufactured and repaired precision instruments. The patent was issued in September 1846, when Howe was just twenty-seven. After winning patent battles against Isaac Singer and Walter Hunt, Howe ultimately earned a royalty payment on every sewing machine sold—making him a very wealthy man.[22]

Another important innovation was the woven hydraulic hose. "A pure rubber hose doesn't have any strength. But if you form the rubber around a woven tube, a tube woven from cotton or some other fabric, then it's like reinforced concrete, only flexible," explains Charles Sullivan.[23] The woven hose was the brainchild of a Dock Square–cum–Kendall Square company founded in 1880 by retired Civil War Colonel Theodore Dodge, called Boston Woven Hose and Rubber. It required inventing a sophisticated loom to weave the material that would be combined with rubber to make the reinforced hose. The original loom had some eighty thousand parts and was far too complicated for production—but Dodge hired a machinist to redesign and simplify the machine. They set up shop in two rooms rented—startup incubator style—from the Curtis Davis Soap Factory on Portland Street.

The loom, and the woven hose it made possible, was a breakthrough and probably did more to put Kendall Square on the innovation map than anything before it, even Charles Davenport's railcars. Basically, all modern fire hoses and garden hoses came from this invention, says Gavin Kleespies at the Massachusetts Historical Society. "Boston Woven Hose revolutionizes how firefighting is treated. Modern firefighting doesn't happen without Boston Woven Hose."[24]

The company soon outgrew its rented digs and in 1887 erected its own three-story brick building at the corner of Portland and Hampshire Streets, near the end of the Broad Canal. In 1893, it expanded into bicycle tires. Initially, these came in two parts—an inner tube and a casing. A couple of years later, Boston Woven Hose came out with the Vim, a single-tube tire with the first nonskid tread. These innovations helped sustain, and possibly enable, the great bicycle craze of the 1890s—the Gay Nineties. The company by 1896 employed 975 workers and its factory covered nearly 250,000 square feet of floor space.

At its height around 1930, Boston Woven Hose had about 1,200 employees working in nineteen buildings sprawled across some fifteen acres (the company also bought land across Portland Street that served as a parking lot; one corner had reportedly been used for years as a major Boston-area cockfighting pit). Production extended to "rubber hoses, rubber rings, rubber fabrics, rubber mats, brass fittings, gaskets, and various tubings. By 1928, the firm was responsible for nine percent of all rubber goods manufactured in the United States."[25] A whistle atop one building blew to announce shift changes, sound factory fire alarms, and as late as the 1950s signal a 9:30 p.m. curfew for residents under sixteen.[26]

Ultimately, the company could not keep up with more modern competitors. Losing money, it was acquired around 1956 by the America Biltrite Rubber Company, which is still operating today as American Biltrite, based in Wellesley Hills, Massachusetts. Two of the Cambridge plant's buildings were razed in 1969 and others renovated when the site was rebuilt as part of a major redevelopment of Kendall Square (see chapter 8).[27] Today the nine-building campus is a multiuse commercial complex known as One Kendall Square. It's home to a brew pub, several restaurants, a pool hall, a dentist's office, and a cinema—in addition to office and lab space. The eight-acre complex was purchased for $725 million in 2016 by Alexandria Real Estate Equities, which opened its biotech startup incubator, Alexandria LaunchLabs, in part of the main building.[28]

*   *   *

A cast of other interesting characters and companies could be found along the streets and in the offices and factories around Kendall Square in the second half of the 1800s.[29] Some would be famous in their day; a few individuals even made their fortunes. One other man was destined to join the ranks of prominent businesspeople and innovators such as Davenport, Howe, and Dodge, and in a way he went on to become much more of a household name than any of them, still famous worldwide. Like many of the other players of that age, he came from humble beginnings—growing up on a farm and never finishing high school before coming to the Boston area to apprentice as a machinist. He was entrepreneurial, cofounding a company in the heart of Dock Square.

Along the way, the man became a deacon in his church and a fierce advocate of temperance. He grew active in politics, serving as a Cambridge alderman and becoming an early member of the Prohibition Party—running unsuccessfully under its auspices for Congress and Massachusetts governor. Despite this zealousness—at least by modern standards—he was well-respected in his day. But in the end, he would be remembered for his name, not his beliefs or his company.

His name was Edward Kendall. He likely would have been shocked at all the bars—not to mention the liquor store—currently populating Kendall Square.[30]

5 KENDALL BECOMES KENDALL

The new subway line between Boston and Cambridge was only days away from making its public debut. A special group of some 150 dignitaries and other VIP guests had been invited for an evening preview ride. The group met in Central Square, and the three-car train eased out of the station at 6:35 p.m. for the run across the West Boston Bridge to Park Street. In Boston, they would attend a celebratory banquet and then ride back. Only one stop awaited between Central and Park in those days, on the Cambridge side of the river. As the train pulled into that station, the guard (each train apparently had a policeman on board) called out its name: "Kendall Square!"

As the *Cambridge Chronicle* reported, "When Kendall station was reached, three cheers were given in honor of the station's sponsor, Deacon Edward Kendall, who was in the party."[1]

The date was Wednesday, March 20, 1912. Over the course of the evening, the dignitaries inspected each of the four stations on the line: Harvard, Central, Kendall, Park.[2] The official opening took place three days later, on Saturday, March 23. Wishing to be part of that historic occasion, several hundred people got up well before dawn to wait at Harvard station to purchase tickets on the first train to leave for Boston at 5:34 a.m. "Was there a rush? No, it was a stampede," reported the *Cambridge Chronicle*. "Cheers rent the air as 306 people try to outdo one another in their efforts to reach the train first."[3] The three-stop run from Harvard Square to Park took just under eight minutes. During rush hour trains ran every two minutes. If anything approaching that speed and efficiency could be consistently achieved today, it would do a lot to address a main priority of the Kendall Square Association—overcoming the difficulty and unreliability of public transportation to the area.

Folks surely appreciated it at the time. "Rapid transit became an actual and accomplished fact," the *Chronicle* crowed. A new commuter line had been first promised back in 1887, only to be held up for a variety of reasons for a quarter century. But the day had arrived. "Cambridge and Boston are now inseparably connected," the paper continued its rave. "It's Harvard, Central, Kendall and Park Street, one, two, three, four and reverse, eight minutes to and as many fro. No more waiting for a car, no more excuse for being late at the office, and no more excuse for being late on the return from the office, at least so far as car service is concerned, for the new subway is the 'cure-all.'"[4]

How did the little slice of former marshland near the Cambridge end of the old West Boston Bridge become known as Kendall Square? The story begins with Kendall's birth on December 3, 1821, on a farm near Holden, Massachusetts, about forty-five miles due west of Boston, to a Puritan family traced back to the founding of Massachusetts Bay Colony. After dropping out of high school, he tried his hand, unsuccessfully, at his own lumber business, then moved to Boston in 1847 to apprentice as a machinist under his brother James at the West Boston Machine Co. Within about a year, he was promoted to foreman of the boiler-making department—and was still in this job when James sold his business a short time later and moved to California, lured by the gold rush. The purchaser of James's business was the iron maker Allen & Endicott, only a few years away from acquiring 700 Main Street in Cambridge when Charles Davenport retired.[5]

Edward Kendall worked as a foreman at Allen & Endicott for more than ten years before leaving to start his own boiler-making business, partnering with a man named John Davis, Jr. In 1861, the two set up shop in highly affordable Dock Square with literally only a roof over their heads; the shop initially had no walls. "This rather crude building served them well, however, and earnings soon enabled them to add walls," reads one account.[6] When Davis retired in 1865, Kendall brought in a new partner, George Roberts. The firm Kendall & Roberts, later known as the Charles River Iron Works, soon became the largest manufacturer of pressure boilers in Massachusetts.

Kendall & Davis established their business on the south side of Main Street, spanning the intersection of Broadway and Main, on what is now the MIT campus. It sat roughly where the Eastgate Apartments, a residence building for MIT families, stands today. (Eastgate is slated to be torn down and replaced by a new building under the Kendall Square Initiative.) At the time, though, only the river lay behind the business, and Dock Square still boasted a row of docks.

Edward Kendall made his mark not only as a successful businessman but as a civic and social leader. He married the daughter of a deacon and served as a deacon himself for over thirty years, mostly with the Pilgrim Church on Magazine Street in Cambridgeport. Kendall and his parish were passionate supporters of emancipation leading up to the Civil War. He was also a temperance advocate and joined the Prohibition Party, possibly at its 1869 inception.

Kendall soon also entered the political arena. In 1871, he was elected to the Board of Aldermen in Cambridge, serving one term. He later served as a state representative from 1875 to 1876. Roughly a decade after that, in 1886, he represented the Prohibition Party in an unsuccessful bid for Congress—and ran again under its auspices for Massachusetts governor in 1893. He lost that bid as well.

Sometime probably in the mid-1880s, he bought out Roberts and handed daily operations of the business over to his two sons, George and James (ironically, or not, the first names of his two former partners). In 1905, the sons sold the business to a firm that relocated it to South Framingham, Massachusetts.

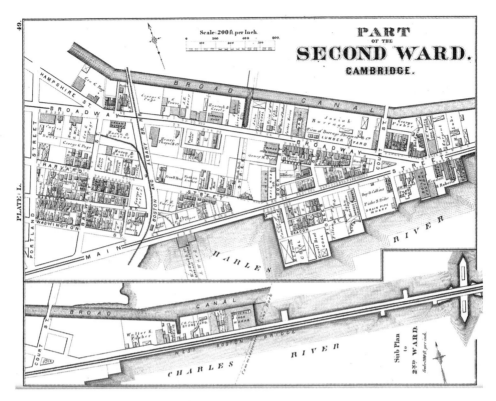

Figure 7 1873 map of Kendall Square/Dock Square. The Kendall & Roberts boiler-makers shop is on the far right, almost where the West Boston Bridge came in. Dock Street runs from Broadway to Main in the middle of the map, connecting about to where Legal Sea Foods is today. The dark patch just to its right, about where Google is today, is a schoolhouse and grounds. Court Street was renamed Third Street. *Source:* Cambridge Historical Commission.

With all of his hats—kingpin of nineteenth-century Cambridge industry, politician, and social leader—Kendall loomed as an outsized presence in the area on the eastern side of Cambridge. According to his biographical sketch, at the "Cambridge 50 Years a City" celebration held in 1896, "the Kendall [company] float was the chief attraction of the parade. Two drays bearing two huge boilers were drawn by six horses. This was indeed an unusual display and was so regarded by the spectators."[7]

So perhaps it was natural that people would gradually begin to refer to the area itself as Kendall Square. According to a short biography of Kendall published around 1920, the name goes "as far back as 1856."[8] That's probably a typo, since 1856 was five years before Kendall started his business in Cambridge. One possibility is 1865, if the last two numbers were reversed.

But whenever people started using the name Kendall Square in place of Dock Square, the new moniker was formalized in signage and people's minds on that March 1912 day when the subway line opened to the public. It's been Kendall Square ever since, even though the name of the subway station was changed to Cambridge Center/

Figure 8 1898 view of Kendall & Sons/Charles River Iron Works at low tide. The Charles was tidal until 1910.
Source: Cambridge Historical Commission.

MIT in 1982 before officials bowed to public opposition and reverted to Kendall/MIT in 1985.

The opening of the station in 1912 was a seminal event in the history of Kendall Square, bringing crucial, modern connectivity essential for growth. "The Kendall Square station will mean the development of the neighborhood," the *Cambridge Chronicle* predicted in 1907 in its coverage of the debate leading to the creation of the station.[9]

Kendall died at age ninety-three in January 1915, almost three years after the VIP crowd hailed him as the train pulled into the newly christened station of Kendall Square.[10] Not only was the nonagenarian on hand for the grand opening of the station bearing his name, he also lived to see the next major milestone in the square's development, one that would cause people around the world to link his name to invention and innovation long after he was gone. That event was the acquisition of land in Kendall Square for the new home of a growing school over in Boston: the Massachusetts Institute of Technology.

6   DAVENPORT'S FAILED DREAM OPENS THE DOOR FOR "NEW TECHNOLOGY"

On March 23, 1912, the very day the new subway connecting Boston and Cambridge opened to the public, another event took place that would change Kendall Square even more profoundly than its new, state-of-the-art transit system. As fate would have it, that was the day when a large swath of property adjacent to the square was formally conveyed to MIT, paving the way for the school's move across the river from the Back Bay.

The Massachusetts Institute of Technology today forms such an essential part of Kendall Square—intertwined with the very definition of what the square is—that it may seem surprising to learn that the institute's arrival was by no means certain. In fact, if Charles Davenport had realized his vision for the marshland on the banks of the Charles River, it's very unlikely that the now-great university would have come to Cambridge at all.

Although Davenport sold his railcar-making operation in 1855, he remained extremely interested in what was going on around Dock Square. His initial base of operations was his country estate in Watertown, which abutted Cambridge to the west—on the opposite side of the city from 700 Main Street and Dock Square. There, on the slopes of a hill facing Cambridge, just below where the Oakley Country Club is today, he had built a two-story Italianate mansion complete with cupola and porches, along with a separate summer house and barn. A fountain, apparently watched over by statues, adorned the grounds—giving rise to the estate's name of Fountain Hill.

High on the list of Davenport's passions was travel. He journeyed throughout the United States, took several trips to Europe, and made at least one sojourn to Cuba. It was during a visit to Havana in the 1850s that Davenport got the inspiration for developing the shoreline on both sides of the Charles River into a vast park and ornamental water body, including a stately residential district occupying much of the area where MIT now stands, to mirror that of Boston's Back Bay. In Cuba's capital, the former wheelwright "saw the small embankment on the bay there, where the people sat under the palms, enjoying the breezes."[1] That made him think of the Charles River and the salt marshes and mud flats that lined it on both sides—much of it on view out the back of 700 Main. He envisioned "a boulevard along each river bank, five miles in length and two hundred feet in width."[2]

Davenport already owned some of the marshland on the Cambridge side of the river. Upon returning to Boston, he began buying up more. He eventually accumulated three-fourths of the shoreline flats between the Cottage Farm and West Boston bridges (now the Boston University and Longfellow bridges, respectively), a roughly 2.5-mile stretch on the Cambridge side of the river. It was, essentially, the backyard of both his own former carriage works and Edward Kendall's boiler-making operation.

The entire section of the Charles leading up to and around what is now Kendall Square had long been unappealing. In the 1800s, the Charles River was dammed upriver for mills, and the bordering marshlands were filled for commercial and residential developments. At low tide, the lower Charles, including the area near Kendall Square, became a settling ground for sewage. By the mid-1800s, several plans had been advanced to fill the mudflats and marshes and make the Charles into a world-class public space and park system, but there had been little traction by the time Davenport got active—and he meant to change that.

Davenport formed the Charles River Embankment Company with some associates in 1880 to pursue his dream of creating Havana-like esplanades on both sides of the river. In Cambridge, his plans included a sea wall or embankment that would protect the wide public esplanade and a line of grand homes to be constructed just inland. All this was imagined for just upriver from Dock Square, almost exactly where MIT now sits.

It almost happened—and there would almost certainly have been no MIT in Cambridge if it had. In 1882, the cities of Cambridge and Boston agreed to build a new bridge across the widest part of the river basin—this became the Harvard Bridge, along what was later named Massachusetts Avenue. The Embankment Company negotiated a deal, giving up the land the city needed for an approach to the bridge and for a two-hundred-foot-wide esplanade in exchange for the postponement of any tax increases on the rest of its land during construction—and permission to develop it. This arrangement, says Charles Sullivan, gave the Embankment Company "the right to build a seawall, construct what's now Memorial Drive, and fill all the land where MIT is now."[3]

Construction of the seawall began in 1883, and the Harvard Bridge was completed in 1892. The Embankment Company hired architect Frederick Viaux to draw up plans for the upscale residential neighborhood he imagined behind the esplanade, extending all the way back to the railroad track embankment that Davenport himself had helped build in the 1850s.[4] Developers would have to abide by certain restrictions—including a twenty-foot setback from the esplanade, a prohibition against industrial or commercial structures, a requirement to use only brick, iron, or stone as building materials, and a minimum height of three stories and maximum height of eight stories.

It all seemed compelling, but things started going haywire. More than 80 percent of the railroad tracks crossing Cambridge had been laid on a high embankment to protect the rails from the marshland below. The embankment had few culverts, severely cutting waterflow from one side of the embankment to the other. That helped dry out the marshes north of the embankment and made them more suitable for development. But the river side of the rail line was still wet—and the marshlands reeked. By around the late 1880s, an aging Davenport had pretty much retired from business life. The Embankment Company he had helped form continued but found it hard to sell any

Figure 9 1875 drawings commissioned by Charles Davenport to promote his vision for the Cambridge Esplanade. In the top image, the Boston Common is on the lower left. Across the river, Davenport and colleagues owned much of the land upriver from the West Boston Bridge. The second bridge is the proposed Harvard Bridge along Massachusetts Avenue that would be built in 1892. He had laid out a tree-shaded esplanade (bottom image, left side), as well as residential streets, where MIT is today.

residential lots on that side. Nor did it help that the presence of the tracks themselves discouraged many buyers.

Then came the devastating panic of 1893, a depression that lasted until 1897 and forced the Embankment Company into bankruptcy. By then, a thousand feet of seawall had been constructed, and essentially all of the marshes and tide flats had been filled—creating solid land for the first time from Harvard Square all the way to East Cambridge. But little development had taken place. Only the Riverbank Court Hotel at 305 Memorial Drive (now an MIT residence called Fariborz Maseeh Hall), the city armory, and a smattering of other buildings were completed during this period.[5] The lots Davenport's group had envisioned for residential development sat unsold for the next twenty years, despite the expansive river and city views that would be so appealing and sought after today.

* * *

The failure of Davenport's plan opened the door for the arrival of MIT. But it took an unlikely series of additional events to bring the already famous school to Cambridge in 1916. Just a few years earlier, in fact, the odds were great that MIT itself would not long exist as an independent entity: the plan had been for it to merge with Harvard as the foundation of a new science and engineering campus across the Charles River in Brighton, about where Harvard Business School is today. Powerful figures, including steel baron Andrew Carnegie, had stakes in making the union a reality.

The Massachusetts Institute of Technology had been chartered as a land-grant school in 1861. Two days after its authorization, the South Carolina militia attacked the US Army at Fort Sumter, starting the Civil War. The war years were spent organizing and readying the school. The first class was welcomed in temporary quarters in 1865, and a more permanent home was established the following year in Boston's Back Bay. By the late 1800s, what many knew as Boston Tech had proven its worth as a top-flight engineering and applied science school, arguably the country's best. Its popularity strained the limits of its tight urban quarters. By the early 1900s, officials were scouting for a new home.

Harvard's president, Charles Eliot, wanted that home to be Harvard. One of MIT's inaugural chemistry professors, Eliot had been chosen as Harvard's president in 1869. He proposed to four successive presidents of Boston Tech that MIT and Harvard join forces. The fourth, Henry Pritchett, took the bait in the 1890s. MIT needed money and space to sustain its growth. Meanwhile, the Lawrence Scientific School, home to Harvard's engineering and applied science program, was overshadowed by the college's liberal arts reputation and struggled to attract enough students to justify its existence. So the union made sense on Harvard's end as well.

In January 1904, the *Boston Daily Advertiser* announced that Boston Tech and Harvard had agreed to merge. The surprise announcement set off a firestorm at MIT. While the agreement specified that the institute would retain its name, charter, organization, and mission, the reality was that MIT would lose its cherished independence and become Harvard's engineering school.⁶ That didn't sit well with faculty, staff, or many alums. As one account summed up the concerns: "Should the Institute now, after nearly forty years' struggle, give up its hard-won independence, sacrifice its fundamental principles, and yield a leadership won the hard way to come under the partial or complete domination of Harvard in the hope of monetary advantage?"⁷

The answer from MIT faculty—by a vote of fifty-six to seven—was an overwhelming "No." A survey sent to graduates of the school showed 2,035 opposed and only 834 in favor. Nevertheless, in June 1905, the school's board—the Technology Corporation—gave the merger the green light by a vote of twenty-three to fifteen.

The union thereby seemed ordained. In anticipation of the decision, a group of wealthy Harvard donors, including Andrew Carnegie and stockbroker Henry Higginson, had already pooled their resources and purchased the tract of riverfront property east of Soldiers Field in Brighton.

But there was one big catch. Under the terms of the agreement, MIT would be on the hook to "erect, furnish, and equip buildings having the capacity of at least its present buildings."⁸ The school intended to sell all or part of its existing land in the Back Bay to fulfill this obligation. But in September 1905, only a few months after the Technology Corporation approved the plan, the Massachusetts Supreme Judicial Court ruled that since MIT had purchased its land with federal land grant funding, it did not have the right to sell it. The union with Harvard fell through. And Pritchett, who resigned as MIT president in 1907, became "perhaps best known for his unsuccessful efforts to negotiate a merger between MIT and Harvard in 1904, defeated by a court decision in 1905."⁹

MIT then resumed its search for a new campus. The university had apparently already considered the Cambridge Esplanade site that Charles Davenport had earlier tried to develop and rejected it. But it was put back on the table under the new president who arrived in 1909, Scottish-born, New Zealand–raised mathematician and lawyer Richard Maclaurin. Maclaurin "saw clearly that his first and most urgent task would be the relocation of MIT and the raising of funds to build the 'New Technology,'"¹⁰ the name given to the envisioned new campus. (The existing one was known simply as Technology.)

Maclaurin's eyes were on the Cambridge Esplanade even before he officially started. On a visit to Boston in April 1909, a few months before he took office, the Scotsman dined at the Beacon Street home of Charles Stone, a founder of the then twenty-year-old Stone & Webster engineering firm (both Stone and cofounder Edwin Webster were MIT grads). They looked out the windows over the Charles to the Esplanade property. The incoming president was enamored. According to one MIT history, "This struck Maclaurin as ideal for size, accessibility, and dignity of setting. A great and noble edifice could here be erected that would be a worthy home for the Institute."¹¹ Stone told how it had been considered and rejected, explaining that Cambridge would likely oppose having another tax-exempt university inside its borders, that Harvard would probably also object, and that several would-be donors—he must have been thinking of

Higginson, Carnegie, and their partners—were unlikely to give generously to support a move to that site after the failed merger.[12]

Maclaurin was not dissuaded. A formal new hunt for an expansion site was instituted. The selection committee's report was submitted to Macluarin that October. The group had four main criteria to fulfill:

1. *Accessibility* for students, faculty, and the public
2. *Affordability*
3. *Space*, with the potential for constructing buildings "worthy of the institute's importance"
4. A *location* "independent of the influence of other institutions"[13]

The report noted the committee had considered at least twenty-four sites, including the possibility of building an island for the campus in the middle of the Charles, where the Harvard Bridge crossed it (this was soon deemed impractical). The one considered best overall was a long stretch in Boston's Fenway/Longwood area near the current Harvard Medical School campus and Simmons College. The Esplanade site, called the Riverbank in the study, was a distant second, with a laundry list of potential problems. The report pointed out that the land was relatively costly, with a lot of owners—thirty-five, as it turned out—with whom to negotiate. In addition, the "encroaching manufacturing district"[14] was deemed worrisome, as was the nearness to Harvard and that school's potential objection to the move. Finally came the worries about losing the school's tax-exempt status.

In addition, when Maclaurin tested the waters for a donation for any new site from Andrew Carnegie, "as Scotsman to Scotsman,"[15] Carnegie turned him down flatly that December and pushed again for the merger with Harvard:

> My dear Mr. Maclaurin:
>
> Ye're no blate.[16] Just think of it, I hav given $3,800,000. towards extending the Pittsburg school,[17] and certainly it has cost as much more to bring it where it is, and you ask me to help Boston, which has received $400,000 from me for the Franklin Institute! I enjoy the joke! Besides, I do not put the Pittsburg school behind even the Massachusetts Institute of Technology. It is a close race and we'll see who is winner by and by.
>
> Harty congratulations upon your success.
>
> Always very truly yours,
>
> Andrew Carnegie
>
> P.S. "If I mistake not, I am a part owner of that ground that my friend Lee Higginson and some of us purchased to unite the two institutions, *which should be done*."[18]

MacLaurin did get a pledge of $500,000 from T. Coleman du Pont, an MIT graduate and future US senator who was then president of his family's namesake chemical business. But it was for a different site, a golf course in Allston. So perhaps in an effort to stir the pot, Maclaurin casually remarked to a newspaper reporter that "Technology might have to pull up stakes and move to someplace where the cost of living is within its means."[19]

Several Massachusetts cities quickly expressed their interest. A group of MIT alumni from Springfield, Massachusetts, offered land in that city, for example. Cities in other states weighed in as well. The *Chicago Evening Post* boasted: "We could support a 'Boston Tech' with our loose change, and we wouldn't, like some cities we know of, have to search all the hinterland roundabout to find the money."[20]

The competition stirred officials in Cambridge to take action. As Maclaurin's biographer wrote, "Restive at being rated as the only city in the state which Technology would never, never consider," the city dropped any objection to MIT retaining its tax-exempt status. By some point early in 1911, a flurry of letters advocating for the institute's move to that city had arrived from prominent officials, residents, and business groups. The Cambridge City Council passed a formal resolution supporting the relocation, which was forwarded to Maclaurin by the mayor with his personal endorsement. "This incident of open diplomacy . . . brought forward the Cambridge land in a new way with the 'No Thoroughfare' sign removed," the biography concluded.[21] That March, Harvard followed suit, notifying Maclaurin it was now also okay with both institutions being in Cambridge.

All this helped vault the Riverbank/Esplanade property to the top of the list—and loosen purse strings. Du Pont amended his pledge to extend it to the Cambridge property, and the state legislature approved a bill authorizing a grant of $100,000 a year to MIT for ten years, so long as the school raised a similar amount itself. By the fall of 1911, negotiations had been completed with all thirty-five owners of the Riverbank area to buy forty-six acres of land bounded by Massachusetts Avenue on the west and Ames Street on the east, Vassar Street inland or north, and the Esplanade. The envisioned campus did not extend east past Ames Street toward Main Street and the Longfellow Bridge at that time—where the MIT Media Lab and the MIT Sloan School of Management, among other buildings, stand today. Similarly, the land southwest of Massachusetts Avenue, now home to Kresge Auditorium, many student dorms (including the converted Riverbank Court Hotel), and the athletic center, was not part of the initial purchase. The price tag for the forty-six acres was $775,000.

Another key piece of the puzzle came together in early 1912, after Maclaurin met with Eastman Kodak founder George Eastman (who was neither an MIT alum nor a Massachusetts resident) at the Hotel Belmont in New York. They reportedly had a warm and earnest conversation in which the MIT president detailed the plans for New Technology. As Maclaurin's wife Alice later related, "The ground was broken so completely in Mr. Eastman's mind that my husband was astonished. When Mr. Eastman was about to leave, he suddenly asked: 'What will it cost to put up the new buildings?' My husband answered that it would cost about two and a half million dollars. Mr. Eastman said, 'I'll send you a draft.'"[22] Eastman's one stipulation was that he remain anonymous—and so he was referred to only as Mr. Smith until his identity was revealed in 1920.[23]

Once Mr. Smith's gift was announced, hundreds of congratulatory letters poured in. One was from Henry Higginson, a stockbroker and Boston Symphony Orchestra founder, whose own land deal forged around expectations of a Harvard-MIT merger had fallen through. His note read in part, "The giver gains more than the receiver, and also stimulates others to do the same—and so we Harvard men are ten times grateful at your success."[24]

Although final conveyance of the property to MIT took place on March 23, 1912, the same day the subway line officially debuted, it would take another four years for the first wave of construction and various and sundry other elements to come together to enable MIT's move. But all the future duck boats were in a row—and a new era visible on the horizon.

* * *

In the end, it wasn't a duck boat that took the dignitaries across the Charles River for the dedication of the New Technology campus. It was a specially designed barge christened *Bucentaur*, after the Venetian state vessels of the same name.

The elaborate ceremony took place in the middle of MIT's commencement in June 1916. During the four years of construction, twenty thousand pilings were implanted in the mud that had been pumped from the bed of the Charles years earlier to support the new buildings. In the meantime, Edward Kendall had passed away at the age of ninety-three, hailed as the city's "grand old man" by the *Cambridge Chronicle*,[25] and World War I had broken out. But none of that was front of mind. A celebration like none before it had been planned.

MIT took over Boston for the occasion, as thousands of alumni and guests descended on the city "to share in the heart-stirring farewell to the Tech that they had known, and to participate in transporting the *lares* and *penates* of the Institute to the new temple of science that had risen on the north bank of the Charles."[26]

Official celebrations ran from Monday, June 12, through Wednesday, June 14. The headquarters was the Copley Plaza Hotel in Boston, close to the original campus. Some five hundred sailed from New York aboard the SS *Bunker Hill*: a correspondent for the school newspaper the *Tech* wired in stories about their progress via the ship's Marconi wireless. When the *Bunker Hill* pulled into Boston Harbor on Monday morning, yachts belonging to Charles Stone and T. Coleman du Pont, among others, escorted her to dock, amid cheers from alumni and undergrads. A twenty-one-gun salute was staged along Commonwealth Pier. Once disembarked, most of the new arrivals then marched to Copley Square, led by the school band.

Early that afternoon, the assemblage reconvened at New Technology to see the cornerstone laid for Walker Memorial, erected in memory of former president Francis A. Walker, who had died in office in 1897. After the dedication, the new campus was opened for inspection. Thousands began walking the halls, drinking tea, connecting with old classmates and professors, and perusing an extensive exhibit called *Fifty Years of Technology*, which presented the institute's contributions to applied science and technical education since its doors had opened in 1865.

Outside, a water fete was being staged on the Charles. Despite intermittent showers, thousands of people gathered on both sides of the river, as well as the two bridges flanking the campus—the Longfellow to the east and Harvard Bridge toward Cambridgeport. An armada of power boats, sailboats, canoes, and racing shells paraded on the Charles. An eighty-one-foot-long Goodyear balloon floated all afternoon over the dome—and a Wright brothers biplane twice took off and landed on the Esplanade itself. At the head of the official reviewing party was an assistant secretary of the Navy named Franklin D. Roosevelt. He was just thirty-four years old and seventeen years away from the presidency.

The next day, Tuesday, was graduation day for the 360 members of the class of 1916. For the gathering alumni, the day brought more festivities and a field trip by steamer to Nantasket, with each class and club parading along the beach. Then it was back to Boston for an extraordinary spectacle.

The pageant started at Technology's Rogers Building on Boylston Street. Undergrads clad in Venetian costumes and armed with halberds provided a guard as some eighty MIT faculty and corporation members adorned in long black gowns, MIT's charter carried among them, marched to the Union Boat Club, almost directly opposite MIT.

There the *Bucentaur* awaited. The barge was captained by alumnus and benefactor Henry Morss, who served on the MIT Corporation board. Morss was also an accomplished sailor: he had won the 1907 Bermuda Race. He piloted the *Bucentaur*, the cardinal and gray MIT flag flying above the stern, dressed as Christopher Columbus.

The barge crossed the Charles to the strains of Grieg's *Land Discovery*. Thousands of spectators had gathered along Harvard bridge and on the Esplanade to watch. An additional ten thousand alumni and guests filled long grandstands set up on both sides of the open area known as the Great Court. Small thrones in the middle of the stands on each side awaited the mayors of Boston and Cambridge, while a larger throne on top of the colonnade was reserved for Massachusetts governor Samuel McCall, who rode in on horseback accompanied by a phalanx of mounted lancers. "In a stand on the eastern side of the open end of the court five hundred chorus singers were in place and an orchestra of a hundred was tuning up. Primitive Men, Mediaeval Students, Nymphs, and Fire Dancers . . . were all in readiness," read one account of the spectacle.[27]

Then a cheer rang out as the *Bucentaur* approached. Searchlights on the pavilions illuminated the scene; rockets and bombs welcomed the vessel. The governor's lancers rode down to greet them and to cheers and music escorted the MIT charter and seal back up the Great Court to their new resting place at New Technology.

* * *

The festivities continued the day after the pageant, when the buildings of the neophyte campus were officially dedicated. Maclaurin closed his dedicatory address by speaking of the genius of the American people for making practical advances, proclaiming: "And so . . . we dedicate these buildings to the great cause of *science, linked with industry*."[28]

His words harkened to MIT's founding mission and foreshadowed the major role the school would play in the decades ahead in the evolution and development of Kendall Square. The West Boston Bridge and the turnpikes feeding into the square had created the connective framework—making the cheap land so near Boston and Harvard accessible by the early 1800s. The canals and railroads built in the decades that followed added to and fortified the connective tissue, as the area became attractive to manufacturers first and foremost, but also to entrepreneurs and innovators who sought to be near the manufacturers and in some cases could use their space or equipment. Now, in the first two decades of the 1900s, modern rapid transit had sped into Kendall Square, followed closely by what was arguably the country's number one school of engineering and applied science.

Some other important ingredients for creating the Kendall Square people would come from far and wide to visit and study at were still lacking—among them formal training for hopeful entrepreneurs and managers, larger pools of experienced executives, and professionally invested capital. All those, though, could be found in the Boston area to some degree—both Harvard and MIT had started programs in business management a few years earlier—so the essence of the future ecosystem was becoming embedded into its fabric.[29] But all that aside, an even more important ingredient remained lacking: the connection between MIT itself and the ecosystem around it.

New Technology's front door was first and foremost about the Charles. Maclaurin had promised his faculty and benefactors "a great white city" along the river.[30] The school eventually selected as its architect an accomplished alum, Wells Bosworth. He had proposed one grand edifice with interconnecting corridors, as opposed to the separate buildings, often with distinct architectural styles, found on so many campuses. Roughly centered among this structure was a giant dome reportedly inspired by the Thomas Jefferson–designed rotunda at the University of Virginia, itself modeled after the Pantheon in Rome. The interweaving thread in Bosworth's design was a center hallway—it became the famous Infinite Corridor—running roughly three hundred feet on either side of the dome. Guarded by colonnades, the dome overlooked the Great Court. A series of descending steps and terraces, never completely built, was supposed to lead to the river. This was to be MIT's face to the world.

The Great Court formed the formal main entrance to the campus but is hardly used today as it connects to almost nothing—and in front of it, Memorial Drive is open to cars. A second main entrance—the chief entry since it was completed in 1938—was through steps and columns along Massachusetts Avenue.[31]

It would be a long wait—more than a hundred years—for MIT's new front door, designed to connect it porously and warmly to Kendall Square (see chapter 1). Indeed, MIT was likely better connected to Boston and the rest of the world than to its own backyard in those early days. When contemplating the move to Cambridge, the school's selection committee had even looked askance at the neighborhood, fretting that the "encroaching manufacturing district" might cramp MIT's style.

In a way, it didn't matter—Kendall Square was booming without MIT.

7 "A CANOPY OF INDUSTRIAL HAZE"

By the 1920s, Kendall Square was in the midst of a remarkable transformation. The "many wooden sheds and storage buildings, frame houses, and large tracts of open land" that defined the area before the subway opened and MIT acquired its land, as described by historian Susan Maycock, had been largely replaced by a mighty manufacturing enterprise.[1] "What one sees today from the Esplanade on the Boston side or from the bridges approaching Cambridge is a forest of factories and warehouses under a canopy of industrial haze," wrote the *Boston Daily Globe* in January 1927.[2]

In the first volume of his *History of Massachusetts Industries*, published in 1930, Orra L. Stone describes how the city, which for centuries had been known worldwide as the educational beacon of America, the home of writers, poets, legislators, and thinkers such as Henry Wadsworth Longfellow and Oliver Wendell Holmes, had been remade: "Today it is its factories, rather than its educational institutions, that render the city famous," proclaimed Stone. "No other New England municipality has grown so fast industrially as has Cambridge in the past decade and a half."[3]

The facts Stone and the earlier *Globe* article laid out were astonishing: 300 percent gain in "manufactures" (this probably means the value of manufactured products) over ten years and, by 1930, more than 375 industrial plants employing almost twenty-five thousand people and turning out products valued at more than $175 million a year. The city, wrote Stone, "has experienced a metamorphosis that characterizes it as as much of an industrial boom town as Akron, Ohio, or Detroit, Michigan."[4]

The center of all this action was Kendall Square. "Within an area of two square miles of Kendall Square, where the greatest manufacturing development has taken place, are located more than 200 plants, whose invested capital exceeds $100,000,000," reported Stone. "Here the searcher of facts finds the homes of the largest manufacturer of soap in the world; the greatest producer of rubber clothing in the world; the largest manufacturer of mechanical rubber goods in the world; the largest plant in the world devoted exclusively to the printing of school and college textbooks; the greatest producer of writing inks, adhesives, carbon papers, and typewriter ribbons in the world; the largest plant in the world devoted exclusively to the manufacturer of confectionery; a branch plant of the largest manufacturer of optical goods and optical machinery in the world, the largest producer of road paving plants in the world; the oldest and largest school supply house in the United States; the only industrial research laboratory of its kind in the country."[5]

Stone's list went on to credit Kendall Square with being the New England, Massachusetts, or Boston-area leader in various additional fields, as well as home to branch plants or operations of nationally famous companies, including Standard Oil. "Between 1910 and 1920 fourscore buildings . . . were erected in that section," he wrote of the building boom that helped enable this growth.[6] All told, within Kendall Square alone Stone identified more than one hundred separate companies employing more than ten thousand people. That marked an almost 500 percent increase in the number of businesses from a decade earlier. Meanwhile, reflecting the surge in jobs, the city's population reached 113,643 in the 1930 census, a more than 60 percent growth since 1890—and actually more than the 105,162 counted in the 2010 census.[7]

Why the huge boom? One major enabler was Kendall Square's accessibility for companies and their workers. While the canals had ceased to be much of a factor (although the local electric utility received fuel shipments by barge along the Broad Canal as late as 1982), the railroads serving the area provided a huge benefit for manufacturers.[8] So did the "uncongested highways" that made Kendall Square and its environs a favorable location for trucking, in contrast to the clogged and narrow streets leading in and out of Boston's wholesale district. Moreover, the subway system that had opened in 1912 in both Kendall Square and nearby Lechmere Square in East Cambridge made it easy and affordable to reach for workers living outside the city. Wrote Stone: "These advantages . . . account in a large measure for the phenomenal industrial development of the last fifteen or twenty years."[9]

Because it was so new to the scene, "MIT was probably a pretty minor factor" in Kendall Square's industrial development in the early part of the century, notes historian Charles Sullivan.[10] Rather, much of the development was driven by companies that had moved to Cambridge or started there in the previous century and continued growing. Boston Woven Hose, for example, was just hitting its peak around this time; it accounted for nine percent of US-produced rubber goods in 1928 and reached its highest level of employment locally around 1930.[11]

Soap was also cleaning up, so to speak. Cambridge had been New England's leading manufacturer of soap for 125 years or more. But the business got a major boost when British-based Lever Brothers, maker of iconic soap brands such as Lux and Lifebuoy, bought out the local Curtis Davis Company in 1898 and established Cambridge as its US headquarters. By 1930, its local workforce had grown from eighty in 1898 to more than a thousand. Its main plant, which included a state-of-the-art laboratory, was a sprawling affair along Broadway crowned by two giant smokestacks—where Draper Laboratory is today.

Lever Brothers would retain a big presence in Kendall Square for another nearly thirty years, finally shutting down its massive factory in 1959 after its technology had become obsolete.[12] The impressive art deco administration building it erected in 1938 at 50 Memorial Drive eventually became home to MIT's economics department and various operations of MIT Sloan, as well as the MIT faculty club on the top floor.[13]

Figure 10 1930 aerial view of the Lever Brothers main Kendall Square factory. Draper Laboratory occupies most of the site today. The road in front of the smokestacks is Harvard Street, which doesn't exist today east of Portland Street. The Grand Junction railway went right up to the factory on the east. The plant's main entrance on Broadway is hidden from view, but Hampshire Street is visible angling up to the left. Just beyond it is the Boston Woven Hose factory at what is now One Kendall Square.
Source: Cambridge Historical Commission.

Another important and innovative industry sweetening the city's tax rolls in this period was candy-making. In 1910, officials counted sixteen confectionery makers in Cambridge. A decade later, their ranks had almost doubled to thirty, and by 1930, the number surpassed forty. The peak came in 1946, when the local directory listed sixty-six candy manufacturers. Iconic products like Necco Wafers, Junior Mints, Sugar Daddies, Charleston Chews, and the Squirrel Nut Zipper, a taffy from the Squirrel Brand Company, also famous for its nuts, were all made at various times in the city. The Fig Newton also traced its roots to Cambridge. The fig paste cookies were first baked in 1891 by the Kennedy Biscuit Company on Franklin Street, only a few blocks from Charles Davenport's original workshop on Massachusetts Avenue.[14]

Most of the candy-making activity took place around the periphery of Kendall Square, in Cambridgeport—but it extended into Kendall Square as well. For a time, Main Street was known as Confectioner's Row, which included the Fox Cross Company, maker of the popular Charleston Chew, at 292 Main Street in the nucleus of the square. This building later housed the MIT Press Bookstore, along with various MIT offices, until renovations under the Kendall Square Initiative forced everyone out. And arguably the most famous candy-making operation, Necco, had a long-lasting impact on

Kendall Square. Creator of the famous Necco Wafers, the company moved to Cambridge in 1927 and opened a state-of-the-art factory at 250 Massachusetts Avenue, in what is generally considered part of the Central Square business district. At the time, it was the world's largest factory devoted entirely to candy-making, and it remained Necco's home until 2003.[15]

Necco spared no expense in erecting the structure, which turned out to be a benefit for the new tenant in 2004, and later the building's owner: Novartis. The drug powerhouse refitted the factory to hold its global research headquarters, called the Novartis Institutes for BioMedical Research (NIBR, pronounced Nibber). The structure was made of concrete, rather than steel frame, making it resistant to vibrations. The floors, between nine and fourteen inches thick, were strong enough to support the trappings of modern drug research: storage tanks, screening apparatuses, and robotic tools. From its base in the old Necco plant, Novartis soon expanded to much of the surrounding area on both sides of Massachusetts Avenue, all the way up to into Kendall Square itself—in the process becoming one of Cambridge's largest employers (more on this key move in chapter 16).[16]

Of all the old Cambridge candy-makers, only the former James O. Welch Company at 810 Main Street was still churning out candies in 2020. The operation is currently owned by Tootsie Roll Industries, but the plant still rolls out the entire global supply of Junior Mints, estimated at more than fifteen million per day. Tootsie Roll also purchased the Charleston Chew recipe and produces those from the site as well, along with Sugar Babies.[17]

In terms of the value of its products, the printing and publishing industry ranked number one at the time of Orra Stone's survey. Cambridge had a rich and storied history in publishing. America's first printing press had been established in the city by Stephen Daye in 1639. The successor of that business, the University Press, was still in existence—employing 225 people, Stone reported. But that was just part of the impressive publishing presence in the city.

Houghton Mifflin and its Riverside Press had continued its reign as the premier publisher in Cambridge. Henry Houghton, who had served as the city's mayor in 1872–1873, passed away in 1895, leaving the business in the hands of partner George Mifflin. The company increased the number of its steel presses from thirty-three in 1889 to sixty in 1905 and added two new buildings to expand its capacity, fueling its growth. It had encountered some troubles when Mifflin passed away in 1921, coupled with a worker's strike that same year. But the business would expand again during World War II and would remain in Cambridge until 1971, when stiff competition and advanced technology finally rendered the Riverside Press unviable.[18]

What became the global brand of Little, Brown and Company began as a Boston bookstore in 1794.[19] By the early 1900s, the company had expanded into a major publisher and was still headquartered in Boston, in an old home of the Cabot family on Beacon Street. But its warehouse and bindery were located next to each other in Cambridgeport, near the Riverside Press. From there, the company published some eight thousand books a day, including a string of Pulitzer Prize–winning novels and plays in the 1920s.

Inside Kendall Square proper, Stone offered a brief account of the Murray Printing Company, which put up a four-story building on Wadsworth Street, very near the subway station. In the 1930s, it produced a wide range of publications—house organs for companies, trade journals, textbooks, and books for other publishers among them—and employed more than one hundred people.[20]

The biggest publishing name in the square, and one of its crown architectural and historic jewels, was the Athenaeum Press owned by schoolbook publishers Ginn & Co. It stood at 215 First Street—next door to Carter's Ink at 245 First Street, for a time the world's largest ink manufacturer and on the other side of the Broad Canal from Murray Printing. The five-story building was completed in 1895, taking up an entire block. The front was meant to evoke an ancient Greek temple, and the top was crowned by a statue of Athena. Ginn's executives undertook an extensive hunt for the location of a new publishing plant. They decided Kendall Square could not be beat. It sat just off the Charles, with an unimpeded view of its muddy waters, and within one mile of the Massachusetts State House. "Here they obtained land at a reasonable price with abundant light, so difficult to secure in the crowded city and so essential to the best quality of work," wrote Arthur Gilman in his history of Cambridge. "Near by are all the great freight stations, affording the best advantages for shipping in all directions. Lines of electric cars bring their employees from any part of Boston or suburbs almost to the door."[21]

The building, now in the National Register of Historic Places, currently houses mostly biotech companies. It is also home to what has long been the only full-service athletic club in Kendall Square, the Cambridge Athletic Club.[22] The Athenaeum Building has had numerous occupants over the years. The manager for the rock band Aerosmith reportedly had offices there in the 1980s. One tenant in the 90s was an MIT spinoff called SensAble Technologies, built around haptics technology that brought the sense of touch to computing applications. SensAble's CEO was Bill Aulet, who later became managing director of the Martin Trust Center for MIT Entrepreneurship. The founder and inventor of most of its pioneering technologies was a young MIT master degree graduate named Thomas Massie. In 2012, Massie, a Kentucky native, was elected as a Republican US Congressman representing the Bluegrass State's Fourth District. Massie's office in Congress is room 314, hearkening to the number for pi.[23]

* * *

In the middle of World War I, as Kendall Square was just entering the growth period described by Orra Stone and MIT was moving to Cambridge, a group of local industrialists founded Manufacturer's National Bank. They established the bank in a two-story building at what is now the corner of Main and Wadsworth Streets. After the armistice, amidst the boom, they expanded the building significantly. About a block wide and five stories tall, it became known as the Kendall Square Building, with the official address of 238 Main Street. At the top, the bank erected a 135-foot clock tower. "It was intended to advertise Kendall Square to Boston," notes historian Charles Sullivan. "It was by far the tallest thing on the horizon anywhere in that part of Cambridge—with a clock you could read from Beacon Hill."[24] The clock tower, which was preserved as part of the Kendall Square Initiative, became an icon for Kendall Square as a place of industrial strength, growth, and innovation.

In the second half of the 1800s and into the twentieth century, that growth was driven mainly by the traditional manufacturing businesses that dominated Kendall Square and East Cambridge. But sweeping change was afoot. "Industries are always maturing and moving to other parts of the country, and so were the smokestack industries," Sullivan explains. "Those were on their way out in the 1920s—and they were being replaced by cleaner industries like printing and binding, furniture-making, warehousing." Such companies had been in the city for a long time—but they were growing. They could take advantage of Kendall Square's location as a distribution center but required far less space than factories as the rising price of real estate made operating big plants in town precipitously more expensive. Many also needed more professional-type workers, not just laborers. "The subway is bringing in skilled workers, as opposed to smokestack industries, foundries, and the like," says Sullivan.[25]

Even more important to the future evolution of Kendall Square, though, was the arrival of companies formed around science and technology, especially the burgeoning fields of radio and electronics. Unlike with the manufacturing boon, these were businesses MIT was much more poised to influence or even to have a direct hand in.

Among the first was a company founded in Boston by Melville Eastham and his partner J. Emory Clapp to make X-ray equipment. Clapp-Eastham moved to 139 Main Street, now a historical landmark, sometime around 1910 (in more recent times, this was the locally famous American Red Cross building) and quickly expanded into making spark coils, variable capacitors, and crystal detectors to supply the radio market then exploding on the scene. One of its early employees, Knut Johnson, told the story of taking the subway to Kendall Square in 1913 in response to the company's ad in the *Boston Post* for a machinist's position. He almost gave up upon seeing a line of several hundred people waiting to apply, but after slogging it out through the interview received an offer letter with a starting wage of thirty-five or forty cents an hour (he couldn't remember which).

In 1915, Eastham raised $9,000 to branch out and form a new company, General Radio, with Johnson as its only other employee. He found space on the third floor of a flatiron building just opposite the Necco plant at the corner of Massachusetts Avenue and Windsor Street. The company would eventually take over the entire building and expand to adjacent space; the complex later became home to the MIT Museum. It employed 440 workers by 1944 and had sales that year of $5.3 million. MIT took equity in the company, and Eastham himself became an influential figure at the institute and nationally before his death in 1964. His legendary company moved to the suburbs and later became known as GenRad. After sales peaked at $344 million in 2000, the bottom fell out and it was acquired the next year by Teradyne.[26]

Two years after General Radio started, in 1917, the trailblazing industrial research laboratory and management consulting company Arthur D. Little (ADL) moved from its Boston base to a newly erected headquarters building at 30 Charles River Road (the roadway's name was changed to Memorial Drive in 1923). The "magnificent three story and basement building,"[27] overlooking the Charles River and adjoining MIT's recently opened campus, was later added to the National Register of Historic Places and is now home to part of the MIT Sloan School of Management. ADL would stay there until 1953, when it moved into a forty-acre industrial park in northwest Cambridge, near the future Alewife T station.[28]

And in 1926, another highly innovative, cutting-edge concern moved its headquarters from Boston to 38 Memorial Drive, next door to Arthur D. Little. This was Warren Brothers, a company founded in 1900 by seven brothers—one of whom, Frederick, had invented bitulithic pavement, a mixture of bitumen and angular stone chips known as aggregate that significantly improved the durability and stability of road pavement. The company followed up on this by designing and building the equipment to lay the pavement. With the automobile just taking off, the timing couldn't have been better. That same year, Warren Brothers established its factory on Potter Street in Kendall Square. The enterprise was a roaring success. By 1930, after continuing to innovate in its pavement, the company employed thousands of workers and had installed more than 135 million square yards of pavement worldwide. One of the older Warren brothers—Stone does not say who, but possibly Frederick—later became a professor of organic chemistry at MIT. After some ups and downs, the company was acquired in 1966 by Ashland Oil & Refining Company.[29]

"The Warren Brothers Company is . . . one of the most conspicuous examples of what Massachusetts brains, capital and inventive genius can produce," wrote Orra Stone. "And so it may be asserted figuratively, and in thousands of instances literally, that all roads lead to Kendall Square because they are directly connected with paving plants made there."[30]

While the impact of MIT on Kendall Square was small in the years before World War II, there was an early harbinger of its future major role with the 1922 launch of the first breakaway startup success with strong ties to the institute since its move to Cambridge. A thirty-two-year-old professor named Vannevar Bush teamed with Laurence Marshall, his undergraduate roommate from Tufts University, and physicist Charles Smith to form the American Appliance Company. The startup was built initially around refrigeration technology, but quickly shifted course—*pivoted*, in modern terminology—to focus on electronics. Its first blockbuster product was a helium rectifier that eliminated the need for expensive batteries by enabling radios to be powered by the alternating currents found in most homes and offices. They called the tube at the heart of their invention *raytheon*, which means *god of life* in Greek. In 1925, reflecting the importance of the tube, the corporate name was changed to the Raytheon Manufacturing Company; eventually it became just Raytheon. The growing startup, originally based on Wadsworth Street, moved out of Kendall Square in 1928 for the suburbs, evolving into a multi-billion-dollar concern and one of the world's largest defense contractors.[31]

Another startup of a different sort, with a much more lasting presence in Kendall Square, got going just after the time of Orra Stone's survey. In 1932, an MIT research associate teaching in the Aeronautical Engineering program while pursuing his doctorate talked the school into founding a lab devoted to aircraft instrumentation. Charles Stark Draper had trained as a pilot himself, reportedly performing daredevil acrobatics to test aircraft capabilities. The MIT Instrument Lab he founded set up shop at 45 Osborn Street, bordering 700 Main Street, where his future friend Edwin Land would create his private lab a decade later. The instrument lab eventually evolved into the Draper Laboratory, which separated from MIT in 1973 as an independent, nonprofit research and development organization. It remains a major Kendall Square presence.[32]

*   *   *

Despite its industrial might and success during the booming 1920s, however, Cambridge did not escape the dark days of the Great Depression—and Kendall Square was irrevocably changed. "In the 1930s, everything went to hell, and then just like every other New England city, the industrial base started to evaporate," says Sullivan.[33] The Depression accelerated changes already underway, driving some companies—many of them older and less efficient—out of business entirely or forcing them to find more affordable locations.

What happened next, just as national productivity and economic performance finally began returning to pre-Depression levels, would have an even bigger and far more enduring impact, not just on Kendall Square, but on the world. The seismic event, of course, was World War II. Kendall Square would be thrust into the national struggle. In the process, the square would be radically transformed from the industrial powerhouse of the 1920s to a virtually unrivaled center of research and innovation.

8 RAD LAB: KENDALL SQUARE'S TIPPING POINT

Early in 1942, not long after the Japanese attack on Pearl Harbor the previous December, Jerry Wiesner was working in the propaganda arm of the Library of Congress, building radio transmitters for what was later called Voice of America. He got a message from an old professor from his University of Michigan days, urging Wiesner to leave Washington to join him at MIT to conduct war research.

"What sort of research?" Wiesner wrote back.

"We are doing important war research," came the terse response from Dutch-born theoretical physicist George Uhlenbeck, his former teacher.[1]

That was all Jerome B. Wiesner needed to hear. At the age of twenty-six, he packed his bags for Cambridge. Other than a few short stints away, including one at Los Alamos and another as science advisor to John F. Kennedy, he never left. He ultimately became president of MIT and later joined with Nicholas Negroponte to cofound the MIT Media Lab, whose I. M. Pei–designed building would be named after him.

The siren call that brought Wiesner to MIT had a hard-to-quantify but nevertheless transformative effect on the future character and culture of Kendall Square, dramatically accelerating the transition from manufacturing to scientific research and technology-based companies. Wiesner was really in the second, or third, wave of gold-plated recruits to what was known as the Radiation Laboratory, dubbed Rad Lab for short.

This unprecedented enterprise had started with just a few scientists in November 1940, more than a year before Pearl Harbor, but with Britain being bombed relentlessly and the writing on the wall for America's entry. By the war's end in 1945, it had swollen to 3,897 staff members, of whom 30 percent, nearly 1,200, were scientists or engineers. That probably made it bigger than any other wartime effort—the Manhattan Project included. Its effects would turn out to be much more pervasive.

Starting from a two-room suite in the main MIT building, the lab spread across fifteen acres, coming to dominate the entire northwest corner of MIT—the space where the Stata Center and the Koch Center for Integrated Cancer Research stand today and the very spot hailed some seventy years later as the world's most innovative intersection. But it might have been even more innovative then—and a lot smarter. At least ten future Nobel Laureates, nine university presidents (two of MIT), four of the first ten presidential science advisors, and four of the initial six Air Force chief scientists were lab alumni—along with many other key figures in industry, government, and academe who together helped shape the postwar technological age. More to the point, the lab's impact on the outcome of the war was vast—arguably far greater than the atomic bomb's.

The Radiation Laboratory was chartered to develop radar systems around a top-secret British invention, the cavity magnetron, which had been shared with the United States in September 1940 as the Battle of Britain raged. It resembled a hockey puck but was capable of churning out high-frequency transmissions on a wavelength of ten centimeters that were variously estimated to be one thousand to three thousand times more powerful than any microwave transmitter then known. This opened up the possibility of breakthroughs across a range of wartime needs.

First on the list was fighting back against Germany's night air raids. The Chain Home network of radar stations the British relied on to warn of German attacks operated on much longer wavelengths of ten to thirteen meters. The stations worked well in daylight, when a pilot's eyes could correct for the several-mile error range inherent in such long wavelengths. But they were ineffective against night attacks. The magnetron's microwave emissions could theoretically pinpoint aircraft positions far more precisely—and the device was small enough to be fitted onto planes, promising to enable pilots to find their quarries even on the cloudiest or darkest nights.

And that was only the beginning. The magnetron held the promise of precisely guiding bombers through cloud cover or enabling patrol planes to spot U-boats in the ocean's vastness. Yet the device was still raw and needed a lot of refinement to be ready for mass production—and no systems had been built around it. The already war-strapped British industry simply did not possess the capability for such endeavors. So Winston Churchill had made the bold decision to share it, among other secrets, with America in hopes of convincing the still-uncommitted US to pick up where British resources left off.

Churchill's gamble worked. On the American side, one of the key figures behind the Rad Lab was none other than Vannevar Bush, the former MIT professor and Raytheon cofounder. After becoming MIT's first dean of engineering in 1932, Bush had left in late 1938 to head the Carnegie Institution of Washington, a respected private research entity in the nation's capital. From there, Bush had shrewdly climbed to the highest echelons of power and influence. "Of the men whose death in the summer of 1940 would have been the greatest calamity for America, the President is first, and Dr. Bush would be second or third," asserted Alfred Loomis, a wealthy investment banker and virtuoso amateur scientist destined to play a central role in the Rad Lab story.[2]

As a young man during World War I, Bush had worked on submarine detection and been startled over the lack of cooperation between the military and civilian scientists. Sensing that the new European conflict would soon involve America, he had conceived of a national committee designed to bridge that gap. He arrived at the White House on June 12, 1940, to meet with Franklin Roosevelt, armed only with a single sheet of paper that laid out his idea in four tight paragraphs. It took less than ten minutes to convince the president. "That's okay. Put 'OK FDR' on it," Roosevelt told Bush. And the National Defense Research Committee (NDRC) was born.[3]

Bush and his NDRC cohorts, including MIT president Karl Compton, had quickly compiled a list of technical projects the upstart civilian group should take over if war came—either because the military had not yet started on them or because the service labs would have to drop them once war began to focus on more near-term priorities. Microwave radar was high on its list—and a special Microwave Committee had been established under Loomis, reporting to Compton. So when the British arrived with their cavity magnetron microwave transmitter, it fell to Bush and his group—and they grasped its potential immediately.

From this point, events rushed forward at a pace unimaginable today. It was decided that a central laboratory should be created under this Microwave Committee to develop radar systems around the magnetron—a lab run by civilians and not subject to military oversight. Loomis wanted the lab at the Carnegie Institution, near the halls of power. Bush didn't agree. "I protested, and we had a hell of an argument that took half the night and a bottle of scotch," he recounted.[4] In the end, Bush prevailed and the job went to MIT. The school already had microwave research ongoing, and it was felt that establishing a new lab there might attract the least attention.

Then recruitment had begun. Headquarters was set up in 4–133 using MIT's designation for the building and room number (it was actually a two-room office that was quickly divided into a suite of connecting rooms that included two labs). Early on, the group tapped Melville Eastham, head of General Radio, as the lab's first business manager. For the technical work, the prime focus was on attracting nuclear physicists already familiar with high-frequency research. To lead the recruiting, Bush roped in E. O. Lawrence, the UC Berkeley physicist who had won the Nobel Prize the year before for inventing the cyclotron. One of his first calls went to a former student, Lee A. DuBridge, then chairman of Rochester University's physics department. DuBridge accepted the director's job, then teamed with Lawrence to beat the nation's bushes for more recruits. I. I. Rabi from Columbia was among the first big names to sign on, leaping at the opportunity to help stop the Nazis. He brought two of his best students—Jerrold Zacharias and Norman Ramsey. Lawrence enticed a pair of his protegees, Luis Alvarez and Ed McMillan. From Harvard came J. Curry Street and a young instructor named Edward Purcell.

This first wave brought in six future Nobelists—and the list of the best and brightest kept growing. By war's end, the staff comprised roughly one-fifth of the country's leading physicists.[5] Alvarez once explained their reaction to the magnetron. "A sudden improvement by a factor of 3000 may not surprise physicists, but it is almost unheard of in engineering. If automobiles had been similarly improved, modern cars would cost about a dollar and go 1000 miles on a gallon of gas. We were correspondingly awed by the cavity magnetron. Suddenly it was clear that microwave radar was there for the asking."[6]

Alvarez was right. All told, the Rad Lab designed nearly half of all radars used in the war. The lab's airborne interception radar remained the Royal Air Force standard until 1957. Its long-range navigation system became the worldwide grid known as Loran; years later, GPS would supplant it. An automatic fire-control radar, the SCR-584, proved crucial to defeating the buzz bomb attacks on Britain in the summer of 1944.

One of its greatest successes involved bombing through overcast skies. Both Europe and Japan were often covered by clouds and bad weather, making visual bombing impossible. The Rad Lab created a bombing radar system called H2X that operated on a wavelength of just three centimeters, more accurate than ten-centimeter systems. It still wasn't very precise: getting within a mile of the target proved extremely rare. But it enabled Allied forces to keep up the pressure. The H2X debuted in November 1943. A communique from the Eighth Army Air Force read: "The availability of this equipment has changed the status of the 8th AAF planes from that of being grounded for all but one or two days per month during the winter to that of being able to bomb at will the most distant enemy objectives through heavy overcast."[7] By the end of 1944, the force depended on H2X for two-thirds of its raids.

Perhaps the greatest triumph came against the U-boat, with air to surface vessel systems that enabled planes to track down the German submarines surfacing to recharge their batteries—even at night. Rad Lab radars laid the foundation for much of the sudden turnabout in spring 1943—when the U-boat went from destroying ninety-five ships in the first twenty days of March against a loss of just twelve of their own to thirty-eight U-boats sunk that May—at a cost of only fifty Allied ships worldwide. That fall, the Germans fared even worse. U-boats claimed only nine of the almost 2,500 merchant ships that sailed in convoy in September and October. Meanwhile, twenty-five U-boats were sunk—19.5 credited to air attacks, with one being shared by an escort vessel. As Captain S. W. Roskill, official British naval historian, wrote: "The centimetric radar set stands out above all other achievements because it enabled us to attack at night and in poor visibility."[8]

A strong case can be made that radar, and especially microwave radar, was by far the most important technology of the war. But when the two atomic bombs fell on Japan in early August 1945 and the war ended soon thereafter, the lab's story was pushed from the limelight. *TIME* magazine, for one, had been planning a cover story on radar and the lab. But the article was cut to three pages and buried deep in the magazine, replaced by a piece on the A-bomb. The Rad Lab was described only as "a great anonymous army of scientists." Not a single name from it was mentioned.

But for the Rad Lab veterans, the mantra became: "The atomic bomb only ended the war; radar won it."

* * *

What did all this have to do with Kendall Square? In the roughly thirty years since MIT had moved to Cambridge, it had had little to do with the square's industrial growth. In fact, with the exception of a few innovations like woven fire hose and bitulithic pavement, virtually all of that once-explosive growth had come from factories churning out common products with little or nothing "high-tech" about them.

That still held true after World War II came to an end. There was almost no well-known company, save Raytheon, that could be confidently stamped: "Launched from MIT." Nor were there many, if any, who might boast of their technology: "Invented at MIT." Even more to the point, when comparing the scene to present day Kendall Square and all the multinational drug and computing companies there to be close to the university, very few—Arthur D. Little might have been the main exception—could say: "Here because of MIT."

But the war laid the foundation for the next phase of Kendall Square's evolution. And more than anything, this metamorphosis traced back to the Rad Lab.

Broadly speaking, the five years of feverish wartime radar work had advanced the knowledge of microwaves and related areas such as crystal theory, radio signal propagation, circuitry, transmitters, receivers, and antennas by an estimated twenty-five years of "normal" development. Nearly a year before the war ended, the Rad Lab commenced the gargantuan task of capturing and disseminating the fruits of all this learning for posterity. Louis Ridenour, a sharp-tongued physicist who had served on the lab's steering committee, spearheaded the effort. It took some 250 Rad Lab staffers, plus an army of stenographers and proofreaders, and bore a final price tag of $495,024.07. But his group produced a twenty-eight-volume (counting an index volume) compendium known as the Radiation Laboratory Series. Published in 1947, it formed "the occupational bible for at least a generation of physicists and engineers studying microwave electronics."[9]

Box 2
How the Rad Lab Transformed MIT—and Shaped the Future of Kendall Square

> Of all the events of World War II, the establishment of the Radiation Laboratory on campus had by far the biggest impact on postwar MIT: its fields of focus and expertise, its ties to government and industry, and the interdisciplinary way it organized itself. While little of this spilled over right away into the Kendall Square neighborhood, it had a major influence on the overall technological scene and laid some key foundational elements that decades later transformed the Kendall Square ecosystem in ways that are still visible.
>
> The Rad Lab's accomplishments and effects include the following:
>
> - Brought many of America's best and brightest physicists and engineers to MIT. Many never left the school or the area.
> - Helped create a breed of scientist and engineer who understood scientific principles *and* knew how to make real-world products based on that science.
> - Set a standard for collaborative, interdisciplinary teamwork. This led directly to the formation of MIT's vaunted Research Laboratory of Electronics and influenced the shape of many other MIT labs and centers.
> - Created deep ties to the military and defense industry; MIT soon became the nation's leading nonindustrial defense contractor.
> - The five years of wartime work advanced the state of microwave electronics by an estimated twenty-five years of normal development. This knowledge played a big role in advancing digital computing and electronics, around which much future Kendall Square development took place.
> - Cultivated a series of dynamic leaders in industry, academe, and government who cemented MIT's standing and connections.

For MIT, the Rad Lab proved transformative as well. A Basic Research Division was formed shortly after the Rad Lab closed its doors to continue its work in fundamental areas such as exploring the electromagnetic properties of matter at various microwave frequencies. By July 1946, this had morphed into the Research Laboratory of Electronics (RLE) and soon included the school's nuclear science program. RLE was directly modeled after the Rad Lab and even housed in building 20, one of the lab's main buildings near the corner of Vassar and Main. Its first director was Rad Lab veteran Julius Stratton, who had worked on Loran during the war and also served in the Pentagon. The endeavor got started with a host of surplus Rad Lab equipment and $600,000 in annual funding from the Joint Services Electronics Program. Twenty-six Rad Lab staffers signed on at RLE, many taking advantage of a special rule that allowed them to work as research associates while attending graduate school.[10] "RLE's establishment catapulted MIT overnight to the forefront of electronics and microwave physics, soon giving rise to a host of important innovations in radar, secure communications, electronic aids to computation, and atomic clocks," I wrote in *The Invention That Changed the World*, which chronicled Rad Lab's story. "By the early 1960s alone, it had awarded some 300 PhDs and 600 master degrees, and helped secure for decades the institute's place as the nation's leading nonindustrial defense contractor."[11]

Even more important, RLE's interdisciplinary model set the tone for how MIT approached a wide variety of other fields in the postwar era—an approach it continues today. RLE's current website describes it as "the first of the Institute's great modern interdepartmental research centers."[12] In the 1960s, the *New Yorker* ran a series of articles about Cambridge. One piece was on the MIT Center for International Studies, then run by economist Max Millikan, son of Nobel Prize–winning physicist and former Caltech president Robert Millikan. "The interdisciplinary approach grew out of the Research Laboratory of Electronics here, whose predecessor—our Radiation Laboratory—began with radar during the war and then moved into communications," Millikan related. "Our Center for Communications Sciences arose from that. Then came other centers, focused around problems rather than disciplines—a Center for Materials Sciences and Engineering, a Center for Space Research, a Center for Earth Sciences, a Center for Life Sciences—each drawing on several disciplines and all trying to re-define the boundaries of learning."[13]

Then there were the Rad Lab veterans themselves, who were searching for new projects. "There was an enormous pent-up set of ideas," recalled Wiesner.[14] Inspiration came on both the scientific and business fronts. A problem with water vapor absorbing the signal of an advanced bombing radar system he was working on late in the war spurred Ed Purcell to begin the experiments that led directly to his discovery of nuclear magnetic resonance when back at Harvard in December 1945. For that discovery, he would share the 1952 Nobel Prize in physics.[15]

Over at Raytheon, which worked closely with the Rad Lab and had been a leading maker of magnetrons during the war, the effect was almost immediate. Late in 1945, Raytheon's star scientist, Percy Spencer, who never finished grammar school, reportedly noticed a candy bar in his pocket melting when he stood close to a magnetron being tested. Intrigued, he grabbed a bag of popcorn kernels and watched them pop when brought near the device. This led directly to the Radarange, the first microwave oven, originally for industrial kitchens. Magnetrons still form the hearts of these appliances.[16]

Meanwhile, Denis Robinson, an English engineer sent to the Rad Lab as a liaison with the British Air Commission, teamed up with an MIT professor and Rad Lab alum named John Trump to launch what may have been the lab's first commercial spinoff. After the war, but before the Rad Lab had officially closed its doors, they found spare space in building 24 and began building Van de Graaff electrostatic generators for commercial sale. To make the machines, the pair drew heavily on what they had learned from radar work—utilizing magnetrons to provide the high energies the generators needed for particle acceleration. They soon corralled Robert J. Van de Graaff, an MIT associate professor and the device's namesake inventor, to launch a company around the work—moving operations to a rented storefront on Mount Auburn Street in West Cambridge. Robinson was named president of the company, called the High Voltage Engineering Corporation. When it launched in 1946, the startup also had the distinction of being one of the first three investments of the pioneering venture capital firm American Research and Development (ARD) under Georges Doriot. ARD, the cofounders of which included MIT president Carl Compton, invested $200,000 in High Voltage Engineering to get it going.[17]

*　　*　　*

Just a few years later, as MIT was building out its new interdisciplinary approach, another critical chain of events got underway that would supercharge the pace of innovation. The catalyst was the detonation of the first Soviet atomic bomb on August 29, 1949. The event set off a flurry of activity to understand the Russian capabilities and defend against a strike. A big effort centered on creating a modern air defense warning system based on radar, which quickly led to MIT and its Rad Lab veterans. George Valley, who had headed the lab's H2X bombing radar, was put in charge of an Air Force committee that included, among others, two fellow Rad Lab veterans and Charles Stark Draper, who had done important war work on gun sites and fire-control systems.[18] Valley's committee concluded that humans would not be able to provide real-time processing of radar signals from the network. That inspired them to wonder about harnessing one of the experimental digital computers just coming online. Valley was musing about the dilemma when he ran into Jerome Wiesner in an MIT hallway and explained their thinking.

It was an early example of the "human collisions" Kendall Square's twenty-first-century developers would try to design for. Wiesner, by then an associate director of RLE, had just finished a review of the sole digital computer at MIT. It was housed in the Barta Building, a redbrick edifice crowned by a minaret and a smokestack a few blocks down Massachusetts Avenue toward Central Square from the MIT campus proper. It actually stood kitty-corner from the Necco plant. The guy behind it all was a reedy engineer from Nebraska named Jay Forrester; he called the machine Whirlwind.

It took almost two years to flesh things out and get all the approvals, but the concept they came up with hinged on MIT creating an entirely new center devoted to developing the air defense network—using Whirlwind as the prototype brain for controlling the system. Louis Ridenour, who would soon be named the Air Force's first chief scientist, advocated passionately for the facility, feeling it would be vital to cementing MIT's standing as a leader in modern electronics. Like RLE, the new laboratory was modeled after the Rad Lab. Initial work took place on campus, mostly in old Rad Lab buildings—classified projects in building 22, unclassified work next door in building 20. By the spring of 1952, a permanent home was ready in Lexington, roughly a fourteen-mile drive from the institute. The operation, which already by then counted 550 personnel, also got a permanent name: Lincoln Laboratory.

The air defense network it spearheaded (IBM was the other main contractor for the computer system) was called Semi-Automatic Ground Environment (SAGE). The effort behind it pushed development of a wide array of computing technology—including graphical interfaces, so airmen could follow the action; modulator-demodulators (modems), to convert digital radar data to analog for transmission along phone lines and then to switch it back to digital for processing; and, especially, computer memories. Forrester's invention of magnetic core memories formed the backbone of digital computers until the rise of the integrated circuit in the 1970s.

Whirlwind and SAGE, in turn, did a lot to spur the digital computing revolution—and to make MIT and the greater Boston area a key part of it. A host of major corporations besides IBM—GE, AT&T, Burroughs, Hughes, Fairchild Camera and Instrument, and others—signed on as SAGE contractors and subcontractors. Others started up or pivoted to support the fervor of activity. IBM led the way in creating core memories. Hundreds, soon thousands, of people were trained as programmers for the new industry.

It wasn't all established players driving this movement, either. New companies were also starting up amid a wave of entrepreneurship. Early in the days when Whirlwind was trying to demonstrate its worth to the Air Force, Forrester's team in the Barta Building had fashioned a smaller test computer to serve as a proof of concept. This effort was led by an MIT graduate student named Ken Olsen, who had served as a second-class Navy petty officer and radio technician during the war. Olsen had joined Lincoln Lab when it was launched. But some five years later, in August 1957, he and colleague Harlan Anderson left to form their own company: Digital Equipment Corporation (DEC). Their first products were small, transistorized lab modules, reminiscent of the test equipment developed for Whirlwind, that helped technical teams evaluate core memories and other computer elements. Two years and change later, the company would produce its first complete computer—the PDP-1 (for programmable data processor). The industry and press took to calling such machines *minicomputers*. Behind the PDP-1, DEC caught fire and soon ranked as one of America's fastest-growing companies. At its peak in 1990, it reported $13 billion in revenue and had 124,000 employees worldwide—claiming its spot as the world's second-largest computer maker, behind only IBM.[19]

DEC had launched with $70,000 from Georges Doriot and American Research and Development, in exchange for 70 percent of the company. While a few years earlier High Voltage Engineering had been ARD's darling, DEC soon stood in a class by itself. The company went public at twenty-two dollars per share in 1966, nine years after it formed. By the fall of 1967, the stock had crossed one hundred dollars per share, and ARD's stake was valued at almost $200 million—more than 2,500 times its initial investment. "Digital Equipment was the venture capital industry's first home run, single-handedly proving that venture capitalists could generate enormous wealth by backing the leader of a hot new business," wrote Spencer E. Ante in *Creative Capital*, his book on Doriot and the early venture industry.[20]

* * *

Yet this explosion of technological innovation and entrepreneurial activity didn't immediately make a big mark on Kendall Square. Even though their origins traced to MIT, neither High Voltage Engineering nor DEC settled in the square. Denis Robinson's company had moved out to Burlington, Massachusetts. Ken Olsen's was always outside Cambridge. Many others were following their leads, with established and startup high-tech companies alike gravitating especially to the Route 128 corridor serving Boston's western suburbs.

Hit by the double whammy of manufacturing decline and the failure of the new high-tech industry to put down roots, Kendall Square seemed almost forgotten—a far cry from its heyday in the 1920s and early 1930s. "After World War II, the industrial development of Cambridge ended. Some industrial buildings were converted to offices and laboratories, but many companies closed or moved to other regions," writes historian Susan Maycock.[21]

Forty to fifty years down the line, large corporations, investors, entrepreneurs, and even governmental bodies from around the world would be beating a path to Kendall Square—and, unlike with the previous boom, MIT would be front and center of the renaissance. But for now, almost everyone just wanted to get away.

SPOTLIGHT: THE F&T—PLACE-MAKING'S FIRST PLACE

One of the most cherished components of any vibrant innovation ecosystem is a place where people of all stripes can gather, hang out, brainstorm, and experience the chance encounters that spark inventions and discoveries, inspire new businesses, or otherwise change lives. In the eyes of many Kendall Square long-timers, that ideal took the form of the F&T Restaurant.

The legendary eatery named after Isaac Fox and Robert Tishman was located on Main Street in the heart of the square from 1924 until forced to close in 1986 to make way for the MBTA station's expansion. At its tables, major advances were reportedly made in how frogs see, detecting cosmic background radiation, and verifying Einstein's prediction of gravitational waves—the last two of which led to Nobel Prizes. But the F&T's clientele ran the gamut from world-famous scientists to politicians, hungry students, and local workers. "People would come in the restaurant, Nobel Prize winners, and they'd talk with a guy who'd come in from sweeping the street or a factory," recalled Fox's son Marvin, who along with Tishman's son Maynard ran the restaurant when their fathers retired. "We used to tell people: no religion and politics."

In a way, leaders in Kendall Square have been longing for something like the F&T ever since its doors shuttered. In January 2010, Cambridge Innovation Center CEO Tim Rowe gathered about thirty entrepreneurs, investors, and others to explore launching a new café-restaurant that might help fill the void left by the F&T. At the time, talk centered on a "third place" (besides home and office) where people could meet and interact, recalls Carrie Stalder, who helped Rowe develop the project. "One of the goals of the cafe-as-restaurant concept was to be that third place for Kendall Square, bringing the F&T back in spirit," she relates.[1] While it met with enthusiastic support, the economics did not work out—though the effort inspired the CIC's Venture Café that will be mentioned in chapter 15, which Stalder cofounded.

Fox and Tishman, both immigrants, first opened their restaurant in Boston's West End in 1919. They moved once to Roxbury before the final move to 304 Main Street in 1924. In later years, a diner was added next door at 310 Main. The partners married two sisters, so their children were cousins. After they retired, Marvin Fox and Maynard Tishman took over and would operate the F&T until its closure.

The F&T was first and foremost a restaurant and deli. It opened seven days a week. During the Depression, the proprietors set out a basket with bread, cheese, and

Figure 11 Restaurant with diner next door, circa 1986.
Source: Cambridge Historical Commission.

bologna—and whoever needed something could grab a bite and pay later, or not at all. "My father said you never refuse anyone," Marvin recalled.[2] The F&T also issued weekly meal tickets so that customers wouldn't need cash each time they came in.

At first, the F&T was frequented mostly by local workers and some families. Over time, it also became a destination spot for MIT students and faculty. Charlie Cooney, later a professor and a cofounder of Genzyme, discovered the F&T in his student days in the late 1960s. "There was a time that was the only place outside a MIT cafeteria you could go," he says with a smile. "They had the best pork and beans. They were open hours that were conducive to the life of a student, and they had pricing that was conducive to a student."[3]

Rainer "Rai" Weiss started going there in the 1950s. He would walk the few blocks from MIT's building 20 near the corner of Main and Vassar, where he worked as an electronics technician in the lab of Jerrold Zacharias, who later became his faculty mentor. An electroplating company sat across the street—with a pickle factory and candy maker along Main as well. The Lever Brothers plant, where animals were rendered to make soap, was only a few blocks away. If the breeze came from that direction, he says, the smell "was disgusting."[4]

Inside the restaurant, a bar sat on the right and a big round table on the left. Then came some smaller, rectangular tables and a few booths toward the back where drinkers could hunker down. You could get pastrami sandwiches and other deli fare or a full meal. "The waitresses were salty," recalls Weiss. "They slung stuff on the table. But they would leave you alone."

Weiss and many of the MIT crowd loved the one round table where five or six people could squeeze in. After he became a professor, he brought visitors and students there regularly. The tables had paper place mats and people would write or draw on the backs. "Scientists love to write stuff down," he says. If you filled up a mat, "you could go to the bar and get a few more of them. A lot of ideas came up in that place."

One regular at the F&T was Jerome Lettvin, an MIT cognitive scientist known for his groundbreaking experiments in how frogs see. Weiss says his colleague developed some of his most important ideas at the F&T.[5] Lettvin, who died in 2011, wrote a poem about the restaurant that he recited at a 2003 commemoration. John Mather partially fleshed out the idea for the Cosmic Background Explorer (COBE) satellite at the restaurant. "I was visiting Rai Weiss in 1974 specifically to have that conversation about COBE," he recalls. "NASA had just announced the opportunity to send proposals, and I knew Rai would be interested." Mather shared the 2006 physics Nobel for this work.[6] Weiss himself advanced some ideas about the antenna for the Laser Interferometer Gravitational-Wave Observatory (LIGO) at the F&T's round table. LIGO led to the first detection of gravitational waves and earned him a share of the 2017 Nobel Prize in Physics.[7] "I think I started thinking about it at the F&T," he relates.

The list of notables who visited the restaurant included politicians like Tip O'Neil and diplomats such as Caroline Kennedy. Marvin Fox says he knew every MIT president who served during the years the diner was open. Another regular was biologist and New York native David Baltimore. "As a Jewish boy, delicatessen food was an important ethnic part of my life. So when I came to MIT in 1960, I was overjoyed to see a delicatessen around the corner," he relates.[8] Baltimore came into the F&T shortly after winning the Nobel Prize in 1975, Marvin Fox once recalled. "I said, 'Congratulations Dr. Baltimore.' He said, 'Marvin, my friends still call me David.'"

The 1970s proved tough, Fox remembered. That was after NASA curtailed plans to develop its electronics center and before Kendall Square's revitalization the following decade. Once construction picked up in the 1980s, the diner was rediscovered by workmen, and experienced boom times at lunch. "That was our business—a beer and a sandwich. We had some of our best years," Fox related.[9]

The F&T closed on Thanksgiving Day in 1986. Some fifteen years later, thanks to an effort organized by Rai Weiss, a historical marker was placed at the site. A small ceremony was held, followed two years later by another one where Lettvin read his poem.

But that was not necessarily the end of the F&T. While advancing his real estate development career, former city councilor David Clem took up an unusual hobby: collecting diners and their memorabilia. After the F&T closed, he secured all the booths from the restaurant, as well as the facade and its signage, and more than two-hundred old menus, signs, posters, and other paraphernalia. He donated two booths to the MIT Museum, but stored the rest. In 2021, his son Chet, also taken by its story, was hunting for a way to somehow resurrect the F&T. Those efforts were progressing but hit a roadblock with the COVID crisis. Perhaps one day the F&T will be back.

9 URBAN MARSHLAND TO URBAN RENEWAL

In the decade following World War II, Kendall Square suddenly felt empty again. The intense work of the Rad Lab was over. Many of the scientists had left, and the flood of visitors had dried up. In many ways, the square had become a new urban marshland, an echo of the salt grass and mudflats of its past, with activity confined to the drumlins rising above the tidal marsh. Now the activity centered around MIT, and much of the rest was a wasteland of vacant and unmaintained factories, storefronts, and other structures—with only relatively few pockets of activity. Echoing the flight from other industrial cities to the suburbs, the Sunbelt, and overseas, companies had fled for cheaper land and labor along Route 128 and beyond. At night, no one wanted to be there; there was little to do anyway.

"Kendall Square was a moribund 19th-century district," remembers Bob Simha, MIT's director of planning emeritus, who headed the school's real estate arm for some forty years from 1960 to 2000. "Companies were sliding away. People were losing jobs. The city was losing income. The few plants that remained, like the vulcanized rubber plant, were smelly and polluted the air."[1]

In a way, the situation was brought into sharp focus by a postwar burst of startup activity from MIT. The pent-up ideas Jerry Wiesner had noticed among the Rad Lab veterans extended to others at the school, who had worked in fields besides radar during the war or had just been stymied in pursuing their own dreams because of the conflict. On the one hand, empty industrial buildings provided cheap space for startups next to campus. On the other, conditions were totally inadequate once the fledgling companies found success and grew. "MIT-spawned electronics companies set up their incubator facilities in old industrial and warehouse spaces in Cambridge abandoned by large companies for state-of-the-art manufacturing plants elsewhere," recalls Simha. "As start-up electronics companies reached the manufacturing stage, they required modern space that they found or developed outside Cambridge along the newly built Route 128."[2]

Already by 1950, it was clear that concrete steps needed to be taken to reinvigorate the Kendall Square area—and that MIT, with a vested interest in improving conditions around its campus, needed to take the lead. The university had previously begun acquiring property next to its original parcel, much of it south of Massachusetts Avenue, where the student center, athletic facilities, Kresge auditorium, and various dormitories stand today. After World War II, the federal government transferred ownership of several buildings it had acquired during the conflict, the Atlantic Richfield Company gave property on Massachusetts Avenue, and Nabisco donated land on Albany Street, toward Central Square. All of this land was intended for the school's future needs. MIT was able to use some of it immediately, but other parcels were leased to commercial businesses until the school needed them.[3]

But MIT officials still worried that the "encroaching manufacturing district" in Kendall Square did not make for the best of neighbors for an elite school. So they also looked for ways to control the destiny of what lay on that side of campus.

A key opportunity to reshape Kendall Square came in 1957. That's when the Cambridge Redevelopment Authority (CRA) acquired a derelict 4.5-acre plot of land just west of the Grand Junction railroad tracks called the Rogers Block, using urban renewal funds provided under the federal Housing Act of 1954.

Once the CRA took possession, the existing wooden tenements and other structures were demolished, and the property placed on the market. However, the city could not initially attract an acceptable developer: the only offer reportedly involved building a mattress factory, which was not appealing. The parcel still sat unsold roughly two years later, when Lever Brothers, which had moved its US headquarters from Cambridge to New York a decade earlier, decided to close its massive, aging soap factory just abutting the property—creating an even more urgent need to fill what was becoming a gaping hole around Kendall Square.

At the time, the City of Cambridge was in dire financial straits, reportedly unable to even float a bond issue to develop the property. Then came a pivotal meeting between Cambridge mayor Ed Crane and former MIT president James Killian, then chairman of the MIT Corporation, at a groundbreaking ceremony for an industrial office park built by Cabot, Cabot & Forbes (CC&F), a major developer. According to Simha, Crane corralled Killian and the CEO of CC&F and asked what they could do to solve the city's problem.

Killian, who wielded immense influence, felt strongly that MIT should play a big role in helping the city improve the environment around campus and could do it while also providing an acceptable return for MIT's endowment. He proposed that MIT acquire the Rogers Block property, and that CC&F develop the site. The men brought the idea to MIT officials, including new president and Rad Lab veteran Julius Stratton—and soon got buy-in from the MIT Corporation.

"What Jim saw was the opportunity to begin to change the character of this area, which was dying rapidly," says Simha. "He had the courage and the imagination and the leadership to convince the committee that it was the wise thing to do."[4]

On January 11, 1960, Killian appeared before the Cambridge Chamber of Commerce to announce plans to create a new commercial office complex focused on technology and research companies on the fourteen acres of the Rogers Block and the Lever Brothers factory. Killian predicted the development would generate more than $800,000 in annual tax revenue, in contrast to the $135,000 the city had garnered from the previous occupants. An MIT professor, Eduardo Catalano, was retained as the project's lead architect. The state-of-the-art development even had a name befitting the future: Technology Square.[5]

*   *   *

Tech Square, as it came to be known, proved a watershed in the history of Kendall Square. "It is in fact the ignition," asserts Simha. "It's where it all started, and it gives you some understanding of how long and complicated, and how difficult, this whole enterprise is. This is half a century in the making. What you see today started in 1959 to 1960."[6]

It was also a big departure for MIT to use its endowment funds to support the purchase of property for commercial development in contrast to future academic needs. "It was a very significant adventure. Universities did not do these kind of things," according to Simha.[7]

The idea was to develop the property in stages. The first phase focused on what is still called Tech Square—its main "face" looking onto Main Street. The second phase, encompassing the old Lever Brothers site, would be developed as demand grew. Catalano's design for Tech Square featured a wide and flat three-story structure that was perched at the front of the property, closest to Main Street. Behind it, three nine-story buildings formed a U shape around an open quad. These structures were set back from the street edge, with the two sides of the U facing each other across the quad and the backmost building looking straight out on Main Street. As Simha once wrote, "At the time, the site was surrounded by unpleasant industrial activities, which led to a building design that did not engage with its neighbors."[8]

The first building in the complex, 545 Technology Square, was completed by the end of 1962. This stands on the right-hand edge of the complex when viewed from Main Street—and is home to Moderna today, with a Fidelity office on the ground floor. The complex was designed as a traditional office development. But the design was flexible enough that buildings could be converted to house labs and R&D activities, which the complex offers now along with conventional office space. "The real problem was to get the first building up and get it occupied—in a sea of old industrial stuff," Simha says. "MIT had to use a lot of chits to get people to come over here and get people to rent space in the building."[9]

The second building, 575 Technology Square, on the left-hand edge when looking from Main Street, opened in 1964 (Area Four restaurant is on the ground floor today). The third and fourth buildings came online in 1966 and 1967.

Project MAC, an MIT research group destined to play a major role in Kendall Square's much-hyped AI Alley days (see chapter 11), moved into 545 Tech Square. The project basically united the institute's various efforts around artificial intelligence (AI)—including studies of vision, language, and mechanical manipulation—with work in time-sharing and other aspects of advanced computing. One line of its ancestry ran directly to the AI Group run by professor and artificial intelligence guru Marvin

Minsky. Another branch traced to the Research Laboratory of Electronics. Still another involved the Tech Model Railroad Club (TMRC) in the basement of building 20, one of the Rad Lab's old quarters. The TMRC became famous in the annals of computing. It essentially had two types of members. On one hand were those who loved to work and play with the model trains. "On the other hand," according to a historical paper by a group of MIT graduate students, "there were those inhabitants of 'the notch'; a small back room where the control system for the tracks and railways was housed. These students were far more interested in the machines and systems that controlled the tracks than they ever were in actual trains. . . . These students of the TMRC were the very first hackers at MIT."[10]

MAC reportedly stood for two things (other names have been suggested as well): *machine-aided cognition* and *multiple-access computer*. Project MAC launched in 1963 under professor Robert Fano with $2.2 million in funding from the federal Advanced Research Projects Agency (ARPA). The award was given by former MIT professor J. C. R. Licklider, then director of information processing techniques for ARPA. Licklider would later return to MIT as Project MAC's second director.

The Project MAC team set up shop on the top two floors of 545 Tech Square. For the most part, offices and labs were on the eighth floor, with computing equipment occupying the ninth floor "penthouse." Many future legends of computing and AI worked there over the years. Besides Fano, Minsky, and Licklider, they included Tom Knight (operating systems), Danny Hillis (parallel computing), John McCarthy (inventor of the LISP programming language), Jack Dennis (computer systems and languages), Richard Stallman (free software and the GNU Project), and Fernando "Corby" Corbato (time-sharing). Minsky, Corbato, and McCarthy would win Turing Awards, the computer field's equivalent of the Nobel.

This was the kind of operation others liked to be near—and it helped Tech Square get going. In early 1964, IBM rented space in the same building for what was initially called the IBM Cambridge Scientific Center. The center would stay there until its closing in 1992. (Project MAC itself was eventually folded into today's CSAIL.)[11]

The remaining space rented more slowly initially, but the development gradually proved its worth, attracting some major names that included Grumman Aircraft Engineering Corporation, as well as other MIT research efforts. The big coup, though, was Polaroid. In 1966, the instant photo juggernaut moved its world headquarters to the three-story structure nearest Main Street, which was soon dubbed the Blockhouse. About as instantly as its film developed pictures, Polaroid became Tech Square's anchor tenant.[12]

Polaroid embodied the postwar future of Kendall Square as a home for science-based technology companies. Well before its headquarters moved to the Blockhouse, Polaroid had established itself as an international powerhouse and cofounder and CEO Edwin Land ranked as one of the most celebrated inventors in American history.

In the late 1920s, the Harvard dropout had created the first inexpensive filters able to polarize light.[13] In 1932, he partnered with one of his Harvard instructors, a postdoctoral student named George Wheelwright, to form the Land-Wheelwright Laboratories to commercialize the filters. The company's name was changed five years later to the Polaroid Corporation.

The partners initially rented a single room in a garage at the corner of Mount Auburn and Dunster Streets in Harvard Square, in part of what is now the funky urban mall known as the Garage. Then after a stint in a dairy barn in Weston, Massachusetts, they moved to Boston, first leasing a noisy, dusty basement on Dartmouth Street and soon finding more expansive quarters at 285 Columbus Avenue, near the Back Bay railway station.[14]

Early success came from polarized sunglasses and camera filters. In 1947, however, Land and Polaroid revolutionized home photography by inventing instant film. The Land Camera, which developed pictures inside the camera in about a minute, was an overnight success when it was introduced the next year.

Land himself had been operating in the square for more than two decades before the headquarters move. In 1942, during World War II, he had set up his private lab less than two blocks away, at 700 Main Street, in part to be close to MIT, where he worked on a variety of wartime projects. This lab, dubbed the Mole's Hole, would serve as his personal sanctuary for many years—and it was there he made a large number of his key inventions. (For more about Land and the Mole's Hole, see chapter 22.)

Polaroid took over the entire complex at 700 Main Street in 1960 and later bought it outright. The company also eventually spread beyond the Blockhouse into every building in Tech Square. It kept some operations there even after moving its headquarters again in 1979—to 784 Memorial Drive, overlooking the Charles River just a half mile away.[15]

Polaroid employed some twenty-one thousand people at its peak and at one point reached $3 billion in annual revenues. But it could not adapt to changing times. The company fatally clung to instant photography even as the writing was on the wall with the advent of digital. Land also made a disastrous bet on an instant 8 mm movie system called Polavision. He was forced to step down as CEO in 1981, remaining chairman of the board—but that didn't last long. "Suddenly Land was fallible, after a spotless 40-year record, and in 1982 he was nudged into retirement," writes Christopher Bonanos in *Instant: The Story of Polaroid*.

A few years later, Steve Jobs, a longtime admirer of Land, called forcing out the Polaroid founder "one of the dumbest things I ever heard of."[16]

*　　*　　*

With the establishment of Project MAC, Polaroid, IBM, and other tenants, Tech Square soon became the most notable tech "drumlin" of the day besides MIT itself. A few other islands of high-tech activity could be found in or near the boundaries of what would be modern Kendall Square—for the most part, standalone from MIT. A case could be made that one was the vaunted Athenaeum Press on First Street near the Broad Canal. Ginn & Company sold the operation in 1950 to Chicago-based John F. Cuneo Press,

Figure 12 The first four buildings of Tech Square, with Polaroid's Blockhouse in front. The modern terraced structure to the right behind them is Draper Laboratory, completed in 1975, with a parking structure opposite.

but the plant, at the time employing 650 people, stayed in operation.[17] In the aftermath of World War II, though, the tech drumlin that came closest to foreshadowing the modern Kendall Square could be found along Memorial Drive, between MIT and the Longfellow Bridge. There, for a brief time right next to each other in what was dubbed Research Row, sat Arthur D. Little, the Warren Brothers of asphalt technology and road-building fame, and the US headquarters of Lever Brothers. Beginning in 1947, they were joined by a soon-to-be-famous enterprise created to develop new products and techniques that could be used by industry, called National Research Corporation (NRC).

The force behind NRC was a Massachusetts businessman, scientist, and inventor named Richard Stetson Morse, who had gotten his undergraduate degree at MIT, then studied physics in Germany. With war coming, he returned to the United States to work for Kodak but left the camera company after five years to found NRC in 1940. After the war, Morse moved his operation into 70 Memorial Drive (now MIT building E-51), right in line with the others—farthest from the bridge. He was already well on his way to the biggest hit of his career.

NRC's first customer had been Oak Ridge National Laboratory in Tennessee, making vacuum pumps for the Manhattan Project. As the war wound down, Morse began applying NRC's vacuum technology to consumer products. He started with freeze-dried coffee—then, around 1945, frozen orange juice. He began selling this product to the nation the following year, offering the first concentrated frozen orange juice on the market. The product was called Minute Maid, to imply how easy it was to make. Despite the catchy name, business faltered. But Morse had learned from the failure of his freeze-dried coffee that he should not neglect marketing. In 1949, two years after moving to Kendall Square, he entered into a deal with Bing Crosby, paying him an undisclosed amount of cash plus twenty-thousand shares of Minute Maid stock to become the company's pitch man. The crooner's popular commercials ran every morning on TV for a time. "The partnership was a smashing success. In 1948, before Bing came on board, the company was selling just shy of $3 million worth of orange juice. Within three years, that amount rose to almost $30 million," according to a historical piece in *TIME* magazine decades later.[18]

Morse would sell his company in 1959, and the operations soon moved out of Cambridge, effectively closing the door on Research Row. But he was not done contributing to the Kendall Square story. In 1961, Morse returned to MIT as a senior lecturer at MIT Sloan, where he started its first entrepreneurship course, New Enterprises. The course has been taught continuously since then and is reportedly the longest-running continuously taught course at MIT. Morse's son, Ken, later became founding managing director of the MIT Entrepreneurship Center, where he served for many years. "In a heartwarming moment, I learned that the MIT Entrepreneurship Center was installed for a year or so in my dad's former office, overlooking the river. My desk was where he had been," Ken Morse reports.[19]

By the time Tech Square got going in the early 1960s, little was left of Research Row.[20] Simha says it was clear to everyone that the process of turning around Kendall Square would be a long one. But sometime probably in late 1963, he got a telephone call that shaped the square's character for decades to come. In the process, it would create an urban myth. The call was from Robert Rowland over in the Boston Redevelopment Authority. He and some colleagues had an idea, and they needed MIT's help—not in the same way the institute had helped with Tech Square but building on that accomplishment. It involved creating a major high-tech center right in Kendall Square next to MIT that would attract a couple thousand jobs directly and serve as a beacon for thousands more. They wanted to bring NASA to town.

10 KENDALL, WE HAVE A PROBLEM

The story of NASA in Kendall Square would be told and retold countless times by locals—even more than fifty years later. People shake their heads and point to the eyesore compound with a gigantic tower building opposite the Marriott Hotel between Broadway and Binney Street. The most common version of the myth runs something like this: The space agency needed a headquarters, and with Massachusetts's own John F. Kennedy serving as president of the United States, he arranged for it to be in Kendall Square. The site was built, NASA moved in, but then, after Kennedy's assassination, new president Lyndon Johnson moved it to his native Texas, where it remains today.

That makes for a nice, neat story—and there is some truth in it. But it misstates what really happened in important ways. Kendall Square was never supposed to have been NASA headquarters. Rather, it was chosen to host the space agency's Electronics Research Center (ERC), which was built. The ERC opened in September 1964 and was closed less than six years later on June 30, 1970. It was given the axe under the Nixon administration, though, not by Lyndon Johnson. And what few likely realize is that the center never came close to reaching the size and scope it was supposed to have achieved. Had it done so, Kendall Square might be on a dramatically different trajectory than it is today.

The space race was in full bloom in the early 1960s. Russia had launched its Sputnik satellite in 1957, shocking the United States with its technological prowess. NASA had been established the following year. In May 1961, barely four months into his presidency, John F. Kennedy had declared to Congress that the United States should set the goal "of landing a man on the Moon and returning him safely to the Earth" before the decade was out. The mission sparked a reorganization of NASA to focus on the manned space imitative. In concert with the restructuring, NASA administrator James Webb and other key officials believed the agency needed to dramatically up its electronics game. "NASA's fundamental dependence on electronics and its need for internal expertise drove the agency to create an entirely new center, the Electronics Research Center," a NASA-commissioned historical paper noted.[1]

As the paper continued, "it is not clear how the Boston area was chosen, or even if NASA considered other locations." However, the long work MIT had done to bolster its ties to the military and other branches of the government served it well. Three of Webb's top advisors—Associate Administrator Robert Seamans; Raymond Bisplinghoff, director of NASA's Office of Advanced Research and Technology; and Director of Electronics and Control Albert Kelley—had direct ties to MIT. Seamans had gotten his doctorate at the institute and been an associate professor there (he would later serve as Air Force secretary and return to MIT as dean of engineering); Bisplinghoff had been a professor; and Kelley was a Boston native who had also gotten his doctorate at MIT. Webb himself served on the board of visitors of the Joint Center on Urban Affairs of Harvard and MIT.

But locating the center near MIT made sense for more objective reasons, given MIT's decision to cultivate deep expertise in electronics and computing after World War II. "Regardless of the politics, Cambridge was the best logical location for an electronics research facility," the same historical paper noted. "The area abounded with electronics resources and talent: MIT and Harvard, the industries along Route 128, the Air Force's Cambridge Research Laboratory and Electronics System Division at Hanscom Field, MIT Lincoln Laboratory, the Mitre Corporation, and the MIT Instrumentation Laboratory (Draper Lab), which already had undertaken responsibility for the Apollo guidance computer."[2]

Then there was Kennedy. The president did apparently take a direct hand in the matter, working with Webb to keep the project out of the NASA budget until after his brother Ted Kennedy's first election to the Senate in November 1962 for fear it might cause problems. "After the President belatedly put the ERC project in the budget process, Congress rebelled," a different NASA paper noted. "In addition to Republican members, Representatives from the Midwest and other regions feeling swindled out of the NASA largesse repeatedly fought efforts to fund the ERC."[3]

In the end, though, the plan to locate the center in the Boston area overcame this opposition. The ERC was to be a world-class facility. For its initial budget NASA asked Congress for $3 million for land acquisition and $2 million for design. The envisioned center would ultimately employ a staff of 2,100—among them nine hundred scientists and engineers, seven hundred technical workers, and five hundred in administrative and support positions. Research would be conducted in five major areas: electronics components; guidance and control; systems; instrumentation and data processing; and electromagnetics. NASA estimated it would need twenty-two acres to house the entire operation.[4]

As part of this process, the Army Corps of Engineers had set up a task force to hunt for the right site for the upstart center. By 1963, it came to Robert Rowland's attention at the Boston Redevelopment Authority, he recalls: "They had at that point investigated I think 165 possible sites in New England, all the way up from Newport up into Maine." The requirements had been openly published, so Rowland and two colleagues took those and launched their own hunt. It was a bootleg project, done on personal time, as they wanted to keep it quiet in case their quest didn't bear fruit. Rowland and his small team took photos of various sites, including some of the traffic patterns around Kendall Square—an area Rowland knew well, as he parked his car there to take the subway to Boston when commuting to his job. They soon concluded that the square, with its aging factories, lagging economy, and location next to MIT and just five minutes by transit to Boston, was the ideal location. The success of the Tech Square development also played a role in their thinking. But when Rowland called Bob Simha, the idea he put forth was that the revitalization project necessary to accommodate NASA should be done differently than the way Tech Square had been handled. His view was that the City of Cambridge should create a much bigger urban renewal project that would be almost completely financed by the federal government. MIT would not have to buy any land, but the school's cooperation was essential, Rowland emphasized. He asked Simha to help bring it all together.

Simha filled in James Killian, who asked for a meeting with various city officials whereby Rowland could present his idea. Allocating a large tract of land for the federal government would take the parcel off the tax rolls. But at the meeting, as Simha related, Rowland "stressed that the project would serve as a catalyst for the economic regeneration of East Cambridge and would create jobs and tax revenue to offset the loss of taxable property devoted to NASA's research center."[5] It would be much easier to develop other land to attract tax-paying enterprises if the NASA center was there as another anchor, he argued.

Under the US Housing Act of 1949, the city could claim land that was classified as blighted or economically depressed and receive federal funding for up to two-thirds of the costs, including relocating and compensating those displaced by the project. That would still leave Cambridge on the hook for the other third, estimated to be more than $5 million—and given that its coffers were still largely depleted, officials were leery. The erosion of the business tax base had shifted much of the tax burden to homeowners, raising an outcry. "It was quite serious, because poorer people were not able to pay the taxes that were being levied because of the shift in the economic circumstances here," says Simha.[6] Some commercial owners having trouble filling space were actually tearing down upper stories of their buildings to reduce their value and lower taxes.

This is where MIT's support could be critical. A special provision of the urban renewal law allowed for much or all of the city's share to be offset by credits MIT had accumulated. More specifically, the value of MIT land and buildings within one mile of the urban renewal project could be credited to the city to meet its share of the net project costs—as long as the school committed to using that land for education, research, and service purposes. Ultimately, MIT presented detailed campus development plans that fulfilled this requirement, and the school was allowed to transfer roughly $6.5 million in federal tax credits to the city, by most accounts covering Cambridge's entire $6,416,500 price tag for its share of the project. As Simha related, MIT supported the plan "with the understanding that the proposed activities in Kendall Square would complement the work MIT had already begun in Technology Square to help re-develop and reinvigorate the city's economy and its residential and community facilities."[7]

In 1964, Cambridge initiated the Kendall Square Urban Renewal Project under which the effort would proceed. This was a seminal step in the future development of

Kendall Square—though once again, things would not play out as envisioned. As an architect of the plan, Rowland was asked to help get it off the ground and in April 1965 took a three-month leave from his Boston job to work for Cambridge. At the end of the first three months, things were on track, so he requested another three months. He never went back to Boston, heading the Cambridge Redevelopment Authority from 1965 until 1983.

Rowland's team finalized the loan and grant application for the feds and had approvals by the end of 1965. In all, twenty-nine acres in the heart of Kendall Square were earmarked for NASA. The area in question was a large tract bordered by Third Street on the east, Binney Street on the north, Broadway on the south, and the railroad right-of-way on the west. The Broad Canal ran through the parcel, and much of its length would have to be filled in. The land was occupied by a fairly large number of mostly small and aging commercial businesses, though very few residents.

A key part of the plan involved the commercial development of an adjacent parcel of land—thirteen acres in the triangle between Broadway and Main Street, with Galileo Galilei Way as its base. This lay between the NASA-designated land and MIT, where the Marriott Hotel and various office buildings stand today, and also the Broad and Whitehead research institutes. At the time, like the NASA parcel, it contained older businesses, but also included a number of working-class homes. This land would stay on the tax rolls and, the hope was, more than make up for what was lost to the federal government as revitalization took place. The vision was to offer a blend of commercial, retail, and residential space that would create a dynamic, almost 24–7 urban neighborhood right next to MIT.

That more than fifty years later planners, university officials, city officials, and residents would still be fighting to fulfill that dream for Kendall Square is a testament to how far off the mark things ultimately went.

* * *

It was all systems go at first. NASA opened the Electronics Research Center on September 1, 1964, moving into temporary quarters in Tech Square while its permanent site could be readied. The following August, the city formally approved a plan that designated the twenty-nine-acre site for the space agency and also allowed for commercial development of the thirteen-acre triangle next to it.

Over the next three years, half the allocated property—14.5 acres—was conveyed to NASA as it became ready. "In this period, approximately 110 businesses were relocated, the existing buildings were razed, and the Broad Canal partially filled," writes historian Susan Maycock.[8] The Cambridge Redevelopment Authority reported that the companies displaced employed more than 2,750 workers. That was significantly more than the 2,100 jobs to be directly created over time by the NASA center, but presumably most of the old jobs were preserved elsewhere and the NASA-created positions would be more modern, longer-lasting, and higher-paying—and that did not count additional jobs created by the expected boon to the neighborhood the ERC would provide.

Figure 13 Early Edward Durrell Stone proposal for Volpe Center.

The space agency's initial design, by noted architect Edward Durell Stone, who had designed New York's Museum of Modern Art, featured three twenty-four-story towers as the core of the complex. These urban behemoths were to be surrounded by courtyards and lower-rise perimeter buildings. A large courtyard, with a circular fountain at its center, served as the main entranceway. "It looked like a Kremlin undertaking," is how Rowland summed it up. A review committee raised objections that resulted in significant modifications that included reducing the height of the proposed towers. But part of the design remained intact, and the first construction work started on the smaller buildings in 1965. NASA seems to have begun transferring operations from Tech Square later that year.

With the project's liftoff going more or less as anticipated, Congress appropriated funding for the Electronics Research Center for the fiscal years 1965, 1966, and 1967. For the next three fiscal years, however, with NASA facing mounting pressure over its skyrocketing budget, no additional funds were approved for construction, although the center continued to add personnel even as other NASA operations were forced to contract. Then, on December 29, 1969, the abort signal came. President Richard Nixon, who had taken office the previous January, issued an executive order without warning to close the center by June 30, 1970.

When the order came through, only one twelve-story tower and five low-rise, concrete perimeter structures had been completed. For the most part, they sat in an urban flatland amid sprawling parking lots. The center itself employed just 850 workers—one hundred of whom held doctorates. They were working on a range of projects that spoke directly to the hopes of transforming Kendall Square into a leading-edge, high-tech center. These included an array of satellite programs, as well as research into nuclear propulsion systems, hybrid computers, holographic displays, and automated landing systems for jet aircraft and the space shuttle.[9]

The closure spurred rumors and conjecture about what had happened. In future years, the story somehow became that Lyndon Johnson put the kibosh on the center to move operations to his native Texas. But at the time, many people figured Nixon had ordered it shut down as a political strike against the Kennedys and Massachusetts, the only state that didn't vote for him in the 1972 general election. "At least that was the general conclusion," says Rowland.[10]

Whatever the motive, Nixon's executive order stunned Cambridge officials. "The closing, bitterly protested by Cambridge as a flagrant breach of contractual obligations,

Box 3
Kendall-NASA Urban Renewal Timeline

In September 1964, NASA opened its Electronics Research Center in Kendall Square. The plan was for a complex that employed 2,100 by 1968. Instead, late in 1969, the Nixon Administration ordered it closed the following June, and peak employment reached only 850. The closure delayed Kendall Square urban renewal and revitalization plans by more than a decade. Meanwhile, the Department of Transportation took over the site. Only recently have plans been laid to revitalize the property again: work started in 2020 and won't be complete for at least a decade.

The following is a brief timeline of key events in the saga:

Early 1964—Cambridge undertakes Kendall Square Urban Renewal Project to bring NASA to town and revitalize the area. Plan involves transferring twenty-nine acres in Kendall Square to NASA.

September 1, 1964—NASA opens its Electronics Research Center in temporary quarters in Technology Square while permanent site is readied. Over the next three years, 14.5 acres is conveyed to NASA. Approximately 110 businesses relocated.

1965–1968—First twelve-story tower building and perimeter structures go up. NASA moves into "permanent" home, growing to peak employment of 850 spread across ten laboratories.

December 29, 1969—President Richard Nixon orders the site closed by June 30, 1970. Urban renewal plans put on hold.

July 1970—Department of Transportation takes over site for the Volpe National Transportation Center.

November 1971—Federal government agrees to declare eleven acres "surplus" and relinquish rights to the parcel so city can develop it commercially.

1975—Cambridge wins $15 million grant from federal Department of Housing and Urban Development to complete urban renewal project.

1977—City finally agrees on mixed-use development of the area around the center and approves necessary zoning.

1982—More than twelve years after the NASA closure decision, agreement is reached for developing the eleven surplus acres. The parcel now includes the headquarters of Biogen and Akamai.

January 2017—Federal government accepts MIT's $750 million bid to purchase and develop Volpe site. The current compound will be razed and government work consolidated in a new, 212-foot-tall facility on roughly four acres. Parking will be moved mostly underground and grounds open to public.

2022 (est.)—Once the new Volpe Center is completed, development will begin to convert remaining ten acres into a mixed-use area to include residential units, lab and commercial space, and retail. Plans call for roughly 1,400 apartments, including 280 permanently subsidized affordable housing units and twenty apartments devoted to middle-income families.*

*Plans were still evolving in early 2021.

necessitated a replanning and reprogramming of the entire renewal project area," the Cambridge Redevelopment Authority sums up.[11] The city, presumably joined by MIT and others, put pressure on the Nixon administration not to totally abandon the site. Nixon's new secretary of transportation was John Volpe, who had just ended his second stint as governor of Massachusetts. Against the advice of some key lieutenants, Rowland says, Volpe paved the way for his department to take over the facility—which was renamed the John A. Volpe National Transportation Systems Center. The Department of Transportation (DOT) took possession on July 1, 1970. Leading up to that point, 611 NASA staffers remained. Of those, 425 transferred to work for the DOT.

But that was just the tip of the iceberg for the troubles the NASA closure caused for Kendall Square revitalization dreams. Unlike NASA, the DOT had no plans to expand onto the additional acreage earmarked for the federal government. Of particular concern were eleven vacant acres on the property's western edge. This is land west of what is now a pedestrian path called Loughrey Walkway that divides the Volpe site from office buildings occupied chiefly by Biogen and Akamai. With NASA's plans curtailed, Cambridge wanted to develop the parcel commercially. But the city couldn't do anything until the federal government released its rights to the acreage.

It took until November of 1971—almost two years after the executive order to close the NASA site—for Uncle Sam to agree to declare the eleven acres "surplus" and relinquish its rights. Even then, things did not proceed simply. The Cambridge Redevelopment Authority's revised Kendall Square development plan faced strong local objections and was rejected by the city council. That led to the creation of a task force—and a long period of meetings, study, and debate.[12] One big step forward came in 1975, when Cambridge won a $15 million grant from the federal Department of Housing and Urban Development for completion of the urban renewal project. In 1977, the city at last agreed on a mixed-use development plan for the area and approved the necessary zoning. The next year, Boston Properties was selected as the main developer of the thirteen-acre triangle (parcels 3 and 4 on the map shown in figure 14). And finally, in 1982—more than twelve years after the NASA closure decision—an agreement was made for the "surplus" federal land (parcel 2). Boston Properties was also selected as developer of that property.

What had seemed like a clear and even inspired path to Kendall Square's revitalization had taken a huge detour. While the NASA center would be described decades later as "the catalyst for the complete redevelopment of Kendall Square,"[13] the short-term effect was that revitalization became stalled for more than a decade in the mess the ERC's closure left behind. The urban marshland would persist for the better part of another generation.

*   *   *

Very little other significant new construction happened in Kendall Square throughout the 1970s as the urban renewal plans were being straightened out. One exception loomed forebodingly across from the Volpe Center toward the river and Boston—on the corner of Third and Broadway. This was an office complex known as Cambridge Gateway. It was developed on land owned by the Badger Corporation, an engineering and development firm that itself had been displaced by NASA after being in Kendall Square since 1936. The capstone of the project: a huge tower of reinforced concrete that was completed in 1970. It loomed as ugly and soulless as the NASA-cum-Volpe tower. Adjoining it was a low-rise curved garage. A twin tower planned for the other side of the garage was never built.[14] Despite its long-standing status as a local eyesore, Cambridge Gateway would at the end of the century become home to the pathbreaking Cambridge Innovation Center that offered affordable, ready-made space for startups, giving it a long and storied role in the Kendall Square story (see chapter 15).

It would take more than a decade after Cambridge Gateway's completion before the first commercial buildings constructed under the urban renewal project finally debuted. Comprising both the triangle (parcels 3 and 4) and the NASA surplus parcel (parcel 2), the master plan called for 2.5 million square feet of new construction, spread across nineteen buildings collectively known as Cambridge Center. All the buildings, not counting the garages, were required to use red brick, to reinforce design cohesion. The center as a whole accommodated a range of uses—laboratory, office, retail, and residential. Parcel 3, at the triangle's widest end, would be dominated in future years by the Whitehead and Broad Institutes. Down at the narrow end in parcel 4 would be a hotel, with another office building at the very tip of the triangle. Parcel 2, the NASA surplus site, was to get low-rise buildings of two to five stories mostly for research and development and light manufacturing.

At first, Boston Properties selected New York–based Davis & Brody as architects. But they were soon replaced by up-and-coming Moshe Safdie & Associates. Safdie had won international fame for designing *Habitat 67* for the 1967 World's Fair in Montreal. *Habitat* was an adaptation of his master thesis at McGill University. In Montreal, he had known Boston Properties cofounder Mort Zuckerman: both men had grown up in that city. When Safdie moved to the Boston area in 1978 to head the urban design program at Harvard's School of Design, Zuckerman had asked for his help on the Cambridge Center project.[15]

The first building completed was Five Cambridge Center in 1981. A thirteen-story office block at the corner of Main and Ames Streets in parcel 4, it offered ground-floor retail—and soon welcomed a Legal Sea Foods restaurant. It was the only building designed by Davis & Brody—the rest were by Safdie. Next online, in 1983, came another office building: the twelve-story Four Cambridge Center, just down Ames at the corner of Broadway. For many years, its ground floor was home to Quantum Books, a technical book store (that space is currently a bar and restaurant called Mead Hall). Between the two buildings sat a parking garage.

The remaining buildings would mostly come online later in the decade. In 1986, the twenty-five-story Marriott Hotel opened near the narrow end of the triangle—the city's largest hotel. The following year, the Kendall Square T station was enlarged and

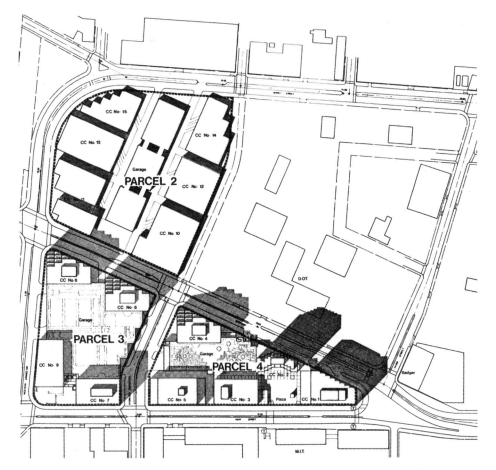

Figure 14 A 1987 map showing completed and proposed Cambridge Center buildings. Unmarked in the upper right is the Volpe Center. Parcel 2 is the eleven "surplus" acres reclaimed by the city. Parcels 3 and 4 account for the thirteen additional acres always slated for commercial development. The first buildings were in parcel 4 in 1981 and 1983. The Marriott Hotel, completed in 1986, is near the narrow end of parcel 4. Amid this parcel is the rooftop garden conceived by Safdie and designed by landscape architecture firm Peter Walker and Partners.

modernized, and the year after that the plaza between the station and the Marriott opened to the public.

Safdie describes seeing Kendall Square soon after he moved to Boston. "It was somewhat like a bombed-out area. I mean it was desolate. There was not a single soul on the streets," he recalls. At the time Zuckerman reached out to him, the real estate mogul was having trouble getting the master plan approved. "He said, 'I'm stuck,'" Safdie remembers.

One of first things Safdie did was to rethink how it all flowed together. "The original plan had a parking garage along Broadway," he says. "So as you came down Broadway, you saw a big goddamn garage. It was very street unfriendly. So I understood that the key to this was to do two things—create an effective piazza that really could become a hub of life and feed into MIT, and secondly, to internalize the garage so its presence from the street would be minimal."

Those insights became core to his plan. He proposed a large piazza facing MIT that "would be the arrival point for the subway." German artist Karl Schlamminger, a friend of Safdie's, came up with the concept of the Persian carpet pattern of paving that characterizes the plaza. To crown it all, and to meet the open space requirements, Safdie moved the parking garage more into the belly of the complex and conceived of a public park atop the structure amid the parcel's cluster of buildings. "A public park on the fourth floor was without precedent," he states. Peter Walker and Partners was the landscape architect. The firm, now known as PWP Landscape Architecture, would codesign the National September 11 Memorial. The Kendall Square rooftop park, which was being expanded and reimagined in 2020, won lots of awards at the time, says Safdie. "But the key thing was it was a precedent, a public park on the fourth floor. I went on in many other places to do it later."

More than two decades after Tech Square, the revitalization effort it had hoped to spark for Kendall Square at last found traction. The effort fell short of its ambitions in some key ways: not only did NASA plans fail to come to fruition, not a single unit of housing was created inside the urban renewal district until 2018. Still, thanks largely to concerted action by a unique collaboration between city, university, and industry officials, the grim conditions of the square were slowly turned around. In the words of Simha, "The important thing is it certainly accomplished one objective, which was to give the city an economic base which now makes it one of the wealthiest, if not *the* wealthiest, in the Commonwealth."[16]

Even as Kendall Square got its long-awaited facelift, though, other questions remained largely unanswered. The most basic was, what kinds of companies would go in the new buildings? If NASA wasn't going to lead the way to the future, then what was? As the urban renewal drama played out, two major advances were steaming forward in science and technology, with MIT playing a big role in both. One centered on artificial intelligence and new frontiers in software and computing. The other centered on the emerging world of gene manipulation—and a field called biotechnology. The first big bets—the bulk of the hype at least—were on software and AI.

11 TECH SURGE: LOTUS TO AI ALLEY

It was a beautiful business plan. The young founder nurtured an idea for what he saw as a stupendous software product—and optimistically conjured up a financial forecast that showed $3 million in sales for the first year. There was just one problem. As is all too often the case with wild-eyed entrepreneurs, the plan proved far from reality. Only in this particular situation, it jackrabbited off in an unusual direction. When the numbers came in, the actual first-year revenues totaled $53 million. "A factor of 1700 forecasting error to the right side," sums up Mitch Kapor.[1]

That was the story behind Kendall Square's biggest locally grown tech star of the 1980s—and its number one software company of all time: Lotus Development Corporation. The company's name was chosen by Kapor, a former teacher of transcendental meditation, as a gesture to the *lotus position*, the cross-legged sitting posture associated with Eastern practices. The first-year results put him and Lotus cofounder Jonathan Sachs in a state of bliss.

A bevy of high-flying information technology company stars arose in Kendall Square during those days. One swath of the square harbored so many "brave new world" artificial intelligence startups that it was known as AI Alley: a billboard touting that moniker loomed over the corner of Broadway and Galileo Galilei Way. Lotus was the big kahuna, though, setting a new standard for office productivity with its combined spreadsheet, graphics, and database product called Lotus 1-2-3. It lost some viability over the years, but sold to IBM in 1995 for $3.5 billion. Kapor even got a tile embedded in the Marriott Hotel plaza as part of the Entrepreneur Walk of Fame that debuted in 2011. The others in that inaugural, and still only, class: Thomas Edison, Bill Gates, Bill Hewlett, David Packard, Bob Swanson, Steve Jobs, and Thomas Edison. Kapor is the only former Kendall Square denizen among them.[2]

Lotus was actually born in Central Square. Kapor, who grew up on Long Island, had gotten an unusual undergraduate degree from Yale University in cybernetics, which combined psychology, computer science, and linguistics. After college, it wasn't exactly a linear path to Lotus for the eclectic Kapor, who already exhibited a passion for flowered shirts and pursued a variety of interests. One of those interests was his college girlfriend, whom he followed to Boston when she got a job at a local public television station. He found he liked life there, especially in Cambridge. "I was certainly influenced by stuff going on around MIT. I did a lot of reading—I made up for my misspent college, when I learned next to nothing, hanging out in bookstores and catching up on [Marxist scholar and philosopher] Herbert Marcuse and all that."[3]

Kapor had spent a couple of years as a programmer and another twelve months or so teaching meditation, gotten his master's degree in counseling psychology, and then worked part-time in a hospital psychiatric unit. He soon left that to go into technical consulting. On the heels of some modest success writing a financial application called Tiny Troll for the Apple II personal computer, he enrolled in the Sloan School's accelerated master program—which compressed the normally two-year program into one intensive year.

Just before entering Sloan in June 1979, Kapor had been hired to adapt a version of Tiny Troll for a California company called Personal Software Inc. (PSI). Much like a record company or book publisher, PSI published software written by different programmers who worked for themselves and paid them royalties based on sales. It had just published VisiCalc, the first spreadsheet for personal computers. VisiCalc was created by two former MIT geeks, Dan Bricklin and Bob Frankston.

Kapor had planned to finish his Tiny Troll revisions while still at Sloan, but it proved impossible to get done while maintaining the intense Sloan workload. He dropped out before completing his degree and moved to Silicon Valley, but returned to Boston less than six months later to devote himself to completing the effort. He called his solo company Micro Finance Systems, and rented a basement office on Franklin Street in Central Square, not far from the VisiCalc team. He ended up making substantial improvements to Tiny Troll, which was ultimately rebranded as VisiPlot and sold starting in early 1981 as a kind of adjunct to VisiCalc, along with a related program for statistics called VisiTrend that Kapor had also created. By that fall, Kapor had earned over $500,000 in royalties, and PSI bought him out for an additional $900,000.

With the influx of money enabling him to buy a house in Cambridgeport and still devote resources to his business, Kapor began to consider new software products he might create. In July 1981, he hired Sachs, a talented programmer who had already developed two spreadsheet programs for minicomputers. Soon thereafter, Micro Finance Systems was reincorporated as Lotus Development Corporation, with Sachs named a cofounder. The partners quickly zeroed in on the idea of creating a program that combined a spreadsheet with graphics capabilities and some word processing features that enabled data to be easily handled and presented. As a Harvard Business School case study by William Sahlman later noted, Kapor had decided the market for such personal productivity tools "was on the verge of exploding."[4]

That would turn out to be putting it mildly. When IBM announced its first personal computer in the fall of 1981, Kapor grew excited about its potential to penetrate the business market. "I saw where the market was going, made a big bet on IBM," he says. By the spring of 1982, Lotus had raised $1 million in a financing round co-led by Ben Rosen at the newly formed Sevin Rosen Partners, which had recently backed Compaq Computer, the first IBM PC–compatible alternative computer. The other lead investor was Kleiner, Perkins, Caulfield & Byers. Lotus soon moved out of its cramped

Figure 15 Mitch Kapor, 1982.
Source: Mitch Kapor.

basement, renting bigger space in West Cambridge near Fresh Pond and the Alewife T station from analytics and consulting firm Abt Associates.

The company released Lotus 1-2-3 on January 26, 1983. But it had started taking preorders the previous November at Comdex, and within a few days advanced sales for the $495 program had already surpassed $1 million—a harbinger of the good times to come. After its first-year total of $53 million, revenues would triple to $157 million in 1984, making Lotus the world's leading personal computer software maker, ahead of Microsoft.

The surging demand forced a hiring spree. With Kapor serving as CEO, Lotus went from about twenty employees in early 1983 to 250 by year's end, and 750 a year after that. To fuel that growth, the company raised another $5 million or so in venture financing. It also filed for an IPO, successfully going public in October 1983 at eighteen dollars per share and garnering another $41 million.

Even by the spring of 1983, packed into increasingly tight quarters at Abt, it was clear Lotus needed more space—a lot of it. "We needed to find a permanent home," says Kapor.

Enter Kendall Square. At that time, Kapor remembers, virtually every big Boston area high-tech company had moved out to Route 128, or even farther west along I-495—led by minicomputer companies like Digital, Data General, Wang, and Prime Computer. The prevailing wisdom was that not only were rents cheaper, affording more spacious offices and pleasant grounds, but the schools were better and houses more affordable and desirable for families.

Kapor didn't buy it. "The issue with all these places out on 128 and 495 was there was nothing out there—there were malls and tract housing and soulless office buildings. I wanted a place that did not require you to have a car to work—good public transit, which would make it easier for many different kinds of people to work there."[5]

Kendall Square fit that bill almost to a tee—close to Boston and its northern suburbs, and on two subway lines: the Red Line running into the heart of the square and the Green Line going to the Lechmere Station in East Cambridge. But there was almost nothing to rent in Kendall Square proper. Only the first building in Cambridge Center—Five Cambridge Center—had opened, and it was fully leased. Beyond that, says Kapor, "There was literally nothing available." A bit farther northeast toward Lechmere, along Binney Street, things opened up. He settled on the former Ashton Valve Company building at the corner of Binney and First Streets. Around the fall of 1983, Lotus took the entire three-story building, which still had glue and assorted junk in its basement. The mess was cleared as part of a full renovation before Lotus moved in. As the company continued to grow, rows of cubicles were set up in the space: employees dubbed the basement offices the Rifle Range.

For Kapor, that building at 161 First Street offered a near-perfect location, almost equidistant between the Lechmere and Kendall Square stations, and also close to MIT, home to interesting research, talks, and potential recruits. "I thought it would send a signal. It would be the place where we could get young, single people to work. I didn't know that we were setting a precedent, but we were," he says.[6]

Lotus also set precedents on the community and employee benefits fronts. As much as he needed funding in early 1982, Kapor was adamant that money-making wasn't all he was about, a fact he stressed in a cover letter sent that January to Ben Rosen along with his business plan:

> While I am keenly aware of the Leo Durocher theory of management ("nice guys finish last"), I am unwilling to operate a priori in an over aggressive, profit–only centered style, nor would I like to see the company as a whole begin to run that way. Rather, I am committed to trying to live with and work out as best as possible the contradictions involved in running a profitable enterprise with my other values, such as the notion that it's better to treat people well as an end in itself, rather than only as a means of improving productivity.[7]

The company practiced what Kapor preached. Lotus is credited with being the first major US company, in 1986, to support a walk to benefit people with AIDS. Six years later, it became the first major corporation in America to offer full benefits to same-sex partners. "The legacy that I'm most proud of is the impact it's had on the people who were there. It was much more long-lasting than the products," says Kapor.[8]

Kapor was never comfortable as CEO of a big company. "I was well-suited to start things and get things to market. It doesn't play to my skill set to be managing at a big company," he relates. Early on, he brought in Jim Manzi from McKinsey & Company as a management consultant. By fall 1984, Manzi had been named president, and one year later, when Kapor decided to leave Lotus's daily operations, Manzi took over as CEO. He led Lotus through the sale to IBM in 1995.[9]

Lotus made a major impact on the Kendall Square and East Cambridge landscape. Less than two years after it moved to 161 First Street, the company expanded across Land Boulevard to a newly completed office building at 55 Cambridge Parkway, overlooking the Charles River. After that, it broke ground on a new headquarters complex at One Rogers Street: the building opened in 1988.

All through the 1980s, the area continued to transform around Lotus. The CambridgeSide mall, a block from the new Lotus complex; Moshe Safdie's Esplanade condominiums along the river; and the River Court condominiums at 10 Rogers Street all went up.[10]

This activity dovetailed with development back in Kendall Square proper—where the Cambridge Center complex was mostly completed in the second half of the 1980s and more office buildings arose down Main Street leading to the river. Along Third Street, though, things got sketchy. The Volpe loomed as a towering blob. Between it and Lotus, you were hard pressed to find thriving buildings. "We really felt apart, distinct from Kendall Square. In between us and it there was nothing," remembers Kapor.

"A lot of parking lots, a lot of vacant lots," adds Reed Sturtevant, a two-time MIT dropout and former Lotus employee who has lived or worked in the Kendall Square area for much of the past forty years.[11]

Lotus served as the corporate anchor on the Lechmere end of Kendall Square. But another tech movement happening in parallel had grabbed nearly as many headlines in the 1980s. The most notable of these companies would set up its headquarters barely two blocks from Lotus on First Street, closer to the Longfellow Bridge. It helped form the vanguard of an emerging branch of software and computing called *artificial intelligence*. The company's name, fittingly enough, was Thinking Machines.

* * *

The nexus of the artificial intelligence emergence was Project MAC at 545 Tech Square. A somewhat forced combination of Marvin Minsky's artificial intelligence group and the hackers from the Research Lab of Electronics and other MIT operations, Project MAC occupied the top two floors—with mostly computers on the ninth floor and offices on the eighth. But MIT had a bigger presence in the building. The cofounder of Minsky's group was Seymour Papert, a pioneer in the Logo programming language. His Logo Lab took up about half the third floor. The other half—known as Suite 304—was occupied by the CIA.[12]

The CIA offices were unmarked. No sign in the lobby, nothing on the doors. In the early 1980s, one of the MIT graduate students connected to the Logo Lab was Danny Hillis. Cherub-faced, already losing his hair in his twenties, Hillis had a wicked sense of humor—and an antiauthoritarian bent. He and other lab kindred spirits liked to have fun at the intelligence agency's expense. Every once in a while, for instance, a visitor would ask where Suite 304 was.

"Oh, who are you looking for?" Hillis would ask innocently.

The person would maybe hem and haw a bit and say something like, "Oh, you know, I'm just looking for Suite 304."

As Hillis recounts, "And then I would say, 'The blue moon jumps over the purple sky,' or some other nonsense line." He would then stare at the visitor as if expecting the other part of a code phrase. "And they would be like, 'Uh . . . ,'" he says.[13]

Hillis loved messing with the CIA. Like several other MIT students, he was a talented lockpicker. "That building had Medeco locks, which are really hard to pick," he remembers. "But we did pick the locks on the CIA just to go in there. And there was nothing except a safe—the desks were completely empty." So Hillis left a note:

"Glad to see you put everything in the safe."

It's hard to imagine college students playing games like that with the CIA these days. But such was the wild and wooly MIT hacker culture at the time. Getting past locked doors wasn't seen as a crime; it meant meeting a challenge. "They went wherever they wanted, entering offices by traveling in the crawl space created by the low-hanging artificial ceiling, removing a ceiling tile, and dropping into their destinations—commandos with pencil-pals in their shirt pockets," Steve Levy wrote in *Hackers*.[14]

Hillis tells two other "battle" stories illustrative of the times. One involved an ongoing issue with the elevators. They always had to go to the lobby before starting a run. That meant if you were on the ninth floor and called an elevator that happened to be on the eighth floor, it would first go to the ground floor before coming back up. "It was a real pain," Hillis says.

It dawned on him that the lab had a lot of computing power at its disposal—and the computers were on ARPANET, the early version of the internet. "I was like, let's just put the elevator on the network," he relates. So that's what he did. "You could hit a key on your keyboard, and it would send a call over the internet to this machine to call the elevator. It was probably the first Internet of Things device. The hard part was I had to put a little relay box in the elevator controller. And I thought, well, they're going to notice that and take it out." His solution lay in a prototype graphic printer Xerox had donated to the lab. "We could print labels that looked like official labels when nobody knew that was a capability," says Hillis. "So I printed this label that said 'Warning, do not remove this device without authorization.'" As long as he was there, no one messed with the box.

Hillis's other battle was with MIT. His master thesis centered on building computers whose processors would work simultaneously rather than sequentially. He saw this "parallel computing" as key to cracking artificial intelligence. To do AI and begin to simulate human thinking, you needed to crunch a lot of data quickly, and that took huge amounts of processing power—far more than was available then through conventional techniques.

The problem with MIT arose when Hillis formed a company around his thesis work. "It turned into this very big project with money coming in from DARPA. It was just getting too big to be at the MIT AI Lab. So I started Thinking Machines and basically the administration came to me and said, well, you can't do that while you're a student."

One issue, according to Hillis, was that MIT officials felt his forming a company created a conflict of interest. "And I said, well, that must be for faculty—because they work for you. I'm a customer." MIT also claimed it owned the intellectual property behind the work. Hillis didn't buy that, either. He offered the school a share, which it never took, he says. In the end, "I just went ahead and did it. MIT eventually decided that the faculty conflict of interest rules didn't apply to me."

Thinking Machines was founded in 1983 when Hillis was twenty-six. Its parallel computer, dubbed the Connection Machine, proved a headline-grabbing achievement. Hillis became one of America's most heralded young inventor-turned-entrepreneurs. The company started in a historic mansion on the Robert Treat Paine estate in Waltham. However, Hillis notes, "As cool as it was, we quickly outgrew it." The natural move was back to Kendall Square. "We did want to be back near MIT because Thinking Machines was mostly MIT students, MIT professors."

In 1984, Thinking Machines moved into the top two floors of 245 First Street, in the old Carter's Ink building. Edwin Land's Rowland Institute, created four years earlier, sat kitty-corner across the street, along the river. Hillis and Land developed an inventors' kinship. "He was working on his theories of color vision and things like that,"

Figure 16 Danny Hillis, center, with Connection Machine prototypes at a going away party for staffer Tamiko Thiel in 1985. From left: Keira Bromberg, John Huffman, Thiel (kneeling), Hillis, Carl Feynman, Arlene Chung. Photo courtesy Tamiko Thiel.

Hillis remembers. "And every time he would finish an experiment, he would give me a call and I would come over and watch."

Thinking Machines got almost instant credibility and buzz. Not only was there the allure of its young founder, but Marvin Minsky was a cofounder and director, and former MIT president Jerome Wiesner became an advisor. Things kept building from there as the company became an East Coast version of Silicon Valley almost before Silicon Valley. It had a dynamic CEO, Sheryl Handler, one of the few women to head a tech company (Handler had previously helped start Genetics Institute, one of Boston's first biotechnology companies). Open spaces were created to spur interaction and creative thought, and couches installed so people could take naps or even spend the night. It also was among the first companies to institute free lunches for employees at a high-end cafeteria. Wiesner and other faculty would come over from MIT, often bringing guests. "I think we served better food than the faculty club," remembers Hillis.[15]

Money flowed as well. A $4.5-million multiyear grant from DARPA had spurred the company's formation. Thinking Machines soon attracted $16 million in venture and angel funding: investors included CBS founder William Paley and CBS CEO Fred Stanton. The company grew up fast, becoming profitable in 1989, when it showed a $700,000 profit on $45 million in sales. It entered into contracts with big companies like American Express and Schlumberger; national laboratories at Los Alamos and Sandia; the US Army; and a host of universities.

The first Connection Machine cost $3 million and harbored 65,536 processors able to perform a stunning nine billion mathematical calculations per second, pioneering a class called *massively parallel supercomputers* and doing battle with the likes of Intel and Cray. The company quickly filled up most of the Carter's Ink building and spread to the tower behind it, on its way to reaching a peak of 425 employees and $90 million in revenues in its top year of 1991.

Some big names in tech and science joined forces with Thinking Machines. Early employees included research director Jill Mesirov, who would later become a professor and associate vice chancellor at UC San Diego's School of Medicine; Greg Papadopoulos, future CTO of Sun Microsystems and later a top-tier venture capitalist; and Brewster Kahle, future founder of the Internet Archive. Signing on as Thinking Machines fellows were names like Stephen Wolfram, Douglas Lenat, and a young mathematician named Eric Lander, who had recently switched gears to go into biology and would soon head to the Whitehead Institute near the other end of Kendall Square. While getting ready for one of its first press conferences to showcase a Connection Machine, Hillis was escorting some science reporters around while workmen scrambled to get ready for the event. One reporter turned to Hillis and said, "Hey, that's funny. That painter up on the ladder looks like Richard Feynman," referring to the Nobel Prize-winning physicist.

Hillis glanced over. "Oh, actually, that is Richard Feynman." The Caltech scientist was another Thinking Machines advisor. "He was, like, helping touch up the paint," says Hillis.

* * *

Thinking Machines became one of the iconic companies of AI Alley. The phrase referred to the swarm of startups focused on artificial intelligence, expert systems, and the supercomputers needed to power them. Notable AI Alley inhabitants included Symbolics, Palladian Software, Allied Expert Systems, Brattle Research, Lisp Machines, Bachman Information Systems, and Gold Hill Computers.

Most of the AI Alley companies had strong ties to MIT, and more specifically, its Artificial Intelligence Laboratory. "Practically every faculty member associated with the AI lab is an equity owner of at least one company," Gerald Barber, a cofounder of Gold Hill Computers once related.[16]

The term *alley* referred to the fact that at least a quintet of these upstart companies set up shop on or near Main Street in Kendall Square. Symbolics, Palladian, and Allied Expert Systems were all in various buildings in Cambridge Center. Thinking Machines and Brattle Research stood on different sides of the Broad Canal. Gold Hill was a bit of an exception—it found space west of Tech Square, in a converted Armenian dance hall on Harvard Street. A host of other Cambridge companies not actually in Kendall Square had also jumped on the AI bandwagon in different ways, among them Bolt Beranek and Newman; Arthur D. Little; and Lisp Machines, another AI Lab spinout that had offices on Massachusetts Avenue closer to Harvard Square. To try and catch the wave, IBM set up an artificial intelligence lab in Kendall Square in 1986.

But AI Alley referred to more than the companies and their physical locations. "It is a subculture containing the myriad of social types: the awkward, wide-eyed hackers who send out for pizza at midnight; consultants who call Japan from their car phones; and venture capitalists looking for a stake in the future. They don't all speak the same language, but they are united by the excitement and risk of riding a largely untried technology," proclaimed the *Boston Globe*.[17]

Despite competition from other parts of the country, most notably Silicon Valley and around Carnegie Mellon University in Pittsburgh, the locus of the burgeoning AI industry was Kendall Square. From virtually nothing early in the decade, the industry counted $660 million in revenues in 1986, and was expected to reach $5 billion in 1991, the article reported. "What you're seeing is the birth of an industry that's going to be greater than the conventional computer industry. What happened on Route 128, and then in Silicon Valley, is going to happen here on AI Alley," predicted one leading consultant.[18]

With its eye on the future, on May 24, 1985, Thinking Machines obtained the third dot-com domain ever issued: think.com. The second domain—bbn.com—was issued exactly a month earlier to Bolt Beranek and Newman, which had done pioneering work on the early internet.

The very first domain—symbolics.com—was registered on March 15, 1985. Symbolics was another MIT spinout and, along with Thinking Machines, the other flagship startup of AI Alley. It wasn't as sexy as Hillis's company, but it rivaled or surpassed its counterpart by some key measures—peaking at $84.5 million in annual sales and just under one thousand employees, and successfully going public. And maybe *spinout* isn't the right descriptor for its relation to the AI lab: some considered *mass exodus* more fitting. Four full-time members of the lab and one part-timer left to join it. That didn't count the company's longtime leader, Russell Noftsker, who had previously been administrator of the AI Group-turned AI Lab but was living on the West Coast when he cofounded Symbolics.

A section in *Hackers* delves into the split inside the AI lab and the newly created Laboratory for Computer Science, which replaced the Project MAC group, around starting a company to make computers capable of running the artificial intelligence language LISP.[19] I won't attempt to sort it all out here. But in the end, Noftsker and his core group could not agree with another key member of the lab, Richard Greenblatt, on such things as the structure of the company, ownership stakes, and whether to seek outside capital. So Noftsker and those who sided with him agreed to wait a year for Greenblatt to pursue things his way before forming a company of their own.

Greenblatt was a legendary programmer. He had among other things created the first tournament-level computer chess program, Mac Hack. In 1977, Bobby Fischer had come to Cambridge and played it, winning all three games.[20] Greenblatt launched Lisp Machines in 1979. In April 1980, just after their stand-in-place agreement had lapsed, says Noftsker, the rival group he was part of incorporated Symbolics. All told, the company listed twenty-one people as cofounders. Sixteen of them had degrees from or direct ties to MIT.[21]

On the strength of $500,000 raised from friends and family, the Symbolics contingent licensed the MIT Lisp machine technology and started building computers in a factory near where Noftsker lived in the San Fernando Valley. Shortly after Noftsker was named Symbolics's president—in effect, its CEO—the company raised an additional $1.5 million in venture capital.[22]

Although manufacturing stayed in California, the company set up an R&D arm in a rented MIT building on Vassar Street. In a rare example of a West Coast entrepreneur moving east, Noftsker relocated back to Cambridge and established the company's headquarters there as well. The group set up a connection to ARPANET between the Vassar Street operation and MIT's AI Lab. That led, a few years later, to Symbolics having a representative on the committee working to spin ARPANET out into the internet. "And that's how we got the first dot-com address," Noftsker relates.

The computing category companies like Lisp Machines and Symbolics helped create was soon dubbed *programming workstations*. These were powerful desktop machines that could crunch financial data and perform graphical and scientific calculations. "At the time we started Symbolics there were no single user workstations on the market," says Noftsker.[23] For Symbolics, growth came quickly. Its initial customers were companies and labs doing AI research: Thinking Machines was one of its biggest. The market expanded significantly when a pot of federal funding became available for artificial intelligence work under Ronald Reagan's Strategic Defense Initiative, commonly known as Star Wars.

When Five Cambridge Center opened in Kendall Square, with Legal Sea Foods on its ground floor, Noftsker moved his office there, along with his accounting, legal, and financing departments. Later, he shifted operations at various times to other center buildings and finally consolidated everything in the four stories of Eleven Cambridge Center, which was built to house the company.

Symbolics went public in 1984, trading on the Nasdaq as SMBX and raising $15 million. A secondary public offering less than a year later garnered roughly $35.9 million more. At its peak in fiscal year 1986, the company reached $84.5 million in sales, with just under $9 million in profits, says Noftsker.[24]

Noftsker resisted the AI label for Symbolics, emphasizing that the company provided workstations that let others do advanced AI research. As such, Symbolics had ties to several other AI Alley companies, at times donating its machines to startups creating software to run on Lisp-capable workstations in order to help them build the industry. One of those it worked with was Palladian Software, which for a time owned close to forty Symbolics workstations housed in air-conditioned rooms.

Palladian didn't fly as high, but it made quite a splash. The driving force behind it was a Sloan fellow—like Noftsker, a pilot—named Phil Cooper who was doing his master thesis on artificial intelligence. Cooper believed that artificial intelligence techniques could be used to create software for fields such as accounting and finance that could in essence incorporate the accumulated best practices and knowledge of experts. By the spring of 1984, he had refined this "expert systems" idea into the concept for Palladian and convinced five professors—two specializing in AI and the rest in aspects of corporate management, finance, and strategy—to sign on as cofounders. Abraham Siegal, Sloan's dean, joined Palladian's board.

Later that spring, Cooper began pitching venture-capital firms. Instead of crafting a conventional business plan he carried along a copy of his 214-page thesis together with a one-page summary of projected financials and a list of the heavyweights who had joined forces with him. By that June, he had raised $1.9 million from a trio of blue-chip firms. They included Venrock and Welsh, Carson, Anderson & Stowe, both of which had invested in Computer Pictures, the company he had founded before going to Sloan. The third was probably the number one VC shop of the times, Kleiner, Perkins, Caulfield, & Byers.[25]

It was Cooper who put up the AI Alley billboard. "There was an explosion of AI-related companies," he relates. "So this was mostly a joke to myself. Just for fun I came up with the idea for that billboard and coined the term AI Alley."[26] The billboard showed a photo of the staffs of Palladian and Bachman Information Systems and compared AI Alley to the Garden of Eden, Plymouth Rock, and other "great beginnings." Bachman was another Kendall Square startup that Cooper had cofounded, which was creating a smart database management system. Its other founder was a bona fide legend, Charles Bachman, who had developed one of the first database management systems while at General Electric in the early 1960s. For this pioneering work he had won the 1973 Turing Award, becoming the first recipient without a doctorate. As Bachman's *New*

York Times obituary in 2017 noted, "behind every product search on Amazon, movie recommendation on Netflix or bid on eBay, there is a flood of digital communications mediated by database management software, which owes an intellectual debt to Mr. Bachman."[27]

Given the immense talent and promise concentrated around the field, there was no shortage of hype around Kendall Square's AI Alley companies. Down the road a few years, mockups of Connection Machines would be visible in the *Jurassic Park* control room. In *Mission: Impossible*, Ving Rhames's Luther Stickwell character, the main sidekick to Tom Cruise's Ethan Hunt, requests "Thinking Machines laptops" (which never existed) to hack into the CIA.[28] At one point, the four fastest supercomputers in the world were Connection Machines—a staggering achievement for a company born from a graduate thesis. It all made for great copy.

The line Danny Hillis came up with, and still delivered with a big smile more than thirty-five years after the company was founded: "I want to build a machine that will be proud of me."

* * *

The good times for AI Alley wouldn't last terribly long. Within about a decade, virtually all the alley companies, including Palladian, Symbolics, and Thinking Machines, would disappear or become shells of their former selves. Along with others around the country, they were victims of the downturn proclaimed the AI Winter. The problems they were tackling were just too hard for the technology of the times. Investors got impatient, customers and money dried up, especially with the demise of the Strategic Defense Initiative in 1993—and at least in some cases battles with competitors also drained resources. "What ultimately led to the bust of the first AI bubble was the computer power necessary to do all this stuff," says Phil Cooper, who later led Goldman Sachs's private equity group and now runs his own private equity firm, Pine Island Capital Partners. "It requires unholy amounts of computer power and we just didn't have the horsepower." Even powerful workstations like those from Symbolics were not as powerful as a modern cell phone, he notes. Only today is AI approaching the vision pursued back then.

Thinking Machines officially filed for bankruptcy in August 1994.[29] Less than a decade earlier, a leading consultant had proclaimed to the *Boston Globe* that AI Alley was home to "the birth of an industry that's going to be greater than the conventional computer industry." Needless to say, that did not happen—at least at the time.

The vacant buildings and parking lots around Thinking Machines and Lotus would be fully occupied in the not-too-distant future. But as bright as their promise had seemed, it wasn't software or AI companies that drove the transformation.

And like MIT coming to Cambridge and so many other elements of the Kendall Square story, it almost didn't happen.

12 THE ORDINANCE AND BIOGEN

A small crowd gathered midway along Binney Street in Kendall Square for the ribbon-cutting. The group included a phalanx of civic leaders, MIT and Harvard officials, and the core local scientific team of Biogen, the company that would occupy the two-story building at 241 Binney, at the corner of Sixth Street. The date was February 23, 1982. It was a historic occasion. The newly renovated building was about to become the research and development headquarters of Biogen—making it the first biotechnology company in Cambridge, and the only one to be operating under the city's landmark ordinance regulating recombinant DNA experiments inside its limits. The company had been founded almost four years earlier in Europe, and its headquarters remained in Geneva, Switzerland. But a key part of its founding team was local, including two top scientists, Wally Gilbert of Harvard and Phil Sharp of MIT. So Biogen was also coming home to its roots in a way.

Of the two local founding scientists, only Gilbert was in the crowd that day—Sharp was traveling. In 1981, not long after winning a share of the 1980 Nobel Prize in chemistry, Gilbert had left his tenured position at Harvard to become Biogen's CEO, called the president at the time. The building, outfitted with then state-of-the art lab space, has since been razed and the entire block redeveloped to incorporate two adjacent historical structures into a new Biogen headquarters christened the Phillip A. Sharp Building. Around it today are a bevy of newer office, lab, and apartment buildings, including an entire complex devoted to Biogen. At the time, the roughly seven-year-old Draper Lab campus lay a few blocks around the Binney Street curve to the southwest. But dead ahead and toward the river, other than the drab tower and slabs of the Volpe Center and the first buildings of the Kendall Square urban renewal project (Five Cambridge Center had been completed and the second structure, Four Cambridge Center, was under construction), the entire area was largely deserted or run-down.

It had taken extraordinary effort to get things to the point where Biogen could operate a laboratory in what in a sense was its own backyard. That was largely because of one of the dignitaries present, Cambridge mayor Al Vellucci. The brash Vellucci had led the charge in questioning the safety of the new recombinant DNA technology. He had orchestrated a series of high-profile, often contentious public hearings and other meetings challenging the scientific elites, as he saw them, to prove the technology was safe—and made it clear the city was not going to roll over just because Harvard and

MIT wanted something. "They may come up with a disease that can't be cured—even a monster. Is this the answer to Frankenstein's dream?" he had once posited to a *New York Times* reporter covering the debate.[1]

Vellucci, who would serve on the city council for thirty-four years, was the quintessential local pol—tough-minded, pugnacious, a thorn in the side to many, a fighter for the people to many others. But on this day, amid the third of four nonconsecutive terms as mayor, he was mostly smiling—though perhaps a bit painfully. The hard times, after all, were behind them—Cambridge had passed the first public ordinance in the country setting out the rules for recombinant DNA (rDNA) experiments—and Biogen was the first company to receive the permit. At the get together, the mayor quipped to the small group gathered there: "It's kind of hard for me to say 'welcome,' but we do welcome your taxes."[2]

And boy, did biotech pay its taxes. No one present could have known then the extent of what was happening that day—that the dedication marked an important early step in making Cambridge and Kendall Square the biotech capital of the world. Only a few decades before that 1982 ribbon-cutting ceremony, factories and other industrial operations were disappearing and Cambridge could not even float a bond to redevelop Tech Square just down the street. In the decades that lay ahead, the taxes received from biotech companies and the big pharma corporations that later set up shop in and around Kendall Square, and their landlords, would make Cambridge one of the wealthiest cities in the nation. For fiscal year 2020, the top three corporate property taxpayers were all biotech-centric real estate landowners and developers: Alexandria Real Estate Equities, BioMed Realty Trust, and Boston Properties. They owned a huge percentage of the lab and office space leased by life science companies. Novartis ranked seventh. Together, those four entities alone were to pay $64.2 million in property taxes, about 15 percent of the city's $438 million total tax levy.[3]

But as people gathered for the Biogen dedication in 1982, the days of stuffed coffers were still far down the road. "I wanted to buy the block," remembers Wally Gilbert, "so it would become clear to everybody that we're serious—we're here to stay."[4] But his board had nixed the idea. Even though Biogen had the cash, the $1 million asking price was too steep, the future too uncertain.

*   *   *

When it was chartered in 1978, Biogen seems to have been just the fourth biotechnology company devoted to commercializing recombinant DNA technology in existence. Genentech had been the trailblazer when it came to employing recombinant DNA techniques: it had been formed in the Bay Area in 1976 by a group that included one of the inventors of rDNA technology, Herb Boyer of the University of California, San Francisco. But Cetus across the bay in Emeryville had been founded even earlier, in 1971. And just ahead of Biogen in 1978, another startup called Genex had gotten going in Rockville, Maryland.

But while biotechnology as an industry began to emerge in the late 1970s, the story behind its rise in Kendall Square in many ways dates back to the 1950s, when Cambridge became a center of molecular biology in large part thanks to the arrival of James Watson and Salvador "Salva" Luria. Watson joined the Harvard biology faculty in 1956. Three years later—the same year Lever Brothers closed down its soap factory and a true changing of the guards period in Kendall Square's history—Luria arrived at MIT. The two were closely coupled. Watson had been Luria's first graduate student at the University of Indiana nearly two decades earlier. Luria had helped nurture his career and connected Watson to an opportunity at the Cavendish Laboratory at the University of Cambridge in England. There, Watson had collaborated with Francis Crick on the work that led to their discovery of the double helix structure of DNA in 1953 that would win them both a share of the Nobel Prize less than a decade later. Luria, for his part, would win the Nobel in 1969 for his discoveries on the nature of genetic mutations.

Watson and Luria made for an extremely powerful combination, with an almost magnetic power to attract other top scientists. At Harvard, Wally Gilbert was an assistant professor of physics. He and Watson had become friends at the University of Cambridge, where Watson worked in the Cavendish lab and Gilbert was getting his doctorate in physics. They reunited at Harvard, and through Watson, Gilbert had gotten so hooked on biology that he switched fields. Other soon-to-be-stars drawn to molecular biology there included Mark Ptashne, Nancy Hopkins, and Tom Maniatis.[5]

Meanwhile, at MIT, Luria attracted rising star biologist David Baltimore in 1968; the young scientist initially came as a visiting professor but soon joined the faculty. The Italian-born Luria "had the job of building molecular biology at MIT," says Baltimore. "There really were no departments of molecular biology anywhere."[6] Luria upped the ante dramatically in 1972 when he won a National Cancer Institute grant, part of Richard Nixon's "war on cancer," to create a new facility devoted to basic cancer research. He found space in an old chocolate factory at the corner of Main and Ames on the northeast edge of campus, and he and Baltimore began hunting for faculty to fill it. By the time the state-of-the-art MIT Center for Cancer Research officially opened in 1974, they had recruited a bumper crop of top researchers that included Robert Weinberg, who was already on the MIT faculty; David Housman from the Ontario Cancer Institute; and Phil Sharp and Nancy Hopkins from the Cold Spring Harbor Laboratory on Long Island.[7] "By summer 1974, one of the most powerful teams of cancer biologists in the world was assembled . . . and they were all set to go with a research program that depended crucially on rDNA technology," wrote science historian and MIT Museum director John Durant.[8]

There was more to what was going on than just attracting world-class scientists. "I want to get across the intense Harvard-MIT rivalry at the forefront of molecular biology," notes Victor McElheny, who worked for Watson at Cold Spring Harbor during the late 1970s and early 1980s and later became his biographer. "Obviously, this type of intense frontier competition is at the heart of innovative pressure."[9] Baltimore concurs:

"When we dedicated the cancer center, we had a little ceremony, and Watson came and gave a talk. And Watson said what you people are doing at MIT is what we should be doing at Harvard. There was a tension about that over many, many years."[10]

That keen competition would ultimately play a major role in turbocharging the biotechnology scene in the area. But in concert with the launch of MIT's new cancer center and the arrival of its top recruits intent on doing rDNA experiments came a major controversy that brought the emerging field of recombinant DNA almost to a standstill. It would take several years for the dust to settle. In the end, though, the debate ended up being crucial to the rise of Kendall Square as a biotechnology hub.

In a way, it started with a meeting in Baltimore's office on April 17, 1974. The confab had been spurred by Paul Berg, who was visiting from Stanford. Just six others were present: James Watson, Sherman Weissman from Yale, Norton Zinder of Rockefeller University, Harvard Medical School's Richard Roblin, Daniel Nathans from Johns Hopkins, and National Science Foundation staffer Herman Lewis. It was an extraordinary group. Watson had already garnered the Nobel Prize. Baltimore would receive his in the following year, while still in his thirties. Nathans's would come in 1978, Berg's two years after that.

They were there to talk about the new rDNA technology and its implications. Herb Boyer of UCSF and Stanford's Stanley Cohen had announced a major breakthrough just the year before, making it possible to splice genes from different organisms—plants, animals, humans—to make hybrid molecules with tailor-made properties, such as the ability to produce insulin. Recombinant DNA technology held world-changing promise. But if something went wrong, might it also cause world-changing harm? As historian Durant once posited, "What if genes conferring resistance to specific antibiotics were recombined into naturally pathogenic bacteria? Or what if genes causing cancer were transferred from viruses into the normally harmless bacteria that inhabit the human gut?"[11] No one knew the answers to such questions for certain.

Berg had been asked by NSF officials to investigate, and that had led to the MIT meeting. As Baltimore later described it, "We sat around for the day and said, 'How bad does the situation look?' And the answer that most of us came up with was that . . . just the simple scenarios that you could write down on paper were frightening enough that, for certain kinds of limited experiments using this technology, we didn't want to see them done at all."[12]

The meeting led to an open letter that June signed by a small group of prominent scientists—with Berg, Baltimore, and Roblin taking point at the associated press conference.[13] Among other things, the letter called for a voluntary moratorium on certain types of rDNA experiments while a program to assess the risks was carried out at a secure facility. For the most part, the moratorium was widely followed. The letter, in turn, spurred a historic conference in February 1975 at Asilomar State Beach on California's Monterey Peninsula that drew 139 top scientists from seventeen countries and further advanced the issue. Among its recommendations, the Asilomar Conference, as it became known, called for classifying rDNA research into levels of risk (P1 to P4), with standards for containment associated with each level. Shortly thereafter, drawing on those ideas, the National Institutes of Health issued draft guidelines for conducting recombinant DNA research. And finally, on June 23, 1976, the NIH issued its final guidelines.[14]

For the scientists back at MIT's new cancer center and their counterparts over at Harvard, the cloud of uncertainty and the voluntary moratorium on rDNA experimentation it sparked had lasted roughly two years. But if these researchers thought they were clear to get on with their science, they were sorely mistaken. On the evening of June 23, 1976, just hours after the final NIH guidelines were released, Cambridge mayor Al Vellucci held a special hearing on recombinant DNA experimentation in city hall.

Historian Durant summed things up: "From the point of view of this nascent MIT research community, things soon got a whole lot worse."[15]

* * *

As the Berg letter and Asilomar played out, Harvard had been advancing plans to build a new laboratory largely to pursue recombinant DNA research—it was to be a P3 facility, the second-highest level of security. The school had filed for a building permit with the city earlier in the year. But the plan was not without controversy, even inside Harvard. Two of its prominent skeptics were George Wald, a Nobel Prize–winning biologist, and his wife Ruth Hubbard Wald, also a Harvard biologist. This led to a university-wide meeting that May. That meeting was attended by two reporters, who published a detailed account of the plans early that June in the alternative newspaper the *Boston Phoenix* under the headline: "Biohazards at Harvard: Scientists Will Create New Life Forms—but How Safe Will They Be?"[16]

The combination of the article, the skepticism of the prominent Walds, and the importance of the general issue caught Vellucci's attention. Sensing a major political opportunity, he had spoken to the *New York Times* the week before the hearing evoking the Frankenstein reference, and the place was packed. "I thought Frankenstein would kind of shake them up," Vellucci later acknowledged. "So I played Frankenstein to the high heavens."[17] Stephen Hall described the meeting in his book *Invisible Frontiers*: "It began with a local high school choir filling the City Council chambers with adolescent strains of 'This Land Is Your Land' and a sign reading 'No Recombination Without Representation.' Onlookers and kibitzers packed the gallery, jammed the balcony, spilled out into the hallway . . . Television lights and cameras, for both television news and archival footage, were trained on the councilors and on the witness table. Having gaveled the meeting to order, Al Vellucci basked in the glow."[18]

The mayor opened the meeting with a warning shot across the bow of Harvard—letting it be known the city would not automatically bow to the prominent institution's wishes: "No one person or group has a monopoly on the interests at stake. Whether this research takes place here or elsewhere, whether it produces good or evil, all of us stand to be affected by the outcome."

Figure 17 The scene at Cambridge's 1977 hearing on recombinant DNA.

Then, as various scientists prepared to share their views with the city council, he pointedly directed: "Refrain from using the alphabet. Most of us in this room, including myself, are lay people. We don't understand your alphabet, so you will spell it out for us so we know exactly what you are talking about because we are here to listen. Thank you."[19]

The parade of speakers represented both sides of the issue. Those on the mostly proexperiment side included Baltimore and Gilbert; two other leading Harvard microbiologists, Mathew Meselson and Mark Ptashne; and Maxine Singer, representing the National Institutes of Health. Others, such as the Walds and Jonathan King, a top MIT scientist representing Science for the People, reportedly spoke more of the potential dangers. Each speaker or "witness" was limited to ten minutes, but the first two, Ptashne and Singer, took more than two hours fielding the barrage of questions, reported Hall.

A second meeting held two weeks later brought in more testimony and led to a three-month moratorium on experiments and the resolution to create a Cambridge Experimentation Review Board to help oversee the matter. This was a citizen's committee to the core. Not a single scientist was named to the nine-member CERB. Instead, its members included a nurse, a nun, a community activist, a physician, and a professor of environmental policy and planning, among others. Perhaps more to the point, says then city councilor David Clem, the board included two former city councilors with strong credibility with major constituencies—one aligned with the city's more progressive voter block, and one tied to more conservative or "independent" voters.[20]

It could have been disastrous. Instead, the members did their homework and due diligence. The group began meeting in late August and convened every Tuesday and Thursday through that fall, with the Tuesday sessions open to the public. A climax of sorts came in a five-hour public debate on November 23, two days before Thanksgiving. Scholars Maryann Feldman and Nichola Lowe, who studied the process some three decades later, wrote: "In total, the committee met for over 100 hours hearing testimony, discussing and deliberating the technology merits. The review board consulted widely and sought to educate themselves rather than set up an adversarial debate. . . . The board also visited biology laboratories at MIT and Harvard and participated in a Cambridge Forum on recombinant DNA."[21]

In the first week of 1977, the review board recommended approving recombinant DNA research in Cambridge, with a few safeguards beyond the NIH guidelines. Most notably, these included setting up a five-person citizen's group called the Cambridge Biohazards Committee that would approve any research and oversee regular site inspections by local public health personnel.[22] Finally, on February 7, 1977, a little more than two and a half years after the Berg letter was published, the City Council unanimously passed the Ordinance for the Use of Recombinant DNA Molecule Technology in the City of Cambridge. The entire document spanned just over two pages, setting out in a series of short paragraphs the specific steps any company or institution would need to take to conduct recombinant DNA experiments within the city limits. It was the nation's first municipal biosafety ordinance.

And with it, the entire arc of Kendall Square changed.

* * *

When the ordinance passed, many pundits and experts predicted its restrictions would deter rDNA research and economic development. As it turned out, the biotechnology industry was at such an early stage—nonexistent in Cambridge at that point, in fact—that it wasn't immediately ready to test that theory. But a few years later, when the nascent industry had evolved a bit, the effect of the ordinance was the opposite of what many had feared: it ignited growth. "What it did at a very critical time was it established rules," says David Clem, a junior city councilor at the time who would go on to become a major Kendall Square real estate developer. "Boston didn't get around to doing anything until over ten years later. Somerville even longer than that. And all these younger startup, early-stage companies trying to find cheap lab space—that was really driving it—they gravitated to Cambridge because they knew what the rules were." MIT and Harvard were of course crucial to starting companies, says Clem. But the ordinance in his view played an even bigger role in influencing where they located. "That's really why Cambridge became the heart of life science."[23]

The effects of the ordinance were immediate on university science, however. At MIT, important experiments and research had been on hold. Now they were unleashed. "Later in 1977," notes Durant, "Sharp and his coworkers announced the discovery that the genes in advanced organisms are not continuous strings but instead contain 'introns' that are deleted through a process of RNA splicing before messages are finally translated." Sharp would win the Nobel Prize for this work sixteen years later. David Baltimore, who already had his Nobel, continued his critical experiments on viruses, cloning complete viral genomes. Robert Weinberg and his group, meanwhile, were making pathbreaking discoveries related to human oncogenes—genes that cause cancer. Some considered it a golden age of cancer research.[24]

For much of this period, the top scientists were almost completely unaware of how industry worked. MIT had a long-standing practice of allowing faculty to devote one day a week to consulting and working in industry. This had been fully embraced by engineering and chemistry faculty, and the researchers in computer science and AI had also jumped on the bandwagon. "An engineer's much like a doctor. If they're not out there solving problems, they're not engineers, right? So they teach and do research, and they're like bees. They go learn here and pollinate, learn there and pollinate," says Phil Sharp. But the practice had not made it to biology at all. "Nobody in science was using their one day a week," recounts Sharp. "They are now," he adds.[25]

Sharp's first introduction to business came via an out-of-the-blue telephone call from a man named Ray Schaefer. An MIT alum, Schaefer had entered the emerging field of venture capital and had spent time learning about the field at Boston-based TA Associates and also at Kleiner Perkins in the San Francisco Bay Area. He was a technology executive at International Nickel Company (Inco), a Toronto-based mining concern. He was calling Sharp because Inco had been offered the chance to invest in a new biotechnology company that would turn out to be Genentech. He wanted to fly Sharp to San Francisco as a consultant to speak with the founders and help Schaefer evaluate the opportunity.

"I didn't know what a venture capital was," Sharp relates, deliberately messing up the grammar to show his ignorance of business at the time.[26] He asked Schaefer who at MIT might be able to vouch for him, and proceeded to check him out internally before accepting. It was his first consulting gig. In San Francisco he met with Genentech's two founders, Herb Boyer of UCSF and Robert Swanson, another MIT grad who had previously been at Kleiner Perkins, and would be the biotech's first CEO. Joining them were two other scientists, Arthur Riggs and Keiichi Itakura. The company's first targets were to genetically engineer human insulin and human growth hormone. The men explained their science and approach, and Sharp listened. Afterward, he told Schaefer: "I don't know if you can make a buck on this. But they're going to do it." Inco subsequently invested $400,000 in Genentech, taking about 13 percent of the company. Genentech went public in 1980 at thirty-five dollars per share, rising as high as eighty-eight dollars a share the first day. Sharp's wife, Anne, jokingly chided him, "You know, you should have asked for a little stock."[27]

Not long after the consulting trip, Schaefer and his Inco colleague Daniel Adams began speaking with Sharp about starting another biotechnology company along the lines of Genentech. The Inco duo had also trawled Europe, meeting with top molecular biologists there they hoped to include—but they had gotten a lukewarm reaction overseas. Sharp was open to the idea, and suggested it would be imperative to involve Wally Gilbert at Harvard. Gilbert was highly skeptical about doing anything commercial, but finally, over a pivotal Chinese dinner in Boston, agreed to explore it further. Eventually, he and Sharp sketched out a dream team of leading European scientists who might join them—a list that included Charles Weissmann at the University of Zurich; University of Edinburgh biologist Kenneth Murray; and Heinz Schaller from the University of Heidelberg. Ultimately, those three plus Sharp and Gilbert would be Biogen cofounders, along with Bernard Mach from the University of Geneva, Peter Hans Hofschneider of the Max Planck Institute of Biochemistry in Martinsried, and Brian Hartley of Imperial College London.

On March 1, 1978, the power-packed group, with Gilbert as the scientific anchor, held a brainstorming meeting to examine the idea at the swank Le Richemond hotel on the shores of Lake Geneva. Joining Adams and Schaefer on the investor side of the ledger were two other venture capitalists interested in molecular biology—Kevin Landry of TA Associates in Boston, and Moshe Alafi, who was based on the West Coast and had a seat on the Cetus board. No commitments were made at the Geneva confab, but attendees remained intrigued enough to schedule a second meeting.[28]

Round two began on March 25 at a hotel in Paris. The entire scientific team was there, with Schaefer and Landry representing the potential investors. So too, for a brief time, was Genentech CEO Robert Swanson. Sharp says Swanson had gotten wind that Inco, one of Genentech's investors, was thinking of starting a rival. "Bob Swanson flew to Paris and sat in the lobby of the hotel just furious with Schaefer, and made Schaefer commit to selling the shares in Genentech," Sharp recounts. "So that separated Biogen from Genentech—and then Biogen went on its way and Genentech went on its way."[29]

Over two days, the group dived in deeper. Finally, Gilbert asked the VCs to leave so the scientists could talk among themselves. As the scientific huddle dragged on, Schaefer and Landry grew more despondent. When Gilbert finally emerged, he had a somber look on his face. As Stephen Hall captured in *Invisible Frontiers*: "Schaefer immediately inferred the worst and asked, 'Well, is it all over?' Gilbert nodded. 'How many are going to come with us?' Schaefer asked resignedly. 'All of us,' said Gilbert." Later, Schaefer brainstormed names with Dan Adams. He scribbled an idea down on a piece of paper and asked Adams for his take. The name was Biogen.[30]

Their vision for the upstart company was pretty basic—Genentech had the West Coast. Biogen would take the East Coast and Europe. "Little did we know that there were going to be thousands of biotech companies," remembered Sharp.[31]

Figure 18 A 1982 meeting of Biogen's founders and scientific advisory board. Charles Weissmann stands on right. To his left, seated, are Peter Hofschneider, Brian Hartley, Danny Wang, Ken Murray, Heinz Schaller, Wally Gilbert, and Bernard Mach. Standing behind them, left to right, are Klaus Mosbach, Julian Davies, Phil Sharp (leaning in), Walter Fiers, and Richard Flavell (at the blackboard).

It was a science-first company. "The biologists were purists. So we all brooded about, well, should we do this, is it the right thing to do?" says Gilbert. On top of that, he adds, "We were very distrustful of the VCs."[32] In the end, they all decided that forming the company was a good idea, but in the negotiations that followed, they imposed some special conditions that the investors ultimately agreed to. One was that while the overall budget would be set by the board of directors, budget allocations would be controlled by the scientific advisory board (SAB). The SAB also determined stock and option allocations.

Biogen was incorporated in Luxembourg on May 5, 1978.[33] The next day, the founding team met in Zurich to ratify the charter and finalize other details, including stock allocations. All told, the investors initially put up $750,000—half in cash, and half in in-kind services from Inco. Just as they were running out of cash, drug giant Schering-Plough (now part of Merck) came in with $8 million for a 20 percent stake in the company. (Gilbert says Biogen was so cash-strapped that Schering gave them a loan while waiting for the investment funds to go through.) That was joined by another $1.25 million from Inco—bringing the total raised in the first year to $10 million, counting the in-kind services. With the infusion, Biogen itself was deemed to be worth $40 million.[34]

In 1979, the company tapped Rob Cawthorn, a former Pfizer executive, to serve as Biogen's first president—giving it professional management. Initial research was carried out in the university labs of the founding scientists, using company funds: the scientists insisted they must be free to openly publish results, though Biogen would be given the early heads-up on any key work so it could file for patent protection. All this work proceeded, with the ironic exception of Sharp at MIT. Despite its being one of the most open universities when it came to professors taking outside work or forming companies, he was told that school policy barred scientists from conducting research for companies they held stock in. Sharp had planned experiments related to cloning bovine growth hormone, but decided to hold off and keep his equity in Biogen.[35]

Within less than a year, the company set up its own laboratory in an old watch factory in Geneva, where its headquarters was also established. Julian Davies, a Welsh microbiologist, was recruited to head the lab.[36]

The first breakthrough came on Christmas Eve in 1979, when researchers in Weissmann's lab at the University of Zurich contacted him during his ski vacation to report their first successful experiment to clone alpha interferon. The news got worldwide attention when they announced it the next month—making the front page of the *New York Times*.[37]

As advances continued and competition increased, it became clear to Biogen's leadership that the company needed to set up a lab in the United States and bring its headquarters closer to key scientific founders and universities from which new staffers could be recruited. Fresh infusions of cash, including $20 million from Monsanto and $10 million from Grand Metropolitan, a British conglomerate that operated a hotel chain and various distilleries, among other businesses,[38] gave them more room to invest in the lab and grow.

Those imperatives had quickly led them back to Cambridge and Kendall Square. As Sharp once recalled, "We were going to have to hire young people out of universities, and they would only come to the organizations close to universities where they could keep their contacts. We wanted to be close to MIT, and felt we had to be."[39] The city's ordinance, which no corporation had yet tested, made it especially attractive. Wrote Feldman and Lowe, "The fact that Cambridge had already addressed the debate and had proposed a reasonable solution, which appealed to both sides of the recombinant DNA debate, significantly reduced the chances of public outcry or a politically motivated research stoppage. . . . This was especially appealing to Biogen's large corporate financiers . . . who were eager to avoid any potential negative press that could arise from public opposition to this emerging technology."[40]

Still, the ordinance had been written to apply only to university research laboratories, so some changes were needed to extend it to commercial enterprises. One critical clarification Biogen sought—and eventually got—was for permits to conduct rDNA research to be granted indefinitely, rather than forcing companies to reapply annually. The whole process took some time, but the amended ordinance was passed by a vote of eight to one in the spring of 1981.[41]

As the Biogen team hunted for space, they walked around the Kendall Square neighborhood. The first building in the redevelopment zone—Five Cambridge Center—had only recently gone up. The area next to the Volpe Center remained a vacant lot. "There was nothing to the east of MIT for several blocks," Sharp recalled.[42] Finally, they identified an existing two-story building at 241 Binney, at the corner of Sixth Street. In one form or another, Biogen has been there ever since—in the process gobbling up a lot of the land around it and becoming the icon of (mostly) homegrown biotechnology in Kendall Square.

13 BEGINNINGS OF GENE TOWN

Others were hard on Biogen's heels. Nationwide, a slew of scientific stars had soon taken the plunge into business—either launching companies or signing on as key advisors. By the end of 1979, a dozen biotech companies had been formed. Another twenty-six got started in 1980, and forty-three the following year.[1] The epicenters of this activity were the San Francisco and Boston areas.

A handful of important biotechnology companies took root around Boston in the 1980s. Biogen was the trailblazer. Then came Genetics Institute (GI) and Immunogen in 1980 and Genzyme in 1981. Vertex would join the list, but didn't get started until the end of the decade, in 1989. These weren't the only biotechs to gear up in the 1980s, but they were the largest and most impactful. All were based in Cambridge at critical points of their early history. Biogen and Genzyme long maintained major presences—including their headquarters—in Kendall Square. Vertex, based in Cambridgeport on the other end of MIT until its 2014 move to Boston's Seaport neighborhood, had a smaller Kendall Square footprint. Both Genetics Institute and Immunogen operated solely in other parts of the city. But in one way or another, chiefly by seeding it with experienced managers and future entrepreneurs, all of these pioneering companies played a role in Kendall Square's rise as a world biotech capitol.[2]

GI was born out of Harvard—and some of the issues and frustrations surrounding the early recombinant DNA controversy. It had reportedly started in Mark Ptashne's apartment in Boston. Ptashne was a Harvard professor and a key figure in the effort to build a P3 lab that had kicked off the whole debate with Vellucci and the city. After an attempt to form a company in conjunction with Harvard had reportedly fallen through due to internal objections, Ptashne had teamed with his former postdoctoral student Tom Maniatis, by then a Harvard professor, to launch Genetics Institute, quickly attracting $6 million from four VC firms.

The company had planned to move to Somerville, next door to Cambridge and cheaper. But strong public outcry in early 1981 led to a moratorium on recombinant DNA research that caused it to withdraw its request for a building permit and lease space that spring in Brigham and Women's Hospital in Boston. Its footprint spilled over parts of five floors and encompassed old delivery and recovery rooms and a former morgue. Public opposition soon arose in Boston as well. During the hunt for a more welcoming environment, Al Vellucci gave Genetics Institute brass a two-hour tour of

Cambridge and "welcomed us with open arms," remembered former Baxter Healthcare executive Gabriel Schmergel, who had been brought onboard as CEO.[3] GI decided to move to West Cambridge, into specially built space near Arthur D. Little and the Alewife T station. The buildings were completed and the move begun in 1984—the whole company shifting to Cambridge by the following year. GI went public in 1986, raising $79 million, and grew to reach a $1 billion market valuation by the early 1990s. In 1992, about a year after losing a major patent fight with Amgen, it sold a majority ownership interest to Wyeth (then called American Home Products), which bought it out completely four years later, when GI had some 1,200 employees.[4]

Immunogen was formed in 1980 when a group of investors led by Ray Schaefer of Inco and Kevin Landry of TA Associates, fresh off their Biogen success, teamed with Baruj Benacerraf, a Nobel Prize–winning immunologist from Harvard Medical School and the Sidney Farber Cancer Institute (now the Dana-Farber Cancer Institute). The idea was to build the company around monoclonal antibodies, which would become part of a new class of drugs, called *drug-conjugated antibodies*, to precisely target cancer cells and kill them. The company initially supported research at Dana-Farber and didn't hire its first employee until 1985, when it brought in Mitch Sayare as CEO. Sayare's first office was a small room in Harvard Square, but the company leased an old warehouse at 148 Sidney Street in Cambridgeport and moved in there after it was refitted as lab space in 1987 (Vertex found space in early 1989 in an old warehouse at the corner of Sidney and Allston, three blocks down Sidney closer to the river). Sayare, who was CEO for some twenty-four years until his retirement in 2009, was told by his investors not to be profligate in his spending on space. "Kendall Square had a level of profligacy that Sidney Street did not, even in those days," he says.[5] The company did five rounds of private financing, followed by an IPO in November 1989, and grew to a peak of around 185 employees. But it was not able to bring a product to market until 2013. It still exists today, based in Waltham.

Genzyme, which in 2010 would count eleven thousand employees and some $4 billion in annual revenue, would eventually dwarf them both, on its way to becoming one of the world's largest biotechs—along with Biogen, Genentech, Amgen, and just a few others. In a few years Genzyme would become almost synonymous with its longtime CEO Henri Termeer, who many thought of as the founder. But the Dutch-born Termeer didn't join the company until 1983—two years after it formed.

The company's principal founders are often listed as Sheridan Snyder, a venture capitalist and entrepreneur; and George Whitesides, an MIT professor who later moved to Harvard and who has since founded a string of companies. But there were many others on the founding team. One was Henry Blair, a Tufts University researcher who joined with Snyder in 1981 to get the startup off the ground and later cofounded Dyax, another Massachusetts biotechnology company. And alongside him were eight MIT professors, Whitesides among them, who made up a consulting company known as Bio-Information Associates (BIA).

The BIA octet was self-selected for its members' different specialties—spanning cell biology, molecular microbiology, bioprocess engineering, and organic and surface chemistry. "We positioned ourselves as a professional consulting organization to advise multinational corporations as to the impact of biotechnology on their core business," says Harvey Lodish, a cell biologist who was still at MIT in 2021. As such, they had consulted on projects involving Stella Artois beer and Evian spring water, among others. One day in 1981, Snyder asked BIA to evaluate candidates for a scientific advisory board he and Blair were assembling for Genzyme.

"Sherry Snyder was putting together this company," Lodish remembers. "And it fell to me to advise him, as part of BIA, whether the advisory board was good. And I'll never forget, I was on a bus coming home from Woods Hole and I'm reading the CVs of these people and I call Sherry and I said, 'Look, you know, these are good people. They're not great people. I think you can do better. And he said, 'Would BIA be the advisory board for Genzyme?'" In essence, BIA thereby became the founding scientific advisory board of a nascent Genzyme.[6] One of their number, Charles Cooney, would serve on Genzyme's board for thirty years.

When Genzyme recruited Termeer in 1983, he was an up-and-coming executive for Baxter Travenol, now Baxter International. Genzyme initially derived revenue by selling enzymes used for diagnostics. But Termeer, who would become CEO in 1985, had ambitions for something much bigger. That soon led to him focusing efforts on rare or "orphan" diseases—paving the way to an IPO in 1986 and to Genzyme's first commercial success, Ceredase, an enzyme replacement therapy for treating Gaucher disease that was approved by the FDA in 1991.

In later years, Genzyme would stand alongside Biogen as a paragon of homegrown biotechnology in Kendall Square—moving its headquarters in 1990 to a refurbished building in the old Boston Woven Hose factory complex at One Kendall Square and from there to a state of the art, LEED platinum certified building opposite what is now called Henri A. Termeer Square. But its first home was nothing quite so glamorous—low-rent space on the fifteenth floor of a building on Boston's Kneeland Street, near Tufts Medical School and on the boundary between Chinatown and the famed "combat zone" of strip clubs, adult bookstores, and X-rated movie houses. The ground floor was home to a costume store and a diner whose odors went up through the air vents to the Genzyme offices, remembers Lodish's daughter, Heidi Steinert, who interned there for two summers in the 1980s. The other floors consisted of rows and rows of discounted, off-the-rack ladies clothing.

"At that time, we still had a real Combat Zone, a real red light district," Termeer later recalled. "So I would get propositioned three times—in the morning—getting to the office."[7]

Adds Harvey Lodish: "I mean, that was a freaky place."

* * *

To many, a sketchy multistory warehouse filled with mannequins and clothing racks adjacent to Boston's combat zone wasn't nearly as freaky as the prospect of cloning antibodies or splicing genes and creating new organisms. But with the arrival of Genzyme, and soon thereafter Termeer, the Boston biotech scene was coming to life in the early 1980s. Of the pioneering companies, Biogen alone had located in Kendall Square. Genetics Institute was spreading out in West Cambridge but drawing on MIT as well as its Harvard roots. Genzyme incubated across the river but would soon occupy a fundamental place in the square.

As much as or more than their physical presence, though, these three pioneering firms created a big part of the momentum that would define Kendall Square forty years later and likely beyond—producing legions of skilled, battle-hardened executives, researchers, quality control experts, business development leaders, and marketers who went on to take key roles at a multitude of other companies or who started biotechs of their own.

As the first biotech companies found their footing, one other cornerstone of Kendall Square's life science future was being put into place. It was neither a company nor a classic university scientific operation like Salva Luria's cancer center. It had close ties to MIT, but was set up as a private research organization with a novel structure that gave it an unprecedented degree of independence from the school—and had a freewheeling budget to match.

It was called the Whitehead Institute for Biomedical Research, and at the center of it was the man who in some ways had kicked off the debate that led to the recombinant DNA ordinance in the first place: David Baltimore.

Although the ordinance had in the end encouraged companies like Biogen to set up shop in Cambridge, Baltimore had come to question signing the Berg letter that resulted from the meeting in his office back in 1974. As he said in an interview later, "We may have so dramatized the issue that we made it possible for a lot of things to happen that wouldn't have happened. I think it's at least arguable that the *Phoenix* article, and the local notoriety, the involvement of Science for the People, maybe even George Wald's involvement at all, might not have occurred had it not been [for] Paul [Berg], Jim Watson, and me and the other people who signed that original letter."[8]

But that was behind him when the new opportunity arose. It came in the form of Edwin C. "Jack" Whitehead, a wealthy entrepreneur and hopeful philanthropist. Whitehead and his father, Edwin Weiskopf, who died in 1968, had founded Technicon, a high-flying lab equipment company. Whitehead had taken the firm public in 1969, making him a paper billionaire at one time. Eleven years later, in 1980, he sold the company to Revlon, becoming its vice chairman and single largest shareholder, with most of his stock in preferred shares that provided nearly $20 million in annual income, according to one estimate.

Whitehead had determined that as his legacy he wanted to do something to help medical science. In 1974, he revealed a plan to endow a new research institute at Duke University, but the agreement soon collapsed: one Duke insider said Whitehead wanted a degree of control over it that was incompatible with academic freedom. Whatever the exact reasons, Whitehead gave Duke a bequest and went back to the drawing board regarding his institute. At some point, says his daughter, Susan Whitehead, he determined that the key to success was finding the right leader. This soon led him to Baltimore.[9]

The introduction to Baltimore was made by Lewis Thomas, a physician and celebrated author whose collection of essays, *The Lives of a Cell: Notes of a Biology Watcher*, had won a 1974 National Book Award. Susan Whitehead, along with her brothers, John and Peter, has been a board member of the Whitehead Institute since its inception. She notes: "My father really did like doing big stuff. He was a flamboyant person. He had a real lust and gusto for life. He was a juicy guy. He was not a great match with MIT, to be honest. MIT knew that. He knew that. But he was a great match with David Baltimore. They really hit it off."[10]

Whitehead's original idea had been for his institute to target specific disease areas in an applied or directed fashion. The rebooted concept was to fund basic biological research. Even then, and despite having Baltimore on his side, it was not an easy path to success. Negotiations spanned several years and involved a multitude of faculty and other meetings.

The internal controversy over the idea "was huge—it was huge," says Harvey Lodish, who would become one of the Whitehead Institute's founding faculty members.[11] A big part of the problem was that some MIT faculty were highly suspicious of Whitehead's motives. Another was that the whole concept was largely unprecedented—certainly for MIT. "Today we take for granted that people understand the difference between philanthropy and investment," adds Baltimore. But that wasn't true at the time, he says. Some faculty continued to question Whitehead's motives, even though "Mr. Whitehead had been absolutely clear that this was a philanthropic gift and he would have no business benefit from it."[12]

In the end, both parties got what they wanted. "He was a self-made man. He was very abrupt, made decisions quickly, was very impressed with himself," says Baltimore. "But he knew his limitations and he knew he wanted to support science—but he hadn't been trained in science at all. And so he left me and the faculty to worry about the science."[13]

The final agreement—called the *affiliation agreement*—was largely written by Baltimore and MIT provost Francis Low. It was announced on December 4, 1981. Under it, Jack Whitehead agreed to pay $35 million to build and outfit the institute, which was ultimately housed at Nine Cambridge Center—across Main Street and down one block from the MIT cancer center at the corner of Main and Galileo Galilei Way. He also guaranteed $5 million a year in operational funding and committed to a substantial endowment that, upon his death in 1992, brought his gift's total to some $135 million. On top of all that, Whitehead provided a separate $7.5 million gift to MIT.[14]

One of the biggest stumbling blocks was over Whitehead faculty and their relationship to MIT. In the end, they were given positions at both institutions. The agreement, says Baltimore, "specified that the faculty would all be faculty of MIT from the point of view of teaching and departmental responsibilities but would all be paid entirely from Whitehead and be independent of MIT." A related concern centered on the hiring of new faculty. Although Whitehead scientists would have joint positions at MIT, many at the school worried that the upstart institute would hold too much sway when it came to hiring, upending MIT traditions and standards. Baltimore says that was settled by agreeing to a joint search and selection process "so that when it came to the point where we offered a position, there was agreement on both sides of the street—literally on both sides of the street."[15]

Whitehead had been right to seek out Baltimore: there could hardly have been a better person to head the new institute. "David is one of the great scientists in the last century," says Lodish. He quickly attracted a who's who of biological research. Lodish and one other MIT professor—Robert Weinberg, an oncology expert—became part of the Whitehead Institute from the onset. They were joined by Rudolf Jaenisch, a renowned expert in transgenics recruited from Germany; and Gerald Fink, a genetics pioneer lured from Cornell University. All four were designated founding faculty members. But that was just the start. A host of scientific heavyweights soon signed on—many of them rising stars selected for a special Whitehead Fellows program. Early newcomers included David Page, who eventually became director in 2005; Richard Young, a gene expression expert; Richard Mulligan, a MacArthur "genius" grant winner who much later went into private equity and served on the Biogen board; Peter Kim, later research director for Merck; George Daley, future head of the Harvard Medical School; and microbiologist and future NASA astronaut Kathleen Rubins.

And there was one other early recruit, a mathematician and Rhodes Scholar who would arguably go on to become the most famous of them all: Eric Lander, a pioneer in decoding the human genome and future founder and director of the Broad Institute next door (see chapter 18).

When the agreement was announced, the *Boston Globe* referred to the envisioned Whitehead Institute as the "Taj Mahal of science," which would be "staffed by a thoroughbred stable of geniuses with MIT titles." Baltimore called Whitehead's support "the most extraordinary gift to biomedical research ever made."[16]

The Whitehead, as it quickly became known, began functioning as a kind of virtual institute scattered around MIT more than two years before its permanent home was ready in 1984. "We secured space wherever we could find it. The major place was in the cancer center," says Baltimore. After the dedicated facility opened at Nine Cambridge Center, it took on a life of its own. The funding Jack Whitehead had ensured went beyond what was normally available to academics, enabling them to scale some experiments and purchase specialized lab equipment. "It allowed people to have research programs that required more resources," says Baltimore.

But the extra financial support wasn't what made the Whitehead special in Baltimore's view. "The qualitatively different thing was the sense of belonging," he says. For the first few years, the institute only had a dozen or so faculty members; today, it still has just nineteen. "The key to it was that it was a small environment—that is, the Whitehead building—that was very interactive. And the faculty became very close socially—much more so than faculty in the [MIT biology] department, who were in a much bigger building with less interaction." Jack Whitehead reinforced that by inviting the entire faculty and their families on annual ski trips to Vail, Colorado, paying for rented houses, lift tickets and equipment rentals, and ski lessons for those who needed it—which was most of them. "Most of us didn't even know how to ski," says Baltimore.

Special excursions like the annual ski trip, coupled with the facility's close-knit environment, "broke down what are a lot of the barriers in larger institutions," says Baltimore. "So we just built up this little world of interacting, with very high-quality people, and it was what a unit in a university should be—it was a place that trained young people, that opened new directions, that provided a home for scientists of all kinds."

According to Wikipedia, within a decade of its founding, "the Whitehead Institute was named the top research institution in the world in molecular biology and genetics, and over a recent 10-year period, papers published by Whitehead scientists were the most cited papers of any biological research institute."[17]

The impact on Kendall Square was immediate. The institute became a key place to be: The subsidized cafeteria serving good food drew in students and other "outsiders" from MIT and beyond, increasing the human collision factor. Events held in the small auditorium attracted other guests, as did regular beer parties. Ultimately, many faculty also turned their attention to biotech—helping found a plethora of companies. As of 2020, Whitehead faculty and staff had founded twenty-six companies, a little more than one a year, according to the institute.[18]

In the 1980s, though, predicting the impact of life sciences on the future of Kendall Square was another matter. With Lotus, Thinking Machines, and many of its fellow AI Alley firms on the rise, computing and software was still the bigger story as Biogen and the Whitehead Institute got going. When the I. M. Pei–designed MIT Media Lab opened in 1985, just a year or so after the Whitehead faculty moved into their new building, it grabbed a lot of headlines as well.

Noubar Afeyan remembers the sense of the unknown around biotech well. He had come to MIT in 1983—one of the first doctoral students under a new program in biotechnology process engineering, studying with Danny Wang, an original member of Biogen's scientific advisory board, and with Charlie Cooney, a Genzyme cofounder. An ethnic Armenian born and raised in Lebanon who later moved to Montreal and attended McGill University there, Afeyan would himself become a leader of the square's biotech scene some two decades down the road. In 2021, his venture capital–cum–venture creation firm ranked among the top biotech startup forces in the world, with a string of hits behind it inside and outside of the square—including Denali, Agios, Quanterix, Indigo Agriculture, and Moderna (see "Spotlight: Flagship Pioneering—Kendall Square Company Creator").

But such a startup wave was not even a gleam in anyone's eye at the time. Looking back, Afeyan can only think of a half dozen or so biotech companies with twenty employees or more as late as 1990—with no real clues to how big the field would become. "In the mid- to late 80s, there was not yet the clear defeat of computing and Route 128," he says. "We were all living under the shadow of computing. Biotech was a bit of a curiosity. Nobody, but nobody, was thinking this will become an industry and dominate this place."[19]

14 BUBBLE DAYS: MEDIA LAB TO AKAMAI

In the late 1980s and early 1990s, much of the framework that would define Kendall Square for the next roughly forty years came together. Most of the buildings in the urban renewal project were completed by 1990—anchored by the Marriott Hotel and its plaza leading to the subway station—although work on additional structures continued until 2004, when the Broad Institute opened next to the Whitehead Institute.

New office structures had gone up along both sides of Main Street toward the Longfellow Bridge and Boston, including the tower at One Memorial Drive that is now home to Microsoft New England Research and Development (NERD). Anchoring the scene toward East Cambridge, paralleling the river, were Edwin Land's Rowland Institute, Thinking Machines, Lotus, and several condo buildings. Meanwhile, over at One Kendall Square, the old Boston Woven Hose factory had been refitted in stages starting around 1982. Nearly forty different companies, including various startups, had set up shop there by the time Genzyme moved to One Kendall around 1990.

Some signs of city life were also sprouting up. Several apartment buildings had opened along the square's perimeter. Around the T station, students and workers could even find a few services—a bank, a post office, a florist, and an Au Bon Pain. Near them were a couple of stores that would be sorely missed a decade or so later: a mini-mart, a pharmacy, and a scaled-down Harvard Coop in the Marriott complex.

The restaurant scene was also sparse. Legal Sea Foods had opened at Five Cambridge Center in 1982: it was typically packed at lunch, when you could run into the technorati and biotechnorati—but not so much at dinner. Down by the river in the Thinking Machines building, Michela's had opened in 1985. Michela Larson's eponymous establishment was the first true destination restaurant in the square. A litany of later-to-be-famous chefs and restaurateurs worked there: Todd English, Jody Adams, Suzanne Goin, Barbara Lynch, Christopher Myers.[1] The Boston Sail Loft, in the lobby of One Memorial, offered fantastic views of the river. The Blue Room, another upscale eatery, opened in One Kendall in 1991. There were not a lot of other choices. And if you wanted a bar for a trendy cocktail or a pub to knock back a few beers after work, you were pretty much out of luck. Night life was virtually nonexistent—although in 1995, a nine-screen theater complex, the Kendall Square Cinema, turned on its lights across Binney Street from One Kendall Square.

The technology fabric of Kendall Square was far more vibrant than its social scene. But it, too, existed in a kind of limbo land. The AI Alley firms had dissipated. Most occupants of Cambridge Center were typical offices and lower-tech companies. Meanwhile, although Biogen, Genzyme, Vertex, and a handful of other biotech companies were growing, the life science surge that would largely define Kendall Square after 2004 or so had not yet kicked in. That left the square—like the entire Boston area in the 1990s as Route 128 also faded—as the leader of not much of anything in the high-tech arena.

But those in and around Kendall Square still had some major cards to play. MIT, of course, remained a world-class force in virtually all aspects of technology. Its Artificial Intelligence Laboratory (AI Lab) and the Laboratory for Computer Science (LCS) were outstanding; they would merge in 2003 to form the Computer Science and Artificial Intelligence Laboratory (CSAIL). And in those days, even they paled in stature against the upstart MIT Media Laboratory.

The Media Lab opened in 1985, just about one year after David Baltimore and colleagues moved into the Whitehead Institute a few blocks away. A silvery, four-story rectangular structure, I. M. Pei's ode to the future of computing offered a pleasant contrast to the red brick of Cambridge Center and the drab MIT buildings around it. "Slick as a corporate logo, with somewhat the look of a modern appliance," is how futurist Stewart Brand described it in his 1987 bestseller *The Media Lab: Inventing the Future at MIT*, which helped catapult the lab to the world stage.[2]

Just as the AI Lab and Project MAC had ignited the artificial intelligence gold rush, the Media Lab inspired dreams for and activity in another aspect of computing—what was then called *new media*, which represented the convergence of things like books, film, and TV and which promised novel ways to collaborate, create, and learn. Inside its walls, noted Brand's dust jacket copy, "are intelligent telephones that can chat with your friends, disembodied faces of real people that gesture and converse, interactive video discs, life-size holograms in midair, television sets that comb the networks and assemble programs that reflect each viewer's interests, and glimpses of computerized 'virtual reality.'" In other words, a lot of what has come to fruition in recent years, as well as some things that haven't yet been realized.

The main visionary behind the place was an MIT professor of architecture, Nicholas Negroponte, who cofounded the lab with former MIT president Jerry Wiesner and would serve as its director for the next fifteen years. He had assembled a world-class group of professors from different disciplines—including Marvin Minsky and Seymour Papert from the AI Lab.

Negroponte, now chairman emeritus, went on to other things, most notably founding the One Laptop per Child (OLPC) foundation to bring low-cost, internet-connected computers to children in developing nations. OLPC was based in Kendall Square from its launch in 2005 until 2010, when it moved to Florida. Negroponte says it's impossible to really assess the Media Lab's impact on Kendall Square. One way to try is by looking at startup companies. As of 2020, the lab listed 135 spinoff companies on its website. That averaged out to a bit over four per year, and some percentage of those set up shop in the square, though that was never tracked.

Negroponte thinks the Media Lab also helped entice big tech companies to establish R&D arms in the square. IBM had opened a research outpost in Tech Square in 1964, to be close to the AI Lab and Project MAC. The Media Lab soon drew in several others. Its unique funding model let corporations sponsor various research projects. Sponsors didn't get proprietary rights to the studies they supported, but they did gain access to the lab, which included interacting with professors and students to get a leg up on the future—and on potential recruits. It was only natural that some would open operations nearby to leverage that access. "Whenever I was asked, 'Should we create a lab?' I would always say, 'Absolutely. Do it right in Kendall Square,'" says Negroponte. "And you never know how many people listen to that, or when they do it if it is really because Nicholas said so or because that is a natural thing to do."[3]

One of the Media Lab's early corporate partners was Apple. That was largely thanks to Alan Kay. A future winner of the Turing Award, Kay had started his corporate career at Xerox's Palo Alto Research Center (Xerox PARC). Negroponte had been best man at his 1983 wedding to screen writer and author Bonnie MacBird, who wrote the science fiction movie *Tron*. Several years before the Media Lab opened, Kay had become Atari's chief scientist and had immediately set up a research lab in Kendall Square to align with Negroponte's Architecture Machine Group, as well as Minsky and Papert in the AI Lab. Atari Research Cambridge, based in Five Cambridge Center, opened in July 1982 but closed less than two years later when Atari's business tanked.[4] But Kay was sold on being in Kendall Square. Not long after joining Apple in 1984, he introduced its CEO, John Sculley, to the Media Lab.

Sculley calls the MIT enclave a "secret weapon." It served as a kind of replacement for the faded Xerox PARC, from which Steve Jobs had famously gotten a lot of the ideas—among them, the graphical user interface (GUI)—behind the Macintosh computer that came out in 1984. Recalls Sculley, "Alan Kay came to me in 1985 after Steve Jobs left Apple and said, 'Next time we won't have PARC . . . We need another source for brilliant inspiration.'"[5] That source of inspiration was the just-founded Media Lab.

A few years later, Apple produced a concept video for the Knowledge Navigator, a depiction of what a tablet computer might look like twenty years down the road. "So many of the ideas in the Knowledge Navigator came from Kendall Square and the MIT Media Lab," Sculley says. "Smart agent, multimedia, 3D animation, importance of simulation, object orient programming, ARM microprocessor, importance of video, Newton OS [operating system], and more. So ironically, lots of what Apple became best known for decades later was inspired at Kendall Square."

In 1989, Apple opened its Advanced Technology Lab (ATL) at 238 Main Street, the clock tower building facing the plaza where Main Street and Broadway converge. Its founding director was a former DEC manager named Ike Nassi who would stay at Apple until 2007, eventually moving to California and rising to senior vice president and head of software. The lab chiefly worked on the operating system for Newton, Apple's early handheld personal digital assistant (PDA). The following year, the ATL moved to more modern space on the seventh floor of One Main Street, with a great view of the Charles River. It grew to around twenty-five or thirty employees, Nassi says.[6]

Apple's outpost marked the beginning of a new wave of tech R&D labs in Kendall. Sometime around 1990, DEC opened its Cambridge Research Laboratory in One Kendall Square—in the same building that housed Genzyme.[7] Two Japanese companies with ties to the Media Lab also set up labs in Kendall Square. Mitsubishi Electric Research Laboratories (MERL) opened in 1991 at 201 Broadway. It would eventually grow to around seventy-five researchers tackling projects in AI, optimization, signal processing, modeling and simulation, and other fields. A few years later, in 1993, Nissan opened a lab at Five Cambridge Center focused on the future of "the driving experience." While MERL continues in 2021, the Nissan lab closed in 2001.

Meanwhile, over toward the CambridgeSide Galleria, one small, homegrown research arm arose. Lotus Research was formed in 1992 under Irene Greif, who in 1975 became the first woman to earn a doctorate in electrical engineering and computer science from MIT. "The group was probably never bigger than two dozen people," says Greif, whose charges focused on adding features to existing products and on creating new products. "We always emphasized collaboration."[8] One of its inventions was the now-standard versioning system for spreadsheets, which provided a way to identify and track different versions of a document. The vast majority of the team, including Greif, became part of IBM Research when it opened an expanded research operation around 2000 in the former Lotus space at One Rogers Street. Big Blue later moved its research branch to 75 Binney Street, a few blocks closer to the square itself, leaving only a few business operations at One Rogers.

These labs were harbingers of much more to come. Corporate research arms of information technology giants would become a hallmark of the Kendall Square scene in the 2000s—with the list extending beyond Apple, IBM, and Mitsubishi to Google, Microsoft, Facebook, and Amazon, among others.

*   *   *

As major tech corporations began these early forays into Kendall Square, a momentous, society-changing technological advance was unfolding: the internet.

MIT and early Boston-area companies—led by Bolt Beranek and Newman (now called BBN Technologies)—played a major role in inventing and developing key aspects of ARPANET, the precursor to the internet, as well as the foundation of the internet itself. And as noted in chapter 11, the first three internet domain names had gone to Boston firms, to BBN and two Kendall Square companies: Symbolics and Thinking Machines. Despite this early advantage, the West Coast had quickly stolen the East Coast's thunder, and by the early 1990s it had achieved internet domination. But the net's roots ran deep in Boston's environs—and the region would prove more of an internet force than a biotech force for a number of years still.

In 1994, Tim Berners-Lee, aka TBL to many, who'd written his seminal paper proposing the World Wide Web just five years earlier, arrived in Cambridge to launch the World Wide Web Consortium to develop standards for the emerging medium. TBL and the W3C were initially based in MIT's Laboratory for Computer Science in Tech Square but would move to the newly built Stata Center with the creation of CSAIL.[9]

Throughout the 1990s, a number of internet-based startups arose in and around Kendall Square and the Boston area. One of the highest flying was Sapient, an IT consulting company started in 1990. Sapient went public in 1996 and put its name on the new office tower at One Memorial Drive, visible to all coming across the Longfellow Bridge into Kendall Square. Revenues peaked at around $500 million a year—then the bottom fell out during the dot-com bust of the early 2000s. But the company held on, and ultimately sold to the French advertising firm Publicis in 2015 in a deal worth some $3.7 billion.

Another high-flyer was Art Technology Group (ATG). As MIT undergrads, its two founders, Joe Chung and Mahendrajeet "Jeet" Singh, had been members of Alpha Delta Phi, where fellow fraternity brothers included Colin Angle, Brad Feld, Eran Egozy, Sameer Gandhi, and John Underkoffler. Angle cofounded iRobot. Egozy cofounded Harmonix, developer of *Guitar Hero*. Feld cofounded Techstars and became a world-famous venture capitalist and blogger. Ghandi became a partner with leading venture firm Accel. And Underkoffler, a specialist in novel computer interfaces such as hand gestures, founded Oblong Industries and served as the futurist for the movie *Minority Report*.

The initial concept behind Art Technology Group was to combine internet savvy and design to create things like interactive museum exhibits, but it morphed into building and maintaining e-commerce software and websites—storefronts for the new online world. ATG's first home after its 1991 launch was in Harvard Square above the WordsWorth independent bookstore. After several other moves, it finally settled at 25 First Street in East Cambridge, on the edge of Kendall Square—the same building where Zipcar would later end up and where internet marketing firm HubSpot is now. The company soared at the height of the internet bubble—going from $16 million in revenue in 1999 to over $150 million the following year. It raised $50 million with its initial public offering in 1999 and took in another roughly $150 million in a secondary offering the next year. Then it, too, fell victim to the dot-com crash. The company hung on through massive layoffs and sold to Oracle in 2010. Chung and Singh had long since cashed in and left—pursuing different paths for a time. But that same year they regrouped and set up Redstar Ventures, a boutique venture creation firm based at One Kendall Square.

As the 1990s drew to a close, virtually no Boston area company rose to the stature of its juggernaut West Coast contemporaries—and while Kendall Square had arguably been the number one source of artificial intelligence companies in the 1980s, the same could not be said of the internet era. Says Chung, "The reality is there are a couple of couple of big successes here of that generation, but it's just a drop in the bucket compared to out West. And one thing that never happened here was the massive plow back of money from these successful exits into new companies. Jeet and I like to think we're doing our part [with Redstar], but on the tech side it's a rounding error compared to Silicon Valley."

One internet company from that era stands as an exception. It didn't sell groceries or pet food online, and it didn't pioneer a social media application. Instead, it went after a super hard challenge facing the proliferating internet: congestion. The company started, grew, and thrived in Kendall Square—and remains its only homegrown, publicly traded internet giant. Its main founders were two mathematical prodigies—one an MIT professor, the other a student and former Israeli Defense Forces officer. Their company had a funny name derived from the Hawaiian word for smart or intelligent: Akamai.

* * *

In a sense, Akamai got started in 1995, with a request for help from Tim Berners-Lee. The internet was growing in popularity—and TBL could see a gigantic traffic problem looming: already people lamented the World Wide Wait. His W3C quarters inside MIT's Laboratory for Computer Science on the third floor of 545 Tech Square was only steps away from a star mathematical theoretician—so it was natural to mention the problem to him.

The mathematician was Frank Thomson "Tom" Leighton, who had become a full professor in 1989 while still in his early thirties. He headed the algorithms group. "Tim was down the hall and he was talking about the challenge. He said we're going to have congestion problems," recalls Leighton.[10] "And he knew our algorithms group was the right group to think about it."

The basic problem was that if enough people wanted to access a website's content at the same time, things got clogged, like the checkout line in a busy grocery store. In internet terms, this was called a *hotspot* or *flash crowd*. And this problem was only going to grow worse as more people used the internet. Intrigued, Leighton attracted some DARPA funding to look at the problem. He also presented the challenge to select students, who started working on algorithms that might reduce bottlenecks.

Then, the following year, a very special student arrived: Daniel "Danny" Lewin. The hard-charging Lewin would be killed in the 9/11 attacks, a passenger on the plane from Boston to Los Angeles that was the first to fly into the World Trade Center. It seems almost certain that he was stabbed and killed before the crash, and many who knew him believe he would have tried to confront the hijackers.

"Danny was an exceptional human being. Very, very smart, very driven. Sense of urgency. Great leadership skills. He was a commander in the elite Israeli Sayeret Matkal, which is kind of like Delta force," remembers Leighton. "He had really creative ideas. He liked tackling very hard problems." It was no surprise then that the charismatic Lewin was captivated by the internet bottleneck problem.

Figure 19 A young professor, Tom Leighton, speaking with Danny Lewin in an early Akamai office in 1998 or 1999.
Source: Akamai Technologies.

Lewin soon emerged as a core contributor to a protocol known as *consistent hashing*—which served as his master thesis a few years later.[11] "Hashing is one of the basic primitives in computer science," explains Leighton. "It's a way of deciding where you store something—it sort of balances where things get stored so you can quickly get them later." Consistent hashing brought this to scale. "[There are] lots of pieces of content out there, and you can't put all of them in every server—you've got to spread them around. And the servers go up and go down, and content increases all the time, so you have to do some rebalancing," Leighton explains. "But you want it to be as little as possible. So consistency says basically things stay where they were as much as possible, so you can keep operating without taking the system down. It was a very elegant and optimal algorithm for doing that."

There was lots of work needed to turn the theories and algorithms they came up with into code. But they soon realized that consistent hashing held business potential. In 1997, Preetish Nijhawan, a Sloan School student and a neighbor of Lewin's in MIT student housing, suggested they team up and enter the MIT 50K Entrepreneurship Competition. This prize competition, now called the $100K, had been founded as the $10K in 1990 by Sloan School professor Ed Roberts and others.

In the precompetition that fall, the team grabbed top honors in software and media—one of ten initial categories.[12] "We won $100, which Preetish was pretty excited about," recalls Leighton. "He said that's a big deal to win the software category." After some additional rounds of judging, they were selected one of six finalists for the main competition the next year. However, they came up short in the finale and out of the money, which was split between the top three finishers.[13]

Failure to bring home a big prize in the $50K hit Lewin especially hard. On his twenty-eighth birthday, just a week after the competition, Lewin emailed a friend:

> Silicon Valley makes money on air.... There are many in this business who become horrendously wealthy AND THEY DON'T EVEN HAVE A PRODUCT OR CLIENT!!!!! ... The main thing you do have to do is dedicate yourself to telling lies for a number of years and to spreading bullshit for many, many years. Some people are comfortable with this as long as the payoff at the end is high enough. I am not. The plan is to become a successful company in the right way. That is: have a product, have a market, and have customers who are buying your product.[14]

"It was a slap in the face. Danny didn't like losing," Leighton once recalled.[15] But in preparing for the competition, they had learned a lot about creating a business. Among other things, they had spoken to potential investors. One was Polaris Venture Partners, now Polaris Partners. It had only formed in 1996, but the founding team had a strong reputation. Polaris also had a new venture partner named George Conrades. The seasoned executive had spent thirty-one years at IBM. He had come to Boston in 1994 to serve as CEO of Cambridge-based BBN, where he had launched BBN Planet, which grew into the country's third-largest internet service provider. Several years later, in 1997, he engineered BBN's sale to GTE for $616 million.[16] As soon as the BBN sale was complete, he was enticed by two Polaris founders, Terry McGuire and Jon Flint, to join the firm to work with early-stage companies and help evaluate new deals.[17]

"Within six months at Polaris, along comes Leighton and Lewin—and they started talking about a whole new way to deliver content on the internet," Conrades recalls. Leighton's reputation preceded him. "Everybody knew Tom," Conrades says. And Lewin, who had been born in Colorado but spent most of his life in Israel, was clearly something special. "An amazing combination of MIT and Israeli commando. I'd never seen anybody so physical in everything he did—the way he talked, the way he kidded around, punched you on the shoulder."[18]

As Conrades listened, he thought, "Oh my god, what a big, big idea. Because if it worked like they said it would work, it would solve the bottlenecks on the internet—the World Wide Wait—which we had faced every day at BBN."

Another local venture firm, Battery Ventures, was also extremely interested—and one of its partners, Todd Dagres, had connected with them after seeing the $50K competition. (Dagres later cofounded Spark Capital, another leading Boston venture firm.) Both firms invested $4 million, joined by Conrades and several other angel investors. At the time, Conrades declined the offer to become CEO, but he agreed to take Polaris's board spot and serve as chairman.

Akamai was housed initially in a series of cubbyholes in One Kendall Square. By that time, Nijhawan had taken a job with McKinsey & Co., but Jonathan Seelig, a friend of Lewin's who was also at the Sloan School, had joined the team: he and California businessman Randall Kaplan, who later became a venture capitalist, were listed as cofounders alongside Lewin and Leighton. Leighton became chief scientist, Lewin chief technology officer, Kaplan VP of business development, and Seelig VP for strategy and corporate development.

The original idea had been to sell their service to carriers like AT&T. But they couldn't convince these companies that the technology would work. So they shifted focus to helping large content providers make sure their websites stayed up. That entailed placing frequently requested content on servers strategically deployed close to various internet service providers all over the world—at the "edge" of the internet. Although the term *edge computing* gained great buzz in 2018 or 2019, Akamai has been talking about the edge virtually its entire lifetime.

Disney signed on as an early customer. As Akamai grew, it would ultimately create its Network Operations Command Center, where massive screens provide round-the-clock monitoring of the world's internet traffic. In those days, though, the NOCC was a small room at 201 Broadway, where the company had moved, with maybe a half-dozen computer screens. Leighton remembers huddling around the displays as they served up a single pixel from the Disney network.

Akamai's big break—actually two huge breaks on the same day—came the next month. One took the company by complete surprise. The other was planned, but its scale and impact were not anticipated. The date was March 11, 1999. The total surprise came first—with the debut of March Madness, the NCAA college basketball tournament. ESPN was covering the tournament. "ESPN was hosted at a company called Infoseek, which was the who's who of the internet back then," recalls Leighton. Akamai had tried to sign Infoseek as a customer—but had been rebuffed.

Figure 20 Tracking internet traffic worldwide in Akamai's NOCC.
Source: Akamai Technologies.

"And then on the first day of March Madness, we get a call from Infoseek saying, 'You know that trial you wanted us to do, we'd like to do it now. Can you handle two thousand requests a second?'" Leighton relates. "Now at this moment we're doing one every few minutes. But we're a startup. We had built this thing to scale, so we said, 'Yeah.'"

To "Akamaize," all a customer had to do was tweak the host name on its URLs to point to Akamai. "So within fifteen minutes they'd done this, and we're seeing three thousand hits a second on our platform," says Leighton. "And then we realized that Infoseek had gone down and ESPN was offline and they didn't even know how much traffic there was because all their systems had been crushed. And then they got back online right away. And everybody noticed that suddenly the ESPN site was six times faster than normal, even under this load."

That evening, with the ESPN situation still playing out, the trailer for *Star Wars Episode I: The Phantom Menace* was released for streaming. Despite its first episode status, *Phantom Menace* was the fourth *Star Wars* movie released. Steve Jobs, who had returned to Apple less than two years earlier, had bought the exclusive rights to stream the trailer online to promote QuickTime, Apple's video player. The Akamai crew didn't know that at the time, however, and in their effort to grow their fledgling company had inked a deal with *Entertainment Tonight* to deliver the trailer for its site—starting at 9:00 p.m. EST.[19]

"We had the whole company there," Leighton remembers. "The VCs were there. Because we're going to deliver this big deal *Star Wars* trailer. And so we did it. We delivered it, and things are going fine. Then we start reading news reports that apple.com was down and the internet was toast around where Apple's website was hosted. And then we start seeing other articles that say all the bootleg sites with the trailer are down except for one. And the only place people could see the trailer was *Entertainment Tonight*. Well, that's when we realized that we had a bootleg copy; *Entertainment Tonight* wasn't supposed to have it."

The resulting media attention proved a boon. "We got great press about that," says Leighton. A few weeks later, on April 1, Akamai president Paul Sagan got a phone call from somebody purporting to be Steve Jobs wanting to buy the company. Sagan, previously a Time Warner executive, had joined the team a few months after its inception to bring it some business management chops. He claimed he'd hung up—and accused Leighton and Lewin of playing an April Fool's joke on him. To this day, Leighton isn't sure if Sagan was playing a joke on them by telling his colleagues he'd terminated the call with Jobs. "In any case," he says, "it was Steve, and he was trying to buy the company." Ultimately, Akamai decided not to sell. "But Apple became a large strategic investor and a big customer, and to this day they're one of our largest customers," says Leighton.

If the term *unicorn* had existed then in business-speak, Akamai might have been a unicorn on steroids. Shortly thereafter, Leighton and Lewin finally convinced Conrades to sign on as their first CEO. With more top customers lining up, Akamai went public on October 29, 1999, at the height of the dot-com bubble. Its sales were nothing to write home about—$1.3 million in the first nine months of the year against a loss of $28.3 million. But no one seemed to care. As the Associated Press wrote, "The red-hot investor reception bestowed yet another staggering valuation on a small, money-losing Internet company."[20]

Akamai's stock had been priced at twenty-six dollars per share. But when trading opened, it was already up to $110—and finished the day just over $145 a share. At the time, that 458 percent gain from its initial price ranked as the fourth-hottest IPO in history and gave the company a market valuation of more than $13 billion. But the craziness wasn't finished. A few months later, the stock hit a still all-time high of $345 a share, briefly giving Akamai a market value of $35 billion. "It was totally insane," marvels Leighton. "We're losing a fortune. Very little revenue. The company went to $35 billion, and Apple's share in us was worth more than the rest of Apple, or at the very least, a big fraction; that shows just how wacky things were."

"We were bigger than General Motors," adds Conrades. "It was just going crazy."

* * *

On September 11, 2001, Danny Lewin rushed to Logan Airport to catch American Airlines flight 11, an 8:00 a.m. nonstop to Los Angeles. He had a seat in first class— where all five hijackers were. According to the 9/11 Commission report, most of what is known about what happened onboard is drawn from two flight attendants who called in from the coach section—Betty Ong and Madeline "Amy" Sweeney. They reported that the plane was being hijacked and that two flight attendants up front had been stabbed—neither fatally—and that one passenger had been killed. That passenger was identified as the person sitting in 9B, Lewin's seat—and the commission concluded he had most likely been stabbed by Satam al-Suqami, who was directly behind him. "Lewin had served four years as an officer in the Israeli military. He may have made an attempt to stop the hijackers in front of him, not realizing that another was sitting behind him," the report stated.[21] He is widely considered the first to die in the attacks.

Leighton was still at home that morning when he heard that two planes had crashed into the World Trade Center towers. He turned on the TV and saw total disaster. "There was a lot of just confusion. But I knew Danny was flying that morning, and then it said a plane on the way from Boston to LA had crashed. And I knew he was flying from Boston to LA, but the phones wouldn't work—they just were all lit up with everybody trying to find out what was going on."

So he drove to Akamai's offices, then in Tech Square, went inside the building, "and saw Danny's wife there in tears."

Akamai as a company had little time to grieve that morning. "That same day was a very tough day on the business in terms of getting all the government and news sites back online because there was just a tremendous demand for content," Leighton says. "Plus there were attacks on the government sites. Now, whether they were coordinated or just inspired by what was going on, I don't know. But they were under attack and so we were able to use our capacity to get them all back online."

"Everybody was working like a maniac to deal with that through the grief. It was brutal." But the company rose to the occasion, and the servers held.

* * *

Akamai has had several ups and downs over the years—but nothing compared to that dark day and its immediate aftermath. The dot-com bubble had burst earlier that year. The company's stock had plummeted, and even before September 11, Akamai had laid off five hundred of its roughly 1,500-person workforce. Things grew worse in 2002, with the stock falling as low as fifty-six cents a share—a far cry from the $345 high-water mark barely two years earlier—and the company slashed another five hundred people.

An inspiration came from Conrades's old company, IBM, where CEO Lou Gerstner said something that made the Akamai head take notice. "He made the comment that the dot-coms were fireflies before the storm," Conrades remembers. "That hit me like a ton of bricks because he meant that established businesses, enterprises, were going to use the internet." The dot-com companies that were now struggling or had already failed had been ahead of the curve; the stodgy established giants would ultimately get there and shore things up.

"It gave me hope," Conrades says. Akamai immediately switched its strategy to focus on attracting enterprise customers rather than internet content companies and service providers. "And fortunately, just great fortune, as the dot-coms went out of business, the enterprises were picking up on the internet, and we were able to say, 'Bye-bye, SurfMonkey.com. Hello, General Motors.'"

As a Harvard Business School case study later reported, "By 2009, Akamai was the market-leading CDN [content delivery network] with approximately 60% market share . . . The company also served approximately one third of Fortune's Global 500, including all 20 top global ecommerce sites, 9 out of 10 top social media sites, and all top Internet portals."

Conrades stepped down as CEO in 2005 and handed the reins to Paul Sagan, but he remained board chairman until his retirement in 2018. Sagan led the company through 2012, when he passed the torch to cofounder Leighton.

Akamai has thrived so far under Leighton. It hasn't become a true internet behemoth like Apple, Microsoft, Google, Amazon, Facebook, and others on the West Coast—but its success has been impressive. Revenues surpassed $3 billion in 2020, with over $800 million in profits. It counted 7,500 employees, and maintained some three hundred forty thousand servers across more than four thousand locations in one-hundred-thirty-five countries. Leighton says Akamai is the only public company to ever survive a more than six-hundred-fold drop in valuation.

The nature of its business was shifting, though. As of 2020, there were three main segments, each bringing in roughly a third of Akamai's revenue: content delivery,

Figure 21 The Akamai headquarters building, opened in November 2019. *Source:* Akamai Technologies.

applications acceleration, and security. "In another couple of years, security will be the largest, and before long it will be the majority of our revenue," Leighton asserts. Hacking and cyberattacks have been coming almost since the internet's inception, he says. "But we're at the point now where you got major governments investing *a lot* in all sorts of cyberattacks of various kinds, and a lot of motivations: economic, preparing for some larger hostilities, manipulation of elections, attacking the fiber of society—fomenting chaos, dissent, and hate. It's rampant, and the attack vectors are morphing at a rapid rate."

Throughout its journey, Akamai has been headquartered in Kendall Square. From its cubbyholes at One Kendall, it moved to 201 Broadway barely a block away, then to 565 Tech Square, in the same complex as the Lab for Computer Science. There, it grew to some 1,500 people, taking over the whole building and spreading to two others. When the dot-com bubble burst, and the company nearly went broke, it used that to get out of the lease. But it only moved a few blocks, to 150 Broadway. As it recovered, Akamai took over that building and spread to parts of five others in and around Cambridge Center before moving again in late 2019 to the nineteen-story tower at 145 Broadway where all Cambridge operations are now consolidated.

Conrades says they never thought about headquartering the company anywhere else. "One of the reasons we have always been located in Kendall Square is because we wanted to be within walking distance of MIT." As they grew, shrank, and hung on, then grew again, the entire landscape changed around them. Low-flung older structures came down, vacant lots and parking areas got filled in, and modern new high-rises went up—as did prices—but it was virtually all for biotechs and pharma, not IT startups like Akamai.

"All of a sudden large office buildings, buildings that had labs in them, grew up around us," says Conrades. "The high-tech was moving to California. We were the only ones still standing."

SPOTLIGHT: LITA NELSEN ON TECHNOLOGY LICENSING AND HOW "CLUSTERS FEED THEMSELVES"

A key part of what drives innovation, company formation, and economic growth involves the licensing of scientific and technological advances from universities.

MIT has long been at the beating heart of technological licensing, and its Technology Licensing Office (TLO) is currently headquartered on the fifth floor of 255 Main Street, adjacent to the Marriott Hotel in the center of Kendall Square. The longtime head of the TLO was Lita Nelsen. She retired in 2016 after thirty years at the office, the last twenty-three as director. The TLO makes roughly ninety to a hundred licensing deals annually and helps launch more than twenty companies a year out of MIT, often with the university taking equity. While those companies could take root anywhere, many are based in the square. "You walk down the street and you say, 'Well, I remember when that one was born, and I remember when *that* one was born,'" Nelsen once noted.[1]

The erstwhile TLO director has traveled the world helping people understand the square's "secret sauce," as she calls it. The story of her years in technological licensing offers some key insights about Kendall Square and much more.

Nelsen grew up in Queens and came to MIT on an undergraduate scholarship in 1960. Women, now up to 46 percent of the study body, made up just 2 percent of her undergraduate class. She and some female friends penned a little ditty: "Outnumbered one in 50, I think it's kind of nifty, with 49 fellas and me."[2]

Despite those odds, Nelsen graduated first in her class in chemical engineering, and earned a bachelor degree and a master degree in the subject. She then joined a professor's startup and worked in industry for some twenty years. But then in 1986, she came back to MIT to take a job in the licensing office. "This was a time when tech transfer was beginning to get important, and Stanford was doing a fabulous job," Nelsen recalls. MIT not so much. The office could be a bureaucratic quagmire, not returning phone calls and the like—and it did not have a good reputation with faculty or companies, she says. Ken Smith, MIT's vice president of research, decided it needed fixing. He pushed for a complete reorganization and convinced Niels Reimers, head of Stanford's vaunted licensing office, to come to MIT for a sabbatical year and lead the effort.

Nelsen applied to MIT after seeing an ad for one of the positions Reimers was looking to fill. She and the man who turned out to be the next TLO director, John

Figure 22 Lita Nelsen snowshoes near her home in 2015, when the Boston area received a record 108 inches of snowfall.
Source: Lita Nelsen.

Preston, started on the same day. She fit in right away. "I was a native. I spoke the language."[3]

She also saw something important happening: biotechnology. The field was in its infancy, but she sensed that it was on the verge of inventing the tools and science to address major medical problems. Nelsen had told Reimers, "I agree to take the job if you give me the biotech portfolio." He agreed.

Things progressed from there. "And so we got our act together, we got our policies together, very conservatively. And we started to take some equity in the companies and allowed the faculty to have equity—but still have exclusive licenses," she once recalled.[4]

Preston became director in 1986, when Reimers went back to Stanford. Six years later, after Preston had been promoted, Nelsen was put in charge of the office. When she had started at MIT, the TLO had totaled eight people. It numbered forty-eight when she left in 2016—forty staffers and eight interns. The number of invention disclosures it handled—from MIT and Lincoln Laboratory—went from about 150 to 850 a year. Nearly half of them are now in the life sciences.

"It's data and paperwork intensive," she says of the job. A licensing agreement is "a living document. It's like a marriage. It goes on for twenty years, which is longer than most marriages." Not only does the office have to keep track of modifications and amendments, but "you're managing eight hundred new inventions that come into the office. And that's only that year. You're still managing the eight hundred from the year before."

Here are some other lessons and insights from Nelsen's storied career:

On the Bayh-Dole Act—This 1980 act was "absolutely" critical to technology licensing, says Nelsen. It essentially allowed universities and other entities to own inventions stemming from federally funded research. This enabled a sea change at many universities, allowing them to license inventions made by professors—not only bringing in royalties, but encouraging spinoff companies. Until the act went into effect, says Nelsen, some universities dealing with a few federal agencies benefited from what were called *institutional patent agreements* (IPAs), which allowed essentially the same thing. Bayh-Dole made this widespread.[5]

How licensing revenue is split—The common perception is that royalties are evenly split between the university, the inventor's department, and the inventor. Close, but not quite right, Nelsen relates. First, MIT takes 15 percent off the top. Then a third of the remaining 85 percent goes to the inventors—28.3 percent of the total. The remaining roughly 57 percent is then divided between the school and the department or research center the invention came from, but not necessarily equally. It is a "complicated formula," says Nelsen. "You need an MIT degree to understand it and even then it bores you." So it's roughly okay to think of it as a three-way split—after MIT takes its 15 percent.

Why MIT takes 15 percent off the top—Overhead. "Who is going to pay for patents and inventions that don't bring in money?" asks Nelsen, noting that slightly more than half of the patents get licensed and a much smaller fraction bring in significant revenue. Without that slice, she notes, a licensing office "will rapidly lose money."

On MIT taking equity in startups—The institute will take equity in startups, sometimes for reduced up-front fees, and only rarely an all-equity deal. But it only takes common stock, not preferred stock. That means the school's share gets diluted if there are further funding rounds. Often, Nelsen says, MIT is diluted down to "homeopathic proportions."

It shouldn't be about the money—There are a lot of stories about universities making a ton of money from key inventions like Gatorade. Don't buy it. "This is not generally a money-making business unless you get lucky," says Nelsen. "But you should feel that if you're a university, you have an obligation to turn taxpayer-funded research into curing disease. If you don't hook up with industry, it ain't going to happen."

How Kendall Square grew into a biotech hub that feeds itself—Nelsen once read a study from the UK about how clusters form. Basically, she says, it found that a large company, such as a multinational, moves in to a region or grows there and becomes the anchor. The cluster develops as suppliers arise to support it and other large companies, and things grow from there. But that isn't what happened around MIT. "If you look at Kendall Square and biopharma, it's exactly upside down—100 percent upside down," she says. It was the biotechs that grew and, along with the universities, attracted big pharma. Part of that stemmed from pharmaceutical giants cutting back on R&D and laying off a lot of people in the 1980s and 1990s. "They looked back over their shoulders, and down at the output end of their pipelines there wasn't anything coming out. Why? Because there wasn't anybody putting anything in." They naturally turned to biotech companies, doing deals and acquiring them—often at enormous prices—and in the case of Kendall Square moving to the square to be closer to biotechs and the universities supplying talent.

Even before big pharma arrived in the early 2000s, Kendall Square's infrastructure started to build up around the biotechs, Nelsen notes. "More VCs came in. Landlords started getting used to licensing to blue-sky biotechs that might not be around in a year. Accountants and law firms started to spend more time learning what a cap table looks like, how you advise starters of new companies. That's a specialty. That wasn't there."

On the importance of strong management to a cluster—"In addition to risk capital, crazy risk capital, you need CEOs who know how to manage science toward product and raise capital. And that is a very, very scarce resource—the scarcest resource. In most places they don't exist," Nelsen says. "Clusters feed themselves," she adds. In Boston, early successes like Biogen, Genzyme, and Genetics Institute begat managers with strong technical or scientific backgrounds that helped them became exceptional CEOs. "In biotech you can really see it—more so than in IT."

On whether MIT should start an incubator—"People say to me, 'Does MIT have an incubator?' And my classic answer has been, 'Yes, it's called the city of Cambridge.'"[6]

15 CAMBRIDGE INNOVATION CENTER: KENDALL SQUARE'S STARTUP HEART

After some four rewarding years traveling the world meeting business clients, Amy and Tim Rowe decided it was time for a change. The couple had met while attending MIT Sloan School of Management, and both had gone into consulting after earning their MBAs in 1995—Amy to McKinsey & Co., Tim to the Boston Consulting Group. A few years later, they married, and by the end of 1998 they had decided the long hours and frequent road trips demanded by their jobs didn't mesh with raising a family.

Both had the startup bug, to boot. Internet frenzy hung in the air, with a few local dot-coms, like Art Technology Group and the much newer Akamai, getting great traction. Tim had even worked one summer while in grad school for Joe Chung and Jeet Singh: he had been ATG's thirteenth employee.

As 1999 got underway, Amy was farther along than Tim with her entrepreneurial plans—working on a company she called TravelFit, an online resource for people who wanted to stay in shape when on the road. To help her, Tim took the lead on finding office space. In that hunt, the seeds were planted for one of the icons of Kendall Square: the Cambridge Innovation Center.

CIC is widely acknowledged as a pioneering high-tech coworking space. Yet at the time, says Tim, sharing an office was something few people did, "and the word *coworking* wouldn't be in the lexicon until six years later."[1] It all started that June, when they leased about three thousand square feet in the clock tower building at 238 Main Street in the very nerve center of Kendall Square. CIC has since spread to three other sites in the square, another in Boston, and opened sister facilities in a half-dozen other cities in the United States and abroad, with plans for more. In its twenty-plus years, CIC claims to have been home to more than five thousand companies that have cumulatively raised more than $10 billion in venture investments. As 2020 opened, more than 2,200 startups and other tenants were living in its spaces—roughly 35 percent of them in Kendall Square.[2] The coronavirus pandemic temporarily forced the closure of all CIC locations worldwide in spring 2020, but partly thanks to some innovative and more flexible funding programs, as well as a new subsidiary called CIC Health that provided COVID testing to local companies and individuals, the CIC was building back its client base and finding greater financial security by that fall. It continued the rebound in 2021, with CIC Health, cofounded by Rowe and the well-known surgeon and medical writer Atul Gawande, expanding its testing centers around New England and adding COVID-19 vaccination to its repertoire.

Rowe is a thoughtful, somewhat boyish-looking Cambridge native who attended Amherst College as an undergrad. Besides looking for something for Amy's venture, he and a business school friend, Andy Olmsted, were pursuing their own, still vague idea for a company, and so also had their eyes open for space they might use once their plans crystallized. Olmsted had been working at yet another consulting house, Arthur D. Little, when he got an email from Rowe listing ideas for a possible venture. Well down on the list was a startup incubator. Olmsted, as it turned out, had done extensive research on the topic for ADL. "I think number 7 is cool," he responded, sending along a roughly one hundred–slide deck on incubators.[3]

The incubator idea stemmed from a long-standing fantasy of Rowe's to create an establishment inspired by Rick's Café Américain nightclub from the film *Casablanca*—a kind of watering hole where people from all backgrounds could come together. Rowe imagined himself in the role of Humphrey Bogart's Rick Blaine, making key connections and helping deals get done. That specific idea was probably too fanciful to be practical, so the pair worked on adapting it into something that might attract funding. Olmsted quit his job and moved into an attic bedroom in Rowe's house to keep costs down. Meanwhile, they fell in love with the space in Kendall Square. It was more than the two neophyte companies needed. Happily, Rowe once remembered, "All of our other friends were doing startups, and they wanted places for their companies too."[4] They took the plunge, getting going by raising $600,000 as a loan from Rowe's father, Dick, an educator and entrepreneur who has served as associate dean of Harvard's graduate school of education, president of the One Laptop per Child Foundation, and CEO of the nonprofit Open Learning Exchange.

Olmsted recalls spending weekends painting the place and scoping out rental furniture. They envisioned the incubator focusing on web companies—but soon learned there was a long wait to get an internet connection. "You can't be an internet incubator without internet, can you?" Olmsted quips. So they reached out to Joe Hadzima Jr., a Sloan School instructor whose own neophyte company, IPVision, had set up shop a few floors above them. With Hadzima's permission, they tapped into his internet connection and ran a line down to their space.

One name Rowe and Olmstead kicked around for their venture was Idea Barn. By that fall, though, they'd settled on Cambridge Incubator. Olmsted says other incubators were trying to get going in the Boston area, and they wanted their name to stake ownership of the Cambridge side of the river and proximity to MIT. They also decided to do something to boost its visibility. The internet bubble seemed ever expanding, and the white-hot business technology magazine *Red Herring* was holding its Red Herring East conference at the Charles Hotel in Harvard Square. Rowe and Olmsted decided to host the after-party. They rented out the nearby John Harvard's Brewery & Ale House. "It was an unofficial after-party, but there was no official after-party," says Rowe. "So we just made it *the* after-party and told everybody—because we didn't have enough money to actually sponsor it."[5]

Their vision still wasn't fully formed, Rowe says. "We were just trying to get cool startups to come hang with us." But he and Olmsted were willing to spend to create some buzz. They had business cards printed and hired a modeling agency to direct traffic and populate the crowd with attractive young people outfitted in black t-shirts with the word "Event" on them: they even printed out talking points so the temps would know what to say about Cambridge Incubator. "Men and women so that we were, you know, equal opportunity," Rowe stresses. And it worked. "It was packed. It was a smash hit, and it just launched us."

One of those attending the party was Jennifer Fonstad, a partner at Draper Fisher Jurvetson (DFJ), one of Silicon Valley's top VC firms. DFJ had been thinking about setting up a startup incubator in the Boston area, and the idea fit the bill. As Rowe recounts, she told him, "We'd like to invest."

Within a few months, they were closing on a roughly $10 million investment led by DFJ. Boston Consulting Group, Rowe's old employer, put in $2 million, and a few individual investors also joined. "It was just a classic, 1999-era rocket ship kind of thing," says Rowe. "If you look at the venture capital investing curve, that's when the dollars invested were insane. So that was actually kind of modest by the standards of the time."

For opening its bankroll, DFJ got a significant, still undisclosed, ownership stake in the business. What DFJ invested in, though, was not yet much like the present-day Cambridge Innovation Center. Rather than being a place for startups to rent space, it specialized in what Rowe calls the *foundry model*. He and Olmstead would work with DFJ to create companies in multiple industries, recruiting the people to run them and providing the legal, financial, and marketing expertise to get them off the ground. Because the startups were wholly owned by Cambridge Incubator, they would reap dividends for the incubator and its investors should they become successful.

By early 2000, the venture infusion had enabled a large upgrade of their space. Rowe and Olmsted leased virtually the entire fourteenth floor of MIT-owned One Broadway, the sixteen-story former Badger Building at the corner of Third Street. Despite their new checkbook and ties to DFJ, MIT would only rent it to them if someone guaranteed the ten-year lease. The only one they could find willing to do it was Rowe's father. They used some of the DFJ money to pay off his $600,000 loan, though.

Armed with their cash and a newly signed lease, Rowe and Olmsted began investing in staff and infrastructure. They were soon joined by Geoffrey Mamlet, an entrepreneur who had cofounded and recently sold a corporate travel management business. Mamlet remains a CIC managing director. Over the next several months, the Cambridge Incubator team hired another fifty-odd full-time personnel in marketing, finance, law, tech support, and more. At the same time, Rowe put up an ad at the Harvard Graduate School of Design offering fifteen dollars an hour for students willing to help create "the perfect working environment." The two he hired, David Hamilton and Sven Schroeter, came up with something not far off of what's there today. Resident companies were housed in "bays," whose flexible walls could be slid open to connect to an adjacent bay as the startup grew—and everyone shared kitchen and conference space.

Rowe and Olmsted quickly launched a half-dozen or more ventures, all focused on digital technology. A few attracted additional outside funding, but none made it big. PeopleStreet worked on creating contact entries that updated automatically as people moved or changed jobs. BrandStamp was developing a way for consumers and manufacturers to stay in touch about recalls, updates, and more. Also growing out of the DFJ-Cambridge Incubator collaboration was a venture fund called DFJ New England. It eventually merged with Draper Atlantic, a DFJ affiliate fund in the Washington, DC, area, to form New Atlantic Ventures. Rowe and Mamlet were associated with both funds for years, but are no longer.

The internet boom continued throughout 2000, with DFJ leading a second investment round that brought Cambridge Incubator an additional roughly $8 million. Then the bubble popped in the spring of 2001. "These companies needed more capital from outside investors, and when the doors shut on capital for tech companies there wasn't really a logic for a foundry anymore," says Rowe. The bust aside, some fundamental cracks were also visible in the model. Probably the most significant, Rowe once related, "was that most of our companies were essentially run by employees. Our structure did not free them to become 'true entrepreneurs' . . . When the startup's CEO is an employee hired to carry out a specific business plan, it is tough for them to become free of the original ideas, be agile and pivot. We found that such start-ups don't work so well."

Cambridge Incubator would likely have had to face another issue even if the dot-com bust hadn't reared its ugly head. "What we also learned—I think all of us—is that no matter how smart you are, you can't actually launch 10 companies simultaneously and make them all successful," says Rowe. "Nobody's been able to do that. Maybe Elon Musk, kind of, with maybe two or three. So serially is the answer, not parallel."

The bust brought an end to Cambridge Incubator before it had to fully face such issues. They laid off 90 percent of the staff, worked a deal to transfer the incubator's equity in the startups still standing to DFJ and other investors, and officially ceased operations. Olmstead left to record a music CD and do some consulting. He eventually joined the executive ranks of Research Dataware, a future CIC tenant that offered software as a service to help companies comply with federal regulations protecting subjects of clinical research.[6] That essentially left just Rowe and Mamlet with the bare bones of an operation—including the space, and a lot of desks and chairs—they still hoped to salvage. But even with much lower labor costs, they weren't out of the woods by a longshot. Overriding everything was the lease. Seven years remained on the agreement—a total value of roughly $6 million, which Rowe's father had guaranteed. "It was looking like we were going to go under, and that was going to be a problem for Dick Rowe, because he didn't have that kind of money," says Tim.

Out of this dire condition, the modern CIC was born. Recounts Rowe, "Just to pay the rent, we went back to what we started with—which was a bunch of MIT friends kind of sharing space. Except now we had nicer space." Their idea was to rent the space they had and provide basic services, such as internet, phone lines, and shared copiers, but without the mentoring, legal, and other professional services that Cambridge Incubator had offered.

Despite the internet bust, they found a lot of interest from what Rowe calls "good-old MIT companies that were not necessarily dot-coms, they were just doing good tech. And they needed space and they wanted to be next to MIT." It was still not easy for such fledgling companies to find affordable quarters in Kendall Square. Leases typically ran for five years or longer—and even if startups were willing to sign on for that long, few landlords would take the chance without a solid guarantor. "We were basically solving the problem that the landlords wanted credit and startups didn't have credit," says Rowe.

Their rebooted entity—rebranded the Cambridge Innovation Center—offered much shorter-term subleases for a flat monthly fee. One of the first startups to sign up was a spinout from the MIT Media Lab called Ember that developed chips and software to power wireless mesh networks. Ember's founders, Andy Wheeler (later a partner at GV, Google's venture arm) and Rob Poor, helped spread the word.[7] "When we dropped being a foundry, we filled up," relates Rowe. "Which was great. Except we were still losing money."

The core problem was the difference between their dot-com-era rent cost and what they could reasonably charge startups. There seemed no way out of that math. "It was looking again like we were going to go under," says Rowe. Then, they caught a big break. Sometime in 2002, Rowe learned that another tenant in the building, Millennium Pharmaceuticals, had moved out to consolidate operations near Central Square. Its space on the ninth and sixteenth floors was sitting empty. He was able to cut a deal to take over the space solely by covering the biotech's operating costs—not its entire rent. Those costs ran just three dollars per square foot, in contrast to the roughly forty dollars per square foot or more he was paying on the fourteenth floor. All of a sudden the math changed. The additional space almost doubled CIC's square footage and cut its average cost per square foot nearly in half. Now, he and Mamlet calculated, if they could rent out the new space, CIC had a sustainable, profitable business model. After making some improvements, the space was ready for action in 2003—and it, too, quickly filled.

From that point on, they gobbled up additional space whenever they could. By its tenth anniversary in 2009, CIC had laid claim to eight entire floors in the building and a portion of another—a total of more than a hundred thousand square feet. In 2014, the company opened a Boston location—now spread over ten floors in the Seaport District. That same year, Rowe and Mamlet did their first expansion outside the Boston area—to St. Louis.[8]

The CIC's Kendall Square footprint also grew. In 2013, it expanded to three floors at 101 Main Street, about a block closer to the river from their flagship building at One Broadway. The latest local expansion was to 245 Main Street, at the convergence of Main, Broadway, and Third Streets. They took over parts of four floors starting in fall

Figure 23 CIC founder Tim Rowe (center, far back) at a Venture Café event in Kendall Square.

2019. All three of CIC's Kendall Square sites—totaling three hundred thousand square feet that cumulatively housed 795 tenants as of April 2020—stand in a row along what was once called AI Alley.

Space isn't cheap. Tenants choose among coworking spaces that vary day to day—so-called hot desks—dedicated desktops, and private offices. As companies grow, they can shift from a dedicated desk to a private office or suite, sometimes without much of a move. In spring 2020, prices started at $460 a month for a hot desk and ran to $1,244 per person a month and up for private offices. Only counting the working space, that comes out to roughly $180 per square foot. But the combination of a flexible, short-term lease; shared conference rooms; one price that includes internet, coffee, snacks, and other amenities; plus the chance to be among other like-minded people make it worth it for many. It isn't just for startups, either—though Rowe says startups make up about 80 percent of their clientele. The rest is made up of law firms, CPAs, and various other service providers—as well as venture firms and some large multinational corporations that open small offices in CIC spaces to be closer to entrepreneurs.

For the first few years, CIC was just about "trying to make real estate work," says Rowe. But along the way, more and more tenants started talking about the great community—and how they met an investor or hatched a new startup idea by interacting with their neighbors and others in the building.

A tragic incident in December 2006 drove home the importance of community. An electrical transformer in the building's basement exploded, killing one worker and sparking a fire that sent some one hundred people from CIC and other tenants to the hospital, mostly to treat smoke inhalation. The fire "shook us to our core," Rowe once said. The CIC was closed for five weeks, leaving some 150-odd tenants officeless.

Shortly after the fire, when the building was still shuttered, Rowe noticed a fire official escort two CIC clients back into One Broadway to retrieve their computers. He soon learned that these tenants had called the mayor of their home town, who in turn reached out to Cambridge's city manager, who then got the fire chief's permission for them to enter. As Rowe recalled, "I suddenly realized that I didn't know the Cambridge city council, I didn't know the city manager, I had been blind to the fact that we lived in a larger community with its own needs and issues and cares."[9]

He set out to change that, inviting councilors for coffee and sometimes to speak to and interact with entrepreneurs there. He also began reaching out to various other businesses in the square, large and small, gauging their interest in banding together to raise its profile and bolster its political clout—from getting cracks fixed in sidewalks to improving the mix of retail, eateries, and entertainment venues to developing a plan for future growth and development. It turned out he was not the first to think along these lines. Rowe learned that the Kendall Square Manufacturing Association had been created back in the 1920s (it later evolved into the Cambridge Chamber of Commerce). A group called the Kendall Square Business Association had also materialized in the 1970s, but had been out of commission for at least a decade.

With Rowe spearheading the renewed effort, the Kendall Square Association was formally chartered in February 2009. Rowe served as its president for five years, until the KSA grew its membership—it claims more than 180 members—and collected enough in annual dues that it could sustain a paid president and small staff. Said Rowe at the time, "If you think about Kendall Square as the product that Cambridge offers to the world—the place where the city generates two-thirds of its tax revenue and all of its foreign exchange, if you will—then you might want to step back and say 'Let's have a product strategy.' But until now, Kendall Square didn't even have its own website."[10]

Another lesson from the fire was about how important CIC had become to its tenants. Rowe and Mamlet set up a "command center" at the Marriott Hotel. They found temporary space for their tenants at the nearby Marriott Residence Inn, various MIT buildings, and a private shared-space operator at One Memorial Drive. But Rowe feared many clients would never return—they all had month-to-month leases, so it was easy to move on to other space. When the Badger Building opened again, though, all but five displaced tenants moved back in.[11] That drove home the realization that CIC had evolved into a kind of community of its own. "It was a lightbulb," he says.[12]

Even while working the larger city relationships, Rowe and Mamlet began taking more steps to foster interactions and camaraderie inside the Badger Building. Rowe became a big fan of urban design advocate Jane Jacobs, who passed away in 2006. "What she said basically is that the places that seem to be working best are incredibly dense

> **Box 4**
> How Kendall Square Got Its Buzz Phrase*
>
> > Early in the KSA's life, Tim Rowe reached out to his old employer, Boston Consulting Group (BCG), and convinced it to undertake a pro bono study of how Kendall Square stacked up against other innovation hubs. The report was unveiled at a meeting at the MIT president's house on March 3, 2010. Guests included Susan Hockfield; Eric Lander from the Broad Institute; Roger Berkowitz, CEO of Legal Sea Foods; Doug Linde, president of Boston Properties; and Genzyme's Henri Termeer.
> >
> > Former state secretary of economic development Ranch Kimball, then a BCG adviser and the study's lead, reported on the results—which entailed a look at various innovation metrics from twenty-two innovation hubs around the world. Slide 32 of his thirty-four-slide presentation was entitled: "Kendall Square—'The Most Innovative Square Mile on Earth.'"
> >
> > Kimball estimates he has since given scores of talks invoking that phrase. "The fun follow-up line I used in speeches was, 'I said on the planet earth, because the work did not include any examination of exobiology innovation.'"
>
> *The story of the meeting where "the most innovative square mile on earth" phrase was unveiled comes from interviews with Rowe and Kimball; an email Rowe sent to KSA board members following the meeting, shared by Rowe; and the presentation deck, shared by Kimball. Kimball would serve as CIC managing director from 2011 to 2015. Also on hand representing BCG that day was Rodrigo Martinez, who in fall 2020 became chief marketing and experience officer at CIC Health.

places," he says. CIC's staff started working more deliberately to foster that density and interaction between inhabitants as a way to increase collaboration, creativity, and, hopefully, inspiration. They threw "block parties" to bring together companies from different floors and offered yoga classes and set up running groups and other fitness programs. They also worked deals that got tenants discounts on a wide variety of local services, among them bike sharing, gym memberships, and Zipcar rentals.

"Kendall Square has always felt special, not just for the proximity to MIT and all that means, but also because of the entrepreneurial culture of sharing insights and ideas with others, and because of 'bump and connect,'" says Mamlet. "What CIC has been able to do is to bring that spirit of sharing, and the tendency towards 'bump and connect,' into our building, first to a single floor and then to additional floors as we added them one by one to the community. Where 'bump and connect' at the street level might mean maybe 1 in 100 people you passed was a fellow entrepreneur who you knew and wanted to talk with, inside our building, it was magnified and accelerated. You can literally sit on a couch in the kitchen for an hour and have several helpful, meaningful conversations with your peers as they come in to get their morning coffee."[13]

More recently, CIC started implementing other strategies to encourage interaction. Aisling Hunt, who worked in Kendall Square and helped lead CIC's opening in Warsaw in late 2020, points out that every floor has long featured a coffee and snack area with scores of items, including fruit, nuts, energy bars, chips, and the like. But in 2013, CIC started making some snacks proprietary to a given floor—things like ice cream bars or avocados. "People are primarily motivated by their stomachs," says Hunt. "And so this actually does increase traffic between floors significantly, especially when there's avocados involved."[14]

Beyond its block parties, the Cambridge Innovation Center also began offering events about topics of interest to the entrepreneurial community. In 2010, it opened the Venture Café, which features talks from notable leaders such as Steve Case, founder of American Online and now a venture capitalist, as well as angel investor office hours and Q&A sessions with lawyers. Tenants can also hold their own events—and before the COVID pandemic, the Venture Café had grown to the point that it attracted approximately two hundred thousand people annually to various gatherings across roughly a dozen locations. About a hundred events took place each month in Kendall Square alone.

"We kind of grew up with Kendall Square," says Rowe, reflecting on the CIC's twenty-plus-year journey. "If you walked in and you saw hundreds of companies, and you could have the conversations that you were looking to have, this was sort of the small version of what Kendall was becoming. And you know, not modestly I suppose, I think that we were the heart."

SPOTLIGHT: INNOVATION SPACE ZONING—HOW KENDALL SQUARE HOPES TO KEEP ITS STARTUP COMMUNITY VIBRANT

Is Kendall Square a victim of its own success? Chapter 1's walking tour highlighted how the square has become a mecca for large companies. One building after another is now dominated by multinationals, and many fear these giants will suck up talent by offering pay and benefits few startups can match. They are also driving up rents, which have soared in some cases to more than $120 a square foot—out of the range of almost any startup.

Even when startups are willing and able to pay market rates, financial incentives work against them. That's because all tenants are not equal. As Cambridge Innovation Center founder and CEO Tim Rowe explains: "If a startup or a startup shared space is willing to pay you a $110 [a square foot], and Microsoft is willing to pay $110, you're always better off taking Microsoft. Because when you sell the building, the buyer will put a different multiple on the Microsoft stream of rent payments than on the startups or an operator like CIC. They'll say, that's guaranteed—so I'm going to give you full credit for the next twenty years. They'll look at CIC and say maybe ten years, and they'll look at the startup and say we'll give you two years. So actually $110 from Microsoft counts for ten times more guaranteed income than the same rent from a startup in the value of the building when you sell it. And even if you don't want to sell it—if you want to finance it and use that money to do other projects—the same math applies."[1]

So, yes, in that sense, Kendall Square is a victim of its own success. "There's a structural problem in successful areas where they will automatically force out the startups," Rowe notes. "It's not because of the rent, it's because the startups don't have credit." Bill Kane, executive vice president for BioMed Realty's East Coast and UK markets, agrees. "Left to its own devices, the market typically favors the larger, more creditworthy tenants. As Kendall became great, and demand for space increased, it became harder and harder for smaller tenants to lease space."[2]

At the same time, pretty much everyone—city and university officials, companies, and real estate developers themselves—sees great value in the vitality brought by startups to an ecosystem like Kendall Square. To help ensure their presence, special zoning rules have led to a significant amount of space being set aside for startups and other smaller players such as nonprofits, with a lot more in the works as of early 2021.

Rowe was a key player in making this happen. Sometime probably in 2012, with a number of new real estate developments being discussed for Kendall Square, the CIC founder gathered several other operators of shared space and wrote to city officials urging that something be done to avoid squeezing out startups. "If you want to have an innovation district, you have to decide first of all whether you want startups," he says. "Half of all the occupancy in Kendall Square at that time was companies that were born in Kendall Square. If you take that out and they're born somewhere else, the Cambridge economy doesn't look so great. In a down market, the big guys move on; it's the local ones that don't. So I argued to the city that their pendulum-dampening companies are their local ones, and it's half of their real estate demand—and they've got to think about whether that's important for their long-term health. If they accept that it is, then they're in peril—because this success that we're having is going to kill that source of stable occupancy and replace it with big leases to Dell or something like that." Kendall Square real estate taxes make up roughly half of Cambridge's property tax income, Rowe points out. Preserving that mattered to the city.

It wasn't the first time the issue had come up—and officials were listening. In its K2 Plan for the future of Kendall Square, released in 2013, a long-standing committee comprising representatives from local businesses (including the CIC), real estate developers, nonprofit groups, and residents made suggestions about the square's future. One zoning recommendation involved setting aside a small percentage of any new commercial development specifically for smaller players. This was adopted in somewhat different forms for various zoning districts touching parts of Kendall Square. But perhaps the best example lies in the zoning ordinance for the Kendall Square Urban Redevelopment Area in the heart of the square. As adopted in December 2015, the ordinance among other things mandates that 10 percent of any new office or lab space development totaling one hundred thousand square feet or greater has to be set aside as *innovation space*. While this term encompasses a wide range of endeavors, such as incubators, coworking spaces, training facilities, product development and fabrication areas, and even investor offices, this was mostly about protecting the square's startup fabric.[3]

Importantly, the innovation space being created doesn't have to be in the new building that's going up, so long as the landlord makes it available somewhere else in Kendall Square—say, by converting older space to that purpose. That helps make the startup space more affordable since newer buildings command premiums. It also allows a company to take over an entire building, as many large companies prefer to do.

Key factors mandated by the innovation space ordinance include the following:

- If more than forty thousand square feet of innovation space is required, the space can be split into different buildings. Otherwise it must be in one structure.
- To encourage and facilitate interaction, 50 percent of an Innovation Space must be shared space. This includes common areas, kitchen space, and conference rooms.
- Licence agreements, as opposed to leases, for use of the innovation space should be for approximately one month at a time (encouraging startups and short-term stays).

- No single company or entity can occupy more than two thousand square feet, or 10 percent of an innovation space, whichever is greater.
- The average size of private office suites cannot exceed two hundred square feet.

Notably, the developer-landlord doesn't have to manage the innovation space itself. It can lease the entire space to another entity, such as Cambridge Innovation Center, which then maintains it according to the ordinance.

The ordinance also includes incentives for developers to effectively double the innovation space they create. Alexandra Levering, a project planner with the Cambridge Redevelopment Authority (CRA), explains that if a developer offers at least one-quarter of the required innovation space at below market rates (there is no set price for this, but the CRA says it's essentially breakeven for the landlord), then the entire innovation space, including the market rate portion, is not counted toward the maximum square footage allowed in their approved development. This enables the developer to build a bigger building and, in theory, increase its profits since the innovation space as a whole is meant to be profitable. If they take advantage of this provision, developers can designate up to 20 percent of a new building as innovation space.

While a for-profit entity such as the CIC offers market rate innovation space, much of the below-market innovation space in the Kendall Square Urban Redevelopment Area is housed in the Link, on the eighth floor of 255 Main Street. The Link includes space similar to the CIC, but for nonprofit efforts. It also offers classrooms and large meeting spaces, and is designed to hold events, workshops, and training sessions that aim to connect businesses with underrepresented job seekers. "We want to make sure that the economic development that is going on in Kendall Square is not only company-size diverse, but also that the workforce opportunities are diverse," says CRA executive director Tom Evans.[4]

How has innovation space come into play? In order to build the nineteen-story tower at 145 Broadway that opened in late 2019 and became Akamai's headquarters, Boston Properties was required to create 60,496 square feet of innovation space. Its separate development at 325 Main Street—an eighteen-story structure mostly for Google scheduled for completion in 2021—dictated another 44,704 square feet. Boston Properties plans to meet the total 105,200-square-foot requirement for both buildings in its existing building at 245–255 Main Street, designating exactly one-fourth of the total footage required (26,300 square feet) as below market rate space. By early 2020, it had delivered all its required market rate space, much of which is managed by the CIC. And it had brought online over 17,500 square feet of below market rate space in the Link.[5]

That is just the portion of the square that falls under the Kendall Square Urban Redevelopment Area overseen by the CRA. All told, another roughly one-hundred eighty thousand square feet of innovation space has been created or is planned as part of developments in other areas of the square. LabCentral, a celebrated biotech startup facility in an MIT building at 700 Main Street, occupies about seventy thousand square feet of innovation space created as part of MIT's development of several buildings along Main Street—in what's called the Osborn Triangle. And LabCentral is expanding to another one hundred thousand square feet at 238 Main Street as part of MIT's Kendall Square Initiative, detailed in chapters 2 and 23. Of that space, at least one-fifth, or twenty thousand square feet, will be designated innovation space; the rest will be offered to larger companies that do not fit the requirements, says LabCentral cofounder and president Johannes Fruehauf. That space was set to be available by late 2021. Meanwhile, Alexandria Real Estate Equities was creating roughly nine thousand square feet of below market rate innovation space as part of its development of 161 First Street, the original Lotus building, scheduled to open in the first half of 2021. And MITIMCo's planned Volpe development, also described in chapters 2 and 23, will ultimately create another roughly eighty-three thousand square feet of innovation space.[6]

The CRA's Evans sees such ordinances as a useful lever municipalities can use to strengthen the diversity of their innovation engines. "I think it is important," he says. "Once we started seeing the big footprints of folks like Microsoft and Google coming in on the tech side, and likewise you started to see consolidation in the biotech industry and the growing presence of big pharma as well, the idea of saving some space for startups was a key policy move."

16   "NIBBER": BIG PHARMA UPS THE ANTE

"BG. You need to remember this day. This is an important day."[1]

The words, or something close to them, were those of MIT president Charles "Chuck" Vest. *BG* was his shorthanded moniker for Barbara Gunderson Stowe, the institute's vice president of resource development, one of his closest advisors. Stowe had worked alongside Vest for years. In all their time together, across many big meetings, often with famous and powerful people, Stowe says, the even-keeled engineer almost never showed much reaction. But this day, most likely in 2001 after a meeting with Novartis executives in Switzerland, proved very different. So she took notice indeed.

The outcome of the Novartis meeting became public the following May, with the stunning announcement that Novartis would move its worldwide research headquarters to Cambridge—leasing lab space from MIT and committing to spending $250 million to get the new operation going. The news, accompanied by a press conference at Vest's MIT residence, made headlines around the world. It came at a crucial time, almost in concert with the dot-com bubble bursting and ongoing efforts to recover from the 9/11 attacks of the previous fall. More importantly, although it was not clearly evident then, Novartis's decision formed the zeitgeist of the next phase in Kendall Square's evolution: its ascent into what many believe is the primary biotech ecosystem in the world.

It wasn't the Novartis move alone that propelled this transformation. A series of threads converged in the early 2000s to chart this new path for the square—the most notable of which was the continued rise of its homegrown biotechs that produced a critical mass of executive and entrepreneurial talent that fed back into the ecosystem and reinforced it (see chapter 17). Novartis, though, was arguably the single biggest step in the early 2000s—and at the least a major catalyst. It kickstarted a movement that over the next fifteen years saw the vast majority of its rival pharmaceutical companies moving key parts of their research operations to the Boston area, usually in Kendall Square. No place in the world comes close to this concentration. Big Pharma's growing presence helped ignite a talent war, turning up the crank on salaries and real estate prices and making architecture and building amenities into a recruiting and retention tool—changing the landscape metaphorically and physically.

Katrine Bosley, who by 2021 would go on to become CEO of three biotech companies and board chair of two others, was a manager in Biogen's commercial group at the time of Novartis's announcement. She had been working in Boston biotech for about twelve years, and had found her first job—as an administrative assistant at a biotech startup—by answering a classified ad in the *Boston Globe*. The effect of Novartis coming to town was profound, she says. "All of a sudden there was just this massive number of new, didn't-exist-before jobs in the marketplace." Besides offering people more employment choices, it had the effect of raising compensation almost across the board. "Somewhere along the line, even young startups had to make bonuses a part of their basic compensation package," Bosley says. "My recollection was there was an increase in salaries as well." She calls Novartis an "equilibrating" presence.[2]

Novartis's arrival "was very big," says Janice Bourque, who at the time had been president of the Massachusetts Biotechnology Council for a decade. Pfizer and Merck had some presence in the area, but Novartis was taking things to another level—making Cambridge the center of the company's worldwide research. "This was further validation from the outside about big pharma finding the value in Boston, and particularly in Kendall Square. They came to learn and to integrate, but they came with tremendous resources."

"And then," she says, "it all just took off from there."[3]

*   *   *

The new research enterprise would formally be called the Novartis Institutes for BioMedical Research—but it soon became known as "Nibber," the shorthand pronunciation of its acronym, NIBR. The plural "s" in Institutes was a typo introduced in the rush to incorporate the new entity, but Novartis quickly rationalized it by lumping other research entities under it in one broad umbrella—a global reorg "done to adjust to the 's,' not the other way around," one insider notes. Glitch aside, it was clear from the start that the Swiss drugmaker was not hedging its bet. The $250 million it committed was just to get Nibber going. The first step involved leasing and equipping 255,000 square feet in Tech Square—including a new building that wasn't yet complete—and hiring an initial cadre of four hundred scientists and technical staff. In early 2001, MIT had repurchased the property, which, with the institute's help, had kickstarted the renewal of Kendall Square nearly forty years earlier.[4] Around the same time, probably a bit later, the Swiss company also took out a forty-five-year lease, with the option to buy, on the former Necco candy factory on Massachusetts Avenue near Central Square. It planned to spend $175 million refurbishing the plant, which would mean adding another roughly five hundred thousand square feet to its Cambridge footprint—and close to another seven hundred scientific staff.

This wasn't just an expansion of Novartis research operations. The company and its CEO, Daniel Vasella, had decided Novartis needed to reinvent how it invented drugs. This in itself might seem counterintuitive. Novartis had received nine new drug approvals in the United States in 2000 and 2001, more than any other drugmaker—and fifteen approvals worldwide in 2001 alone. It was riding high on the strength of its hypertension drug Diovan, which had turned into a blockbuster. In 2001, it had also won FDA approval of Gleevec, its anticancer drug, which would also become a multi-billion-dollar product—and its pipeline was also considered among the industry's strongest.[5]

But for several core reasons, the company didn't see business as usual as the way forward. For one thing, its sales base was shifting. In the few years leading up to the decision, the United States had eclipsed Europe as Novartis's largest market—rising to 43 percent of total sales, against less than one-third in Europe—and the gap was widening. It made sense for research to reflect that reality. Perhaps even more important, the genomics era was emerging, bringing with it a plethora of potential new drug targets and computational tools, and dictating a new breed of scientist. Plus, recalls Vasella, "Frankly, I was not totally happy with our productivity in R&D."[6] This was especially true for the research side of the equation, he says.

A pivotal discussion took place at an offsite meeting with the executive committee, probably in 2001, says Vasella. "We were discussing our research and what was working or not working so well, and how we could improve the productivity, the output, the innovation level, and where we could invest additionally. The principle was relatively simple: if you invest more, it doesn't necessarily improve productivity, but more is more likely to deliver more. And then the second question is, of course, well, investing is okay, but where do you invest?"

Expanding existing operations in Basel was the most economic, says Vasella, "because you could just add to existing structures and processes and people." But a strong case could be made for the United States. They looked at options in New York, New Jersey, North Carolina, California, and Massachusetts. But as executives considered that panorama, it was clear that Novartis's most effective relationships were working with academia. "We had it in Switzerland, with ETH," Vasella says, referring to Eidgenössische Technische Hochschule—the public research university in Zürich where Einstein studied. "The interaction with their scientists was always a very good one—an important one historically. And we concluded that the most exciting academic environment [in the United States] was in Cambridge. And maybe we wouldn't gain anything additional being in the Basel region and just being bigger."

As another official later put it, the company wasn't just looking to be closer to the latest advances—it wanted to be close to where the science would be thirty years and more down the road.[7]

To underscore its commitment to this new path, Novartis recruited a renowned scientist to head the effort. He was Mark Fishman, a professor at Harvard Medical School and chief of cardiology and cardiovascular research at Massachusetts General Hospital. Fishman had no real business experience: he was a scientist, who had pioneered the use of zebrafish as a model for the human circulatory system. Yet the company was starting essentially from scratch in Boston, explains Vasella, so finding a respected leader was key to attracting the other talented researchers Novartis would need. Fishman was named president of NIBR, with the authority to reshape worldwide research. He also joined Novartis's executive committee.

Vasella, Fishman, and Vest were accompanied by US Senator Ted Kennedy, among others, at the big press conference kicking things off. "Why Cambridge?" posited Vasella in his remarks. "Analysis shows that it is more and more difficult to attract and retain scientific talent, so we have to go where the talent is. Cambridge is a pool of scientific talent not found elsewhere in the world." Recapping the announcement, MIT's news office reported that in addition to Harvard, MIT, and nonprofit research centers like the Whitehead Institute, Cambridge was then home to sixty biotech companies—fifty-two of them within one mile of campus, in Kendall Square, East Cambridge, and Central Square.[8]

Fishman and a few others initially set up offices on two floors of 400 Tech Square, almost next door to Project MAC and the AI Lab. Their new dedicated building at 100 Tech Square was already under construction and would be ready in March 2003. It held offices and lab space for roughly four hundred scientists and technical staff. Novartis moved 130 employees to Cambridge by the summer of 2003—all but ten of them from its existing US research operation in New Jersey, with the rest coming from Europe—and made a big push to fill out its ranks with new hires. "We were the first Pharma company in the neighborhood at all, as far as I am aware," Fishman relates.[9]

Novartis's announcement was a shot heard around the life sciences world. "I would say *shock* is a good word for it," recalls Tom Hughes, who headed Novartis's cardiovascular and metabolic disease area in New Jersey and was among the first to transfer to Boston.[10] He had gotten a hint of what was coming at a management retreat that January in Interlaken, Switzerland, while having a drink and a cigar with Vasella in their hotel lobby. The Novartis CEO had revealed something of the new plans for research—without getting deep into details. "Dan had this vision that we had to do it differently," Hughes recalls. "It was just one of these moments where you realize here is somebody who has access to the billions of dollars that it's going to take to take that step and make it meaningful. It was very bold."

The shock was both internal to Novartis and industry-wide. Many Basel-based researchers feared their lab might be closed or that they might be ordered to uproot to the United States. It also meant a whole new way of doing things, with all the angst that comes with that. "You were taking a company that had built itself on the Rhine in Switzerland and putting this massive infrastructure in place in the United States and putting a US scientist and academic in charge of it," sums up Hughes. For the industry, the move signaled a new day in several key ways. Labs had largely been sequestered, isolated—for their academic feel but also for secrecy. Hughes had previously worked at a Sandoz lab surrounded by deer-populated woods, but also barbed wire. Now they would have a city facility, within hailing distance of two world-renowned universities. Novartis was upping the ante on not only the way it approached research, but also the competition for talent and science.

Hughes served on the advance team—moving in August 2002 so his kids could attend school when it started. He had done his doctoral work at Tufts in nutritional biochemistry in the mid-1980s, and often would ride his bike from his Somerville apartment to the Tufts Medical Center in downtown Boston. His route took him through Kendall Square, right past Tech Square, where Novartis's first offices would be. "It was really a kind of scary place," he said of Kendall Square back then. Things were a lot better in 2002—with much of the urban renewal plans complete. But the area was still chockablock with old factories and vacant lots, and he realized Novartis would be part of an important step to continue the transformation.

As the Tech Square facility was coming online, the Necco plant's refurbishment got underway. Selecting the old candy factory as NIBR's eventual headquarters was largely down to Fishman. He'd been scouting locations with a small team from Switzerland. Their car pulled up to the Necco plant, but one look at the aging facility was enough to scare off the visitors. One turned to him and said, "We're not getting out. You can go look." Fishman did, and loved what he saw.[11]

A bit later, Hughes was part of an early reconnaissance mission to the old plant. Rain poured down outside and water seeped into the leaky building. "They were making Mary Janes," Hughes recalls, referring to Necco's peanut butter-and-molasses candy. "They had tarps up to divert the water from the manufacturing line. It was horrific and we were horrified that we would be moving there. Mark was right, though."[12]

The building proved almost ideal for Novartis's purposes, with thick floors able to support modern-day equipment and a concrete frame that resisted vibrations. As part of the makeover, the plant was placed on the National Register of Historic Places, earning its owner, the DSF Group, a tax credit totaling 20 percent of the renovation cost. To qualify, the company preserved the factory's exterior, including the water tower on top that had been painted to resemble a giant-sized roll of Necco wafers. Novartis launched a contest for an original design that reflected the life sciences research that would be going on inside, offering $500 toward art supplies for any Cambridge school that submitted an idea: every school did. The winning design, though, came from a Harvard astrophysics student: a strand of DNA in which the bonded pairs of the double helix were painted the colors of a Necco wafer.[13]

The newly refurbished facility opened in the summer of 2004. A ground-floor loading dock was transformed into a winter garden and main lobby area, where a sunlit atrium acted as a gathering spot. The former power plant became a cafeteria and auditorium. Glass proliferated—as windows and walls—and many of the state-of-the-art labs were positioned on the outer edges of the building, offering researchers a view. "In its day, the Necco building was a demonstration of manufacturing," architect Ed Tsoi told the *Boston Globe*. "In our day, it will be a demonstration of research."[14]

Novartis soon far outstripped its original plan. Despite signing a forty-five-year lease, the company exercised the option to buy just a few years later. It subsequently expanded its footprint in the area dramatically—buying out adjacent properties that included a former gay club (the Paradise), Budget rental car office, and an antique dealer—and then moving across Massachusetts Avenue into a three-building complex previously occupied by semiconductor maker Analog Devices. In 2020, the company boasted a massive presence that employed 2,300-plus and extended back into Kendall

a)

Figure 24 (a) 250 Massachusetts Avenue, a former candy factory owned by the New England Confectionery Company (Necco), became the headquarters of the Novartis Institutes for BioMedical Research (NIBR) in 2004. Novartis eventually repainted the building's landmark water tower (b), transforming a roll of Necco Wafers into an equally vibrant DNA helix (c).

Square itself, including one wing of Charles Davenport's old building at 700 Main Street. Novartis never relinquished its presence in Tech Square, either, where it maintained labs and offices spread across three buildings.[15]

The expanding campus in essence fused part of previously old and funky Central Square with the sparser, higher-tech scene of Kendall Square. "We extended Kendall Square," says Jeff Lockwood, global head of communications for NIBR. "Kendall extends to Central, but the reason that happens is Novartis puts that anchor at 220–250 Massachusetts Avenue [the Necco plant], then things start filling in backwards."[16]

It would take a while, but a wave of pharma companies followed in Novartis's footsteps. Within about a decade, Takeda, Johnson & Johnson, Pfizer, Sanofi, Merck, Eli Lilly, Ipsen, AstraZeneca, and AbbVie all had local presences, with Bayer, Bristol Myers Squibb, and Servier arriving in their wakes (see chapter 19).

Everyone wanted to get closer to the cutting edge of science and form the relationships that often proved essential to future success. But how it would all play out was just an unknown when Novartis made its announcement in May 2002. A major international drug maker was trying to disrupt itself before the market disrupted it—and while a lot of thought had gone into the decision, it was still a gamble. Tom Hughes helped lead a Novartis job fair that took place in early 2003 at the University Park Hotel (today's Le Meridien), just a couple blocks from the Necco plant that was being refurbished. NIBR needed to hire like mad, but Hughes vividly remembers the sense of uncertainty and even dread that hovered over his colleagues on the eve of the fair as they set up tables around the various disease areas they were recruiting for. "I didn't even know if people would want to work for Novartis," he recalls.

But when he arrived the next morning, Hughes was stunned. "It was literally like lines were out the door and around the corner." All told, more than 1,800 people turned out that day. This was going to work, he remembers thinking. "It was clear from the very moment we showed up."

SPOTLIGHT: BOB LANGER—PERSONIFICATION OF KENDALL SQUARE'S SECRET SAUCE

Key to Kendall Square's ascent as an innovation ecosystem is faculty entrepreneurship—and an environment where scientists are applauded and rewarded for starting companies, rather than frowned upon or even ostracized. MIT, focused on engineering and applied science, was always more business-oriented than Harvard and virtually every other research university. It had a long-standing tradition of allowing faculty to spend 20 percent of their time, or one day a week, consulting for industry. The school also permitted professors to form companies, as Vannevar Bush did with Raytheon in 1922. But until around 1980, it was relatively rare for faculty to truly embrace entrepreneurship.

All that has changed, of course. Today, the Boston area, like many other regions, has seen large numbers of faculty—most visibly from MIT, but also from Harvard, Tufts, Boston University, Northeastern, and others—form companies themselves or become founding scientists in startups. More than a few academic scientists—George Whitesides, Timothy Springer, and George Church at Harvard; David Walt, previously at Tufts University and now at Harvard's Wyss Institute; and Sharp, Harvey Lodish, Jim Collins (previously at BU), and Sangeeta Bhatia at MIT, to name some of the most notable—have become serial entrepreneurs.

It's safe to say, though, that there is no one like MIT professor Bob Langer, whose Koch Institute lab counts more than one hundred researchers. The chemical engineer, an upstate New York native, earned his doctorate at MIT in 1974. As of early 2021, his 1,500-plus research papers had been cited more than three hundred thousand times, and he counted nine-hundred-plus patents issued, with another five-hundred-odd pending. His prodigious output helped earn him the nickname "the Edison of medicine."[1]

Langer landed his first industrial consulting gig in 1979 with Genentech. He didn't found his first company until 1987—a drug-delivery startup called Enzytech—but he soon picked up the pace. Heading into 2021, Langer had founded or cofounded forty-one companies, including one, Teal Bio, that was born in the pandemic to develop advanced mask technology. "Bob by himself has created more companies than most states have," says former Massachusetts secretary of economic development Ranch Kimball.[2]

The soft-spoken Langer remembers well the reaction from some MIT faculty around the time Enzytech was founded. "A lot of faculty members were very negative

Figure 25 Bob Langer in the Koch Institute lobby.
Source: Langer Lab.

about it. A couple of them didn't want me to get tenure," he relates.³ That outlook has been almost totally vanquished now. Rather than being an impediment to tenure, company formation can be an asset to it—and a badge of honor for faculty members. There wasn't a pivotal moment when that became true, says Langer. "I think it just became gradually more accepted, because more and more people did it."

Langer's philosophy is first and foremost to swing for the fences—to tackle problems that if solved can have the biggest impact on improving people's lives.

His lab focuses on several core areas: biomaterials, drug development and delivery, and tissue engineering. Believing that different backgrounds and expertise can streamline the road to solutions, Langer has built a multidisciplinary research team that includes biologists; chemical, electrical, and mechanical engineers; medical doctors; materials scientists; and even veterinarians and physicists.

Another aspect of Langer's philosophy is that he often encourages his postdocs and researchers to join and create startups if they want to. "One of the reasons a lot of the companies have done well is that the champions have been our students who've gone to them," he once noted. "They really believed in what they did in the lab and wanted to make it a reality."⁴

Looking back on the companies he has helped launch, "I would say the great majority started in Kendall Square," says Langer. Some, like Moderna, formed in 2010, have stayed in the square their whole lives. Others, such as hair care company Living Proof, started in Kendall Square but were acquired and moved out: its headquarters on Binney Street was long adorned with a gigantic poster of spokesperson Jennifer Aniston. A recent startup, Vivtex, which is developing drugs targeting gastrointestinal disorders, is based in LabCentral at 700 Main Street. Says Langer of the square: "I think it's the greatest ecosystem in the world. Part of what makes it what it is are the great universities and hospitals in the Boston area. It's hard to find any place that's got colleges of that quality within two miles of each other. The students that get trained are outstanding—and people want to stay. That's the nucleus, and with that comes of course everything else that's important, like venture capital, lawyers, executives, and so forth."

And, he relates, "I can walk there."

Langer's companies:
1987: Enzytech—Microsphere drug delivery (merged with Alkermes)
1987: Opta Foods—Food ingredients (originally part of Enzytech, acquired by Sun Foods)
1988: Neomorphics—Biocompatible materials for tissue growth (acquired by Advanced Tissue Sciences and Smith & Nephew)
1992: Focal—Biodegradable materials for prevention of surgical adhesion (acquired by Genzyme)
1993: Acusphere—Imaging agents with porous microsphere technology
1993: EnzyMed—Combinatorial pharmaceuticals (acquired by Albany Molecular Research)
1997: Advanced Inhalation Research—Pulmonary drug delivery (acquired by Alkermes; Civitas Therapeutics, an Alkermes spin-off based on AIR technology, later acquired by Acorda)
1998: Reprogenesis—Scaffolds for tissue growth (merged with Creative Biomolecules and Ontogeny to form Curis)
1998: Sontra Medical—Transdermal drug delivery (acquired by Echo Therapeutics)
1999: Transform Pharmaceuticals—Polymorph crystallization (acquired by Johnson & Johnson)
1999: MicroCHIPS—Silicon-chip-based drug delivery
2000: Combinent Biomedical Systems—Transvaginal drug delivery (acquired by Juniper)
2001: Momenta Pharmaceuticals—Complex-sugar-based therapeutics
2003: Pulmatrix—Inhaled therapeutics
2004: Pervasis—Therapeutics for vascular healing (acquired by Shire)
2005: Living Proof—Hair care products (acquired by Unilever)
2005: Arsenal Medical—Nanofiber-based drug delivery (now Arsenal Vascular and 480 Biomedical—Bioresorbable scaffolds)
2005: PureTech Health—New ways to treat GI and immune disorders
2005: In Vivo Therapeutics—Scaffolds for spinal cord therapy
2006: T2 Biosystems—Nanoparticle-based diagnostics
2006: Semprus BioSciences—Medical device coatings (acquired by Teleflex)
2006: BIND Biosciences—Targeted nanoparticle-based therapeutics (acquired by Pfizer)
2007: Selecta Biosciences—Targeted nanoparticles

2008: Seventh Sense Biosystems—Microneedle blood collection technology
2008: Taris BioMedical—Urological drug delivery
2009: Kala Pharmaceuticals—Mucosal drug delivery
2010: Moderna—Modified messenger RNA delivery
2011: Blend Therapeutics—Combination medicines
2012: Xtuit Pharmaceuticals—Novel cancer treatments
2013: SQZ—Delivery to cells
2013: Gecko Biomedical—Medical adhesives
2014: Arsia—Antibody formulations (acquired by Eagle Pharmaceuticals)
2015: Lyndra—Super long acting oral delivery
2015: Frequency Therapeutics—Molecules to restore hearing and other tissues
2015: Olivo Labs—Novel skin care (acquired by Shisheido)
2016: Siglion—Super biocompatible materials/cell encapsulation
2017: Suono Bio—Gastrointestinal delivery
2017: Seer Bio—Nantechnology for proteomic analysis
2018: Vivtex—GI tract on a chip
2019: Lyra Pharmaceuticals—Nasal drug delivery (spun off Arsenal)
2020: Teal Bio—Advanced mask technology in response to COVID-19 pandemic

17 HOMEGROWN BIOTECHS MAKE THEIR MARK

It was a scalding summer's night around Boston—still 87 degrees at 6:00 p.m. the evening of August 21, 2003. But it wasn't as hot as the local biotech industry. More than 850 people had registered for an outdoor event set up in the courtyard between the MIT Media Lab and the school's medical center. Actually, it was more than an event. According to the invitation and program, it was a celebration: the Celebration of Biotechnology in Kendall Square.

The event was spearheaded by Ken Morse, managing director of the MIT Entrepreneurship Center, and cohosted by the Harvard-MIT Division of Health Sciences and Technology, which offered a popular PhD-MD degree. It was the second year of the celebration, which would be held annually for the next several years. But year 2 was when it really took off. "We are gathered together this evening to celebrate the biotech revolution and bear witness to the phenomenon of Kendall Square," Morse told the crowd. Of the 850 registered guests, 150 were founders or board members of biotech companies, he related. "One key take away is that Robert Frost's maxim 'good fences make good neighbors' does not apply to the biotech community in the 21st-century. Rather, all of us at MIT believe in active connections and teamwork with our friends and neighbors here in Kendall Square and beyond. Biotechnology and entrepreneurship go together like Rogers and Hammerstein . . . Boyer and Swanson . . . MIT and the biotech community. Each needs the other to achieve its full potential to be great to make a difference in the world."[1]

After Morse, a series of short remarks were made by some local heavyweights: Genzyme CEO Henri Termeer; Eric Lander from the Whitehead Institute, who was just embarking on a bold adventure to start a new research institute; Janice Bourque, president of the Massachusetts Biotechnology Council; and MIT president Chuck Vest. The night was really for networking—to help ensure fences stayed down.

While the long lines for Novartis's job fair foreshadowed big pharma's big presence, the local biotechnology industry provided another driving force for the surging life science ecosystem around Kendall Square. One California-based biotech, Amgen, had joined the fun—opening a research center near Biogen in 2001 that grew to employ roughly five hundred. But the city's homegrown companies fueled virtually all the industry's local growth.

A quartet of early stars had emerged, reshaping Kendall Square's character. They included two that had helped pioneer the biotechnology field: Biogen and Genzyme. Both had gotten key drugs through the FDA approval process and risen as global forces, vastly increasing their footprints in the square—with Biogen expanding into multiple structures along Binney Street and Genzyme opening a stunning LEED platinum certified headquarters. In their vanguard came slightly more recent startups, led by Vertex and Millennium Pharmaceuticals, both of which were headquartered on the other side of MIT nearer Central Square, but expanded into Kendall. And behind them came a wave of newer startups destined to become future leaders, including a second Phil Sharp–related company called Alnylam Pharmaceuticals.

In contrast to what had taken place in computing and defense electronics, as these and other companies grew, the vast majority didn't flee to Route 128 and lower-cost or more family-friendly suburbs; the growth of biotech proved a city phenomenon. Companies wanted to stay close to the same forces that had attracted Novartis: top universities that provided scientific expertise and future talent, along with a world-class hospital system and medical community across the Charles River. Young startups also wanted to be within easy reach of their scientific founders, often MIT and Harvard professors. The ecosystem was further enriched by an increasingly sophisticated venture capital community geared toward life sciences, leading business schools; its own trade group that worked to build ties to government and demystify the industry in the eyes of a skeptical public; and a strong service provider network that included lawyers, accountants, and real estate developers.

The San Francisco Bay Area had been viewed as the clear leader in biotech since the industry began in the 1970s. Millennium founder and CEO Mark Levin had been a venture capitalist in San Francisco when he conceived the genomics company in the early 1990s. "We in San Francisco looked on Boston and Cambridge as kind of, 'My God, who are these people that came over on the Mayflower?' Essentially very conservative, not much going on—it was Biogen, GI, and a few other companies. But San Francisco was where it was for biotech."[2]

Only the scene was changing dramatically, Levin realized to his surprise. Genomics represented an "enormous revolution, which I thought was much stronger in the Cambridge area because the biology was stronger." What's more, the Bay Area's biotech scene had gotten incredibly diffuse. Some companies, like early pioneers Cetus and Chiron, took root across the bay from San Francisco in Berkeley and Emeryville—while others lined the peninsula by Stanford and farther south. Startups had arisen around Genentech in South San Francisco, but that area was mostly a soulless industrial park. "The Bay Area is extraordinary, obviously," says Levin. "But it never reached that critical mass where everybody's bumping into each other in restaurants and cafes and really creating this energy. It wasn't Harvard and MIT and all these new institutes all within a mile or two of each other. Literally, you could feel the difference once Boston and Cambridge started to take off."

* * *

Levin had sensed the transformation when he founded Millennium in 1993. By a decade later, liftoff had occurred. The leader of the rapidly expanding pack of homegrown companies remained Biogen. In 2003, when Novartis opened its facility, the pioneering biotech was riding high on profits from its multiple sclerosis drug Avonex, which had been introduced seven years earlier. That same year, the Kendall Square company merged with San Diego–based Idec Pharmaceuticals to form Biogen Idec, the world's third-largest biotech after Genentech and Amgen. Biogen would return to using its original name in 2015. It also took over as the largest corporate employer in Cambridge, with 1,467 employees: it retained the top spot until passed by Novartis in 2008 and has been among the leaders ever since, with over 2,300 local employees in 2020.

It had not been an easy path to success, however. In the early 2000s, Biogen was on its fourth CEO—fifth depending on how you counted one, Jim Vincent, who had served two nonconsecutive stints. Indeed, things had started to go wrong almost as soon as the company moved to the square. By 1984, Biogen's research-heavy overhead had racked up an estimated $100 million in losses, with the jury out on whether it could survive. In his years at the helm, Wally Gilbert had relied on licensing or selling Biogen's patents to keep things going, but without a steady revenue stream from its own products, the company could not keep up with costs.

Vincent was brought in to right the ship. The hard-charging former executive of Abbott and Allied Health and Scientific Products—he drove his Porsche to work, an anomaly in non-ostentatious Kendall Square—took the reins in 1985. He immediately orchestrated an aggressive restructuring plan that involved cutting costs, renegotiating patent licenses, recruiting a more experienced operational management team, and nurturing Biogen's own drug pipeline.[3]

"I came here to build an operating company," Vincent once recounted. The science was incredibly strong, if unfocused—but Biogen was especially lacking on the non-scientific side, he related. "If you look at the founders, you would expect them to hire great scientists. It was also pretty clear that the staffing outside of science was not of the caliber needed to create an operating company."[4]

Biogen's original ambitions had spanned human health, agriculture, veterinary science, and even the environment. Vincent let all those go except human health, and even there he narrowed the focus. He also shuttered various European locations, including selling the company's flagship lab in Geneva to Glaxo. Overall headcount was cut from 500 to 225. "The insecurity level here was so high," Vincent said. "You had so many talented people here; they had been winners all their lives, and this was really their first experience with losing."[5]

On the licensing front, Vincent worked to restructure existing deals—or buy back patents altogether. Perhaps most notably, he brought Biogen's beta interferon technology, which Schering had largely not pursued, back into the fold. And in 1986, not long after Vincent arrived, the FDA sanctioned alpha interferon for use against a form of leukemia, making it the first genetically engineered interferon drug approved in the

United States and providing a significant source of royalties for Biogen. Vincent also inked a spate of new deals.[6]

It took a while for the changes to take root. Biogen reported losses eclipsing $70 million between 1985 and 1988. But in 1989, the company reported its first profit—of $3.2 million on total sales of $28.5 million. All told, Biogen's royalty stream, just $1.7 million in 1986, soared to $150 million by 1996.

During this time, the company focused on developing two homegrown drugs. One was Hirulog, a blood thinner invented by a young University of Chicago–trained scientist named John Maraganore, who had joined the company in 1987. It aimed at a huge market—preventing clotting in heart patients. The other centered on the newly reacquired beta interferon technology, which Biogen's scientists thought could provide better treatment of hepatitis B.

Neither went in a straight line. Hirulog encountered a big setback in 1994, when a trial indicated it was safer than the current treatment involving heparin, but no more effective. Although likely to win FDA approval, it would not be the home run Biogen sought. Vincent and new president and chief operating officer (COO) James Tobin, who arrived from Baxter International, halted Biogen's work on the drug and licensed it to another firm. "Hirulog was the favored son; to put a bullet in that was emotionally the most difficult thing the company had faced," Tobin once said.[7]

Beta interferon, meanwhile, had been investigated as a treatment for hepatitis, but just as with Hirulog, trials indicated it would likely only offer a modest improvement over the industry standard: alpha interferon. But outside research showed beta interferon might be effective in treating multiple sclerosis—and the company pivoted to focus on that. Its drug, Avonex, won FDA approval in 1996.[8] Overall sales soared from about $135 million in 1995 to $259.7 million in 1996, fueling a profit of $40.5 million. By the end of the decade, Biogen recorded a net income of $138.7 million on revenues of $557.6 million. Sales of Avonex reached $394 million, just over 70 percent of the total. Avonex was Kendall Square's first blockbuster.

Vincent stepped down as CEO in 1997 but remained on the board. Tobin took over. The new chief executive resigned in 1998, however, after intense disagreements with Vincent: he later became CEO of medical device maker Boston Scientific. John Maraganore, who had moved into business development after his baby Hirulog was licensed away, summed up their differences: "Jim Vincent was an Abbott guy, and Jim Tobin was a Baxter guy. And I can guarantee you that Abbott and Baxter people are like oil and water together at the end of the day. The Abbott leaders are strategic. They're highly competitive individuals, their view of winning is you decimate your competitor. And Baxter was operationally excellent. They were collaborative people. They were a lot nicer than the Abbott people in many ways as individuals. But you would never get an Abbott person and a Baxter person to really work well together."[9]

After Tobin's departure, Vincent returned as interim CEO. A third Jim, Jim Mullen, took the reins in 2000. He had joined Biogen in the early 1990s and unlike his predecessors had risen through the ranks. He would stay in the top role until 2010.

In the late 1980s, following Vincent's downsizing, Biogen occupied its original building on Binney Street and a two-story red brick lab facility just across from it. Employees parked in a gravel lot next to the second structure: one day, someone backed into Vincent's Porsche and smashed it up pretty good. By the early 2000s, Biogen had expanded to a series of newly built lab and office buildings between Binney and Broadway.

Biogen dominated its part of Kendall Square, which featured slick modern buildings erected as extensions to Cambridge Center. But down Binney Street, past the eyesore Volpe complex between Third and First Streets, the square seemed largely stuck in its 1980s state, characterized mostly by a smattering of aging buildings and parking lots.

As the life science era blossomed, that area represented one of the square's last unfilled holes. But that, too, was changing. Anchoring the overhaul effort was the region's other early local biotech success story—Genzyme. The development's mastermind was a well-known figure who harkened back to recombinant DNA ordinance days: former city councilor turned real estate developer David Clem.

*   *   *

Back when David Clem had been an MIT graduate student in urban studies and planning in the early 1970s—he withdrew to run for city councilor, later reentered the doctoral program, but never finished his dissertation—he had attended an event where representatives of McDonald's and Burger King shared how their companies chose locations for new restaurants. The McDonald's man had described a sophisticated location strategy—down to the exact spot selected. "I couldn't believe how much research they did. I was blown away," recalls Clem. And then the Burger King guy's turn came. He just said: "We go within five hundred yards of McDonald's."[10]

The event sparked an epiphany of sorts for Clem. Although he told the story in humor, he modified it to form a cornerstone of his own approach to Cambridge real estate development: "I located within five hundred yards of MIT."

That philosophy had been incorporated into the strategy of the Athenaeum Group where Clem had become a partner. The firm had bought and redeveloped the old Boston Woven Hose factory now known as One Kendall Square.

In the early 2000s, though, Clem was well into his next endeavor. He had sold his Athenaeum Group stake in 1993, figuring to retire and relocating to Vermont. But he couldn't stand being idle and after only three weeks had formed a new company, Lyme Properties, to develop life science space. His first purchase was a building along Sidney Street in Cambridgeport, just a few blocks from MIT's athletic fields. Fast-growing Vertex became the first tenant and later moved its headquarters to another Lyme property nearby.

The big prize Clem had set his sights on, though, was a ten-acre parcel between Second and Third Streets—roughly bounded by the Broad Canal on the south and Linskey Way to the north—that had once held a coal gasification plant. When Clem

Figure 26 1995 aerial view of Kendall Square and the ten-acre parcel Lyme bought. Biogen's growing campus is in the middle left. After the Genzyme Center opened in 2004, filling a longstanding hole, other commercial, residential, and retail developments soon followed.

and Lyme bought the property in 1998, it had been mostly used for parking lots for a few decades—a big part of Kendall Square's wasteland image.

Previous rezoning efforts for the brownfield area had failed. But Clem broke the logjam thanks to artful maneuvering and a plan for economically cleaning up the old site. He secured a permit to develop the parcel into a mixed-use property offering lab and office space, condos and apartments, a possible hotel, a performing arts center, and a touristy kayak launch from the remaining stub of the Broad Canal. "Because I had a few political skills and little bit of history on the council, and I understood zoning as well as anyone in the community development department did at the time, we were able to get a special permit in record time," he relates.

Lyme's concept heralded a new era for the square—offering life beyond the Marriott while also addressing growing concerns about the lack of residential and community spaces. The centerpiece was a trio of large office and biotech lab buildings flanking a small park that featured a skating rink in the winter and a place for summer concerts and open-air eating in warmer months. Clem was allowed to break the rest of the site into parcels zoned for various uses. He then sold off parcels not intended for life science use to other developers, enabling Lyme to stick to its expertise. These included a one-acre lot bordering Third Street intended for the performing arts center and two plots next to it earmarked as residential.[11]

As Clem worked those deals, he hunted for an anchor tenant for the lab and office complex he dubbed Cambridge Research Park. That quest led him straight to Genzyme and its CEO, Henri Termeer. Second only to Biogen on the local biotech scene in the early 2000s, Genzyme had followed a less rocky path than its Kendall Square counterpart. In contrast to Biogen's string of CEOs, it had had only one boss since 1985—Termeer. Genzyme, with its unique focus on orphan drugs, had not endured Biogen's near-death experience either. The company had gone public in 1986. Five years later, Ceredase, a treatment for Gaucher disease, received FDA approval. That had been followed in 1994 by a genetically engineered successor, Cerezyme. And in 2003, Genzyme would secure FDA approval for Fabrazyme, an enzyme replacement drug for those with Fabry disease, a rare genetic disorder. In that pivotal year, when Novartis was holding its job fair and Biogen was merging with Idec, Genzyme stood as the world's fourth-largest biotech and third-largest Cambridge corporate employer, with just over one thousand employees.[12]

As the 1990s came to an end, Genzyme was expanding rapidly and looking for a new headquarters. Back in his Athenaeum Group days, Clem had negotiated the ten-year lease, with options to extend, that brought Genzyme to One Kendall Square at the start of the decade. He hoped history would repeat itself. Termeer's initial response proved discouraging, however. He told Clem his sights were set on Allston, next to Harvard Business School—near where Genzyme had already built a manufacturing plant. It was only after it became clear that the location presented a zoning and negotiations nightmare—at the confluence of several local, state, and federal jurisdictions—that Termeer decided to stay in Kendall Square.[13]

Lyme launched an international competition to find an architect for the project, and Clem gave Termeer a spot on the design jury. Six finalists were chosen, whittled down from more than two hundred entries. In the end, they agreed on German architect Stefan Behnisch. His design imagined a glass-walled building that leveraged natural light and featured large shared areas to facilitate collaboration. A vast atrium would extend the entire twelve-story height of the building, which would include a library, auditorium, employee cafeteria, training rooms, and ground-floor retail space. It would also harbor eighteen indoor gardens, at least one on every floor, as well as a green roof dotted with various plantings. Rooftop mirrors would follow the sun and reflect sunlight throughout the interior.

As Clem relates, Behnisch made a big impression on both himself and Termeer with the way he summed up the design. "He said, 'We want to create a building that you think about from the inside out. It's a space you'll be happy to work in, as opposed to a space you drive by and look at for a brief moment.'"

A talent war was erupting in the Boston life sciences scene. Biogen and Genzyme were growing, but so were Vertex and Millennium, with a multitude of startups in their wake. And then, of course, Novartis had launched a massive hiring spree.

Behnisch's words "clicked with Henri," Clem remembers. "He said, 'You know, it's all about talent. I want to be able to look out at all of these other buildings when I'm trying to attract talent to Genzyme and say we've got greenhouses. We have solar panels that track the sun.'"

Figure 27 Model of Genzyme Center.
Source: Sanofi Genzyme.

Genzyme Center opened in late 2003.[14] It was an office building only: Genzyme maintained lab space some twenty miles way in Framingham, Massachusetts. Termeer set up on the top floor. The only other thing up there was the cafeteria.

At one point it looked like the building would be certified as LEED silver or possibly gold. Termeer insisted they go for platinum—with Genzyme covering the added costs to make that happen. "The difference between meeting the highest standard and a good standard such as Silver or Gold is very large for a company of this kind," Termeer once said. "Setting a good standard simply falls shorts and sends the wrong signal to potential partners, regulators, our employees, and our patients."[15]

The building, which did win LEED platinum status, garnered a slew of awards, including the 2008 Harleston Parker Medal, acknowledging the most beautiful work of architecture in the Boston area.

Lyme held similar competitions for other buildings in its complex. California-based Steven Ehrlich was chosen to design 675 West Kendall, a six-story lab and office space across Athenaeum Street from the Genzyme Center. It featured a full-height atrium with skylights and ground-floor retail, including a restaurant that looked out on the park/skating rink.

Vertex, Lyme's big tenant in Cambridgeport, leased the entire building. The biotech had been cofounded in 1989 by Joshua Boger, a Harvard-trained organic chemist who had become a senior manager inside Merck's research organization in New Jersey, and San Diego–based venture capitalist and MIT grad Kevin Kinsella. They settled on Cambridge after Boger visited other leading life science centers in San Francisco, San Diego, and Research Triangle Park; they were won over by the strength of Cambridge's molecular biology and computational chemistry expertise, in close proximity to an outstanding medical community. "Nothing else offered that mix," Boger says.[16] The company went public just two years later, riding the fever for "rational drug design" that promised to employ intense computing resources to increase the odds of discovering and developing drugs. Boger boasted to Wall Street that he could be ten times more efficient than big pharma. "Not being too math-oriented, they didn't realize I meant instead of failing 99 percent of the time, Vertex would only fail 90 percent of the time," he says.

Vertex's first drug, Agenerase, a treatment for HIV/AIDS codeveloped with GlaxoSmithKline, won FDA approval in 1999. "That proved we could actually do what we said we would do," Boger says. It also helped pave the way to a massive agreement with Novartis to discover and develop drugs to inhibit one or more members of a family of enzymes called kinases. The compact was worth up to $800 million, with $200 million guaranteed. After that deal, says Boger, "people were crawling all over us. We were getting intense interest from all the major pharma companies. They wanted to do a Novartis-type deal in another human gene family."

The prospect of a second deal spurred Vertex to snap up the lease for Lyme's new building. Looking back, says Boger, "We got out over our skis." Even before the building was ready to open in 2003, it became clear a big deal would not soon materialize. In the end, Vertex moved into the first three floors, using it for groups in medicinal chemistry, computational chemistry, and drug metabolism and pharmacokinetics, among a few other things. It sublet the rest—to Genzyme and startup Momenta Pharmaceuticals.[17]

Even without another megapartnership deal, Vertex continued to grow. It cracked the top ten of Cambridge corporate employers in 2006, with 836 workers. At one point, says Boger, the company was spread across thirteen different buildings. "We had thirteen buildings, thirteen receptionists, thirteen loading docks." The scattered operations took a toll on collaboration and productivity, so the company decided to try to put everything in one place. Here, the limitations of Cambridge's rapidly swelling ecosystem manifested themselves. Finding the right space—with at least some room to grow further—in increasingly crowded and expensive Kendall and Central Squares proved impossible, says Boger. In July 2008, the company announced plans to consolidate operations in Boston's Seaport District—although due primarily to the economic climate, the move would not take place until 2014.[18]

Meanwhile, the parcel's third major lab and office building, a six-story structure at 650 East Kendall, arose straight across the park from the Vertex building. Only a hole had been dug at the site when Lyme sold its Cambridge Research Park properties to BioMed Realty as part of a $531 million deal in August 2005: it would eventually sell its entire portfolio to BioMed through two additional transactions, with the three deals totaling $2.14 billion. BioMed completed the building in 2006, but the first core tenant, Aveo Pharmaceuticals (later called Aveo Oncology), experienced a setback with its drug approval shortly after it moved in. Aveo's space was soon scooped up by Baxter.[19]

The other parcels on the former brownfield site were largely developed by 2010, though some took longer. The two Watermark apartment blocks added more than 450 units to the area. The first, a twenty-four-story complex later known as Watermark Kendall West, was completed in 2006 (its seventeen-story sister building, Watermark Kendall East, opened some seven years later). Watermark West also became home to Evoo, an upscale ground-floor restaurant with a connected pizzeria called Za. In coming years, Canal Way bordering the Watermark would sport a series of additional restaurants and the square's first liquor store.

The Lyme-BioMed efforts foretold of future development in the area. Several years after Genzyme moved in, two more apartment complexes opened on the Volpe side of Third Street. One had been intended as condos for MIT- and Harvard-affiliated people, a plan that never fully materialized.[20] A further retail strip—with restaurants, coffee shops, a coworking space, and a gift shop (now a bank), among other things—appeared around them on both sides of Third. Virtually all the restaurants, bars, and coffee shops offered outdoor seating areas in warm weather. On a nice spring or summer day, the area would often be packed in the early evenings—mostly with the after-work crowd.

Box 5
Bringing Life to Kendall's Ground Floor

> Cambridge was intent on increasing the neighborhood feel around the Genzyme Center.* But success long proved elusive. "There was really no neighborhood. No one wanted to be there," says Jesse Baerkahn, a principle figure in changing that.**
>
> In 2005, Baerkahn and Mahmood Firouzbakht, friends attending Northeastern University law school, set up a residential real estate office in Kendall Square to earn money for school. They began marketing the still under construction Watermark apartments to affluent young professionals. Then, in 2007, Baerkahn created a partnership with developer Twining Properties to lease retail space in the Watermark and a few nearby properties. The business eventually expanded beyond Kendall Square, and in 2011, Baerkahn took full ownership, rebranding it as Graffito SP—an urban retail real estate development and brokerage business focused on "activating" the ground floors of mixed-use buildings.
>
> When the business launched, leasing retail space in Kendall Square was a tough proposition, says Baerkahn. But there was a case to be made: lots of daytime activity due to the exploding business scene, well-heeled customers, and proximity to MIT. "No one had ever marketed Kendall as a destination for retail."
>
> Recruiting Evoo, a high-end New American restaurant then in mid-Cambridge, to move to Kendall Square in 2010 marked the first big success. Fuji, a Japanese restaurant, then opened on Third Street, as did Tatte, a trendy bakery and cafe. Baerkahn and his team also helped attract a day care center and other restaurants and shops. Landlords typically offered recruits deals that tied monthly rents to a percentage of sales. That greatly reduced the risk of moving to Kendall Square. "The early deals felt a lot like partnerships," says Baerkahn. "Landlords recognized commercial tenants wanted to be in neighborhoods that were unique and different—and part of it was the kind of retail."
>
> "In this very organic way," he adds, "everybody kind of relatively quickly shifted their focus to we can actually do this."

*In January 2000, the City Council approved, with a few exemptions, a residents' petition to impose an eighteen-month moratorium on new residential development over twenty units and commercial development over twenty-thousand square feet in portions of East Cambridge that included parts of Kendall Square. The petition cited several concerns, including increased traffic congestion, rising real estate prices, lack of public open space, and encroachment of business development into residential areas. The city manager established the Eastern Cambridge Planning Study Committee, which included city staff, residents, business leaders, and outside experts. As a result of its study, the city required office and lab buildings to include ground-floor retail space. Some details from various Simha emails and comments. See Cambridge Community Development Department, "Eastern Cambridge"; Dabek, "City Council Approves New Development Ban."
**Baerkahn interview, June 24, 2020. All Baerkahn quotes are from this interview.

It wasn't Boston—and Saturday and Sunday nights remained almost completely dead. But it marked a new chapter for Kendall Square: the part of the square that had arguably been the dullest and least active had become perhaps its most diverse and vibrant.

* * *

The fourth powerhouse local biotech was Millennium Pharmaceuticals. Its path was reminiscent of Vertex's, in that it was the brainchild of people outside Boston who zeroed in on the area around MIT as the best place to start their company.

The company had been founded in 1993, most visibly by Levin. After shining at Genentech in the 1980s, he'd joined the Mayfield Fund, a leading Bay Area venture firm, where he had cofounded its biotech arm and helped launched seven companies, sometimes serving as interim CEO while they got going.[21]

That had been the plan with Millennium as well. In plotting the genomics startup, Levin had identified a cadre of leading scientists to build it around. Top of his list was Eric Lander, who had recently launched the Whitehead Institute/MIT Center for Genome Research, which had quickly emerged as a leading genome center. Two others were in New York: Rockefeller University's Jeffrey Friedman and Raju Kucherlapati of the Albert Einstein College of Medicine. The fourth, Daniel Cohen, was an ocean away—in France. All four were listed with Levin as Millennium cofounders.

None would join Millennium full-time, but Lander's home turf made sense as a central location to establish the company. Plus, Levin had concluded, it was strongest scientifically. Millennium launched with $8.5 million in funding, led by Mayfield and joined by some blue-chip venture firms. Levin moved east to serve as interim CEO, intending to return to San Francisco. But after interviewing a host of candidates, they only made one offer: to Steve Gillis, a founder of Immunex, then Seattle's most successful biotech, who declined to leave Washington State. Finally, Levin gave up looking. His interim role lasted twelve years.[22]

Exhibiting a penchant for Hawaiian shirts and a wardrobe of colorful shoes and socks, Levin blew into the largely staid East Coast biotech community like a fresh wind. He famously showed up at the annual Millennium Halloween party one year dressed as Cher, with his wife as Sonny Bono, and their daughter as Sonny and Cher's daughter Chastity (now Chaz Bono).

Fun aside, Levin was a dealmaker—taking Millennium public in 1996 and ultimately inking close to $2 billion in partnership agreements with a litany of big pharma companies. Millennium's six-year, $465 million alliance with Bayer, signed in 1998, ranked as the largest-ever deal between a biotech and a pharmaceutical maker.

Millennium's first home was One Kendall Square, a few blocks from Lander's Whitehead Institute digs. "Eric was there all the time early on," relates Levin. After about a year and a half, Millennium took over an MIT-owned building near the Hyatt Regency on Memorial Drive. A few years later, the company expanded closer to Central Square—eventually occupying four new buildings along Sidney and Landsdowne Streets. Its core lab operations and headquarters would remain near Central Square.

But in the late 1990s, Millennium made two forays back into Kendall Square. While waiting for the new buildings, it leased several floors in One Broadway, the old Badger Building. "We had a lot of the business people there," says Levin. The company pulled back from One Broadway when the new sites were ready in the early 2000s, around the time of the dot-com crash. That was why it sublet the space at bargain-basement prices to the struggling Cambridge Innovation Center—ultimately providing the operating margin Tim Rowe and Geoff Mamlet needed to create a profitable business (see chapter 15).

The other expansion into Kendall Square was part of a critical move for Millennium. By the late 1990s, on the strength of its scientific prowess and unparalleled partnerships, its stock market valuation had soared. "The company was valued at like $18 billion without any drugs in the clinic," says Levin. "So what we said in 1999 was that we needed to go out and use this $18 billion in value to acquire companies that were much farther along. We have to move downstream rapidly before the shit hits the fan."

That led to several mergers and acquisitions. Later that year, Millennium acquired LeukoSite in an all-stock deal that gave LeukoSite shareholders 35 percent of Millennium—worth some $655 million at the time, but within a year or so more than $3 billion. Based in the historic Athenaeum building at 215 First Street, LeukoSite had a drug candidate in late-stage trials. Millennium kept its operations in the Athenaeum Building. Then in late 2001, Millennium also bought South San Francisco–based COR Therapeutics for almost $2 billion. Those two acquisitions proved crucial. COR's anti-clotting drug Integrilin was already on the market and doing close to $250 million in sales annually—giving Millennium a much-needed revenue stream. LeukoSite's drug, a treatment for multiple myeloma called Velcade, won FDA approval in 2003 and turned into a blockbuster. "The shit did hit the fan, around 2002," says Levin. "Everything crashed, but by the time it crashed we had acquired those companies."

Levin stepped down as CEO in 2005, handing the reins to Deborah Dunsire, who had been recruited from Novartis in New Jersey: Dunsire and Sherri Oberg of Lexington, Massachusetts–based Acusphere were probably the only Boston-area women to lead publicly traded biotech companies at the time. Levin remained on the board and Dunsire served as CEO and president through Millennium's 2008 sale for $8.8 billion to Japanese pharmaceutical company Takeda. It was the first of several acquisitions Takeda made of Boston biotechs (more on Takeda's presence in chapter 19).[23]

* * *

Millennium never did bring its own drug from inception through to approval. "We were, frankly, a decade early," says Levin. "Almost every genomics company in the 1990s and early 2000s failed. They had no products."

The solid exit makes Millennium one of the region's success stories. But its biggest success may lie in its impact on the Boston life sciences scene, which it seeded with research talent, experienced executives, investors, and company creators. Indeed, wrote industry commentator John Carroll in 2016, Millennium ranked as "probably the biggest single biotech executive mill in the big and rapidly expanding hub city."[24] Among other things, it spawned Third Rock Ventures (see box 6).

Another prime example of Millennium's legacy is John Maraganore. After Hirulog was sold, Maraganore led Biogen's business development for about three years. In 1997, he joined Millennium to head its biotherapeutics subsidiary. A few years later, Maraganore transitioned to more central business operations, where he engineered the COR Therapeutics deal. In this role, he moved back to Kendall Square, based at One Broadway.

Box 6
The Third Rock Bet

> In October 2006, John Maraganore, Mark Levin, and Kevin Starr sojourned to Las Vegas to play blackjack. The three friends still make the trip almost every year. "We do that for seventy-two straight hours with a couple hours for break," says Levin.
>
> Levin was then Millennium's chairman. Starr retired as Millennium's CFO in 2003 to try producing movies. "Kevin became a transformed individual. He was a blue suit, white shirt, red tie executive from Biogen who became this rock star individual," says Maraganore, smiling.
>
> At the blackjack table, the trio jawed about how the venture capital industry had not been doing a good job in biotech—with lackluster returns for almost a decade. "They were bringing a lot of me-too products forward—or taking products that had been tried in some areas and trying to use them in different areas," says Levin. "It wasn't new biology that would change medicine, it wasn't a new product engine."
>
> And thus Third Rock Ventures had been conceived. Maraganore was all-in with a startup called Alnylam. But Levin and Starr brought in Bob Tepper, who had long headed Millennium's R&D. And in September 2007, Third Rock launched with a $378 million fund. Tepper had heard a radio show mention *3rd Rock from the Sun*, a sitcom in which aliens visit Earth. He said, "'What about Third Rock?' And we loved it!!" relates Levin.* Two other Millennium alums—Nick Leschly, project lead on Velcade; and former VP of metabolic diseases Lou Tartaglia—joined as partners.
>
> Levin found his juices flowing. "It was really exciting. We were talking about epigenetics, gene therapy, genetic disorders, personalized medicine, and all kinds of new areas for the future."
>
> The idea behind Third Rock was to conceive, develop, and launch companies itself—rather than fund entrepreneurs pitching to them. "The original vision was we're going to build a venture firm that's not really a venture firm," relates Levin. "We said we're going to go out and hire people like a biotech company. So we had CEOs, we had chief operating officers, we had heads of R&D. The idea was to have experts who actually had experience, and then a lot of young people from academia, and build three or four new companies a year."
>
> Third Rock soon established itself as a top-tier firm. From its Back Bay headquarters, it seeded companies on both coasts—but mostly in Cambridge. Third Rock listed fifteen public portfolio companies in mid-2020. All were in the Boston area, eleven in Cambridge—six in Kendall Square or East Cambridge and five near Millennium's Cambridgeport haunts. The largest was Bluebird Bio. Its CEO was Leschly, who took Bluebird's helm in 2010. Based on Binney Street, the company ranked in the top twenty-five of corporate employers in Cambridge in 2020 with 843 employees.**

*Levin email, June 5, 2020.
**Bluebird was founded in 1992 as Genetix Pharmaceuticals. Levin met the CEO in 2009, and decided to invest. Joined by Genzyme, Third Rock rebooted and rebranded the company and Leschly became CEO (Levin email, June 5, 2020).

Maraganore was still there in 2002, when he was contacted by two well-known figures on the biotech scene. One was Christoph Westphal, a rising star at Polaris Ventures. The other was Phil Sharp. They were part of a small group of leading scientists and investors forming a startup around another emerging field of science—RNA interference, which employed RNA molecules to block the expression of disease-causing genes.[25] The men wanted Maraganore as CEO.

That led to a crucial meeting in summer 2002 in Sharp's cramped office at the MIT cancer center. The startup's name was Alnylam, a twist on the spelling of Alnilam, the brightest star in Orion's belt, and the Arabic word for a string of pearls.[26] "We talked about Alnylam and RNAi and where it was going," recalls Maraganore. "So I started doing a lot of reading about the science and thinking a lot about the applications. I was very happy at Millennium, but over time I realized that this RNA thing could be transformative for medicine. I decided to jump ship."

Alnylam launched later that year with Maraganore as founding CEO. Just as Millennium had experienced with genomics, it proved far harder than expected to commercialize RNA interference. Alnylam refers to its "dark ages," when many lost faith in the technology. As mentioned in chapter 1, it would take until 2018 for the company to receive FDA approval for its first drug—the first in the RNA interference field—a treatment for hereditary transthyretin amyloidosis, a rare, deadly disease. The following year, it gained approval for a medicine combatting acute hepatic porphyria (AHP), an inherited metabolic disorder that causes a buildup of toxic enzymes. In November 2020, it won its third approval in three years—for a drug treating a rare kidney disease known as *primary hyperoxaluria type 1*. Alnylam's medicine was the first of any type approved in the United States to treat the disease in both adults and children. The following month, the European Union approved a cholesterol-fighting drug called *inclisiran* that Novartis had developed using its technology, for which Alnylam received royalties.

Early in 2021, Alnylam was awaiting FDA approval for inclisiran and had four other drugs of its own in late-stage trials.[27] It had also begun preclinical studies around an antiviral drug designed to fight COVID-19. Alnylam's idea was to develop two separate interfering RNA molecules that target different regions of COVID-19. This dual-targeting protects against the virus's ability to develop resistance. Alnylam would develop the drug as an aerosol so that it could be given noninvasively. Because vaccines aren't 100 percent effective, one idea would be to offer it as enhanced protection, especially for elderly or high-risk people. It could also serve as an alternative prophylactic for prevention of infection. Finally, it might be given after infection to lessen the virus's effects. "One of the things that we want to make sure of is that there's a really clear place for this approach in the pharmacopeia of therapies that are emerging," Maraganore says. But if there seemed to be, and the data continued to look good, Alnylam might have been able to start clinical trials later in 2021.[28]

On its long road to market, Alnylam became a major figure in a fresh crop of Kendall Square life science startups. The company's first home was in the newly created

Science Hotel on Memorial Drive, about midway between MIT and Harvard. This development, offering small suites and short-term leases, was geared to life science startups. It marked the initial foray into Cambridge by Alexandria Real Estate Equities, which would soon become the city's largest corporate taxpayer and largest owner of life science space—with over $1.5 billion in assessed real estate value in 2020.

Even before going public in 2005, a growing Alnylam needed a more permanent home. It found one at Third and Binney Streets, a stone's throw from the still under construction building Vertex had leased. Alnylam's digs, later bought by Alexandria, offered a compelling illustration of the tipping point the square had reached. The building had been constructed for Palm, the Bay Area–based maker of the Palm Pilot personal digital assistant and Treo smartphone. But the bursting of the dot-com bubble and other hardships had curtailed expansion plans, and Palm never moved in. With other tech companies also strapped, Palm paid for the building to be overhauled to include lab space so that it could sublet to biotech companies and recoup some costs. Alnylam and two other biotechs became the first tenants.[29]

Maraganore had always wanted Alnylam's home to be in Kendall Square. "I had no doubt that for the type of science-based, innovation-focused company like Alnylam to not be in Kendall Square would be potentially a fatal decision—because of the types of people that you can recruit, the types of people you run into and interact with all the time, even for what it says about your company," he says. "Look at the community of companies that are there, that have been born there and grown there; I think it speaks for itself."

*   *   *

Local biotechs were on the rise, employing thousands and at long last bringing drugs to market. The biotech industry had also matured, growing in sophistication and improving its relations with government and a skeptical public—aided by a trade organization that worked to advance the industry. Like so much of Boston's growing biotech scene, it also called Kendall Square home.

The Massachusetts Biotechnology Council, or MassBio, as it became known, was formed in 1985. Its first full-time president, starting in 1992, was Janice Bourque. She would serve in that role for a dozen years, initially setting up shop in the same One Kendall Square building as Genzyme.[30] Everything had been a big unknown. "No one knew what biotech was—they couldn't even spell it," says Bourque. Even the biggest local biotechs like Biogen were small fry compared to pharma, which mostly ignored them; Novartis's move to Cambridge was a decade away. Academia still largely looked down on the field, she says.

Bourque worked with local biotech leaders to host networking and educational events and to build ties to government, especially the FDA. A watershed came when BIO, the world's leading biotech trade group, chose Boston for its 2000 annual convention. Four days of events and daily news articles about the industry helped put Boston biotech on the map—and increase understanding of the field. "We were above the fold on the front page for all four days. I would say that is what really launched us in the public arena," Bourque says.

The growth of the area's biotech industry dramatically enhanced Kendall Square's standing as a life science ecosystem. Soon, Novartis would announce its plans to base worldwide research in Cambridge—adding to the city's reputation.

One other major new element appeared on the scene in the early 2000s, perhaps tipping the scales completely. The person behind it was Eric Lander, the scientist whose presence had helped convince Mark Levin to launch Millennium in Cambridge in the first place. Lander envisioned a scientific institute focused on weaving the area's strong biological and biomedical research with the still-emerging field of genomics—in a bigger way than anything yet attempted. When he spoke at the biotech celebration that hot August night in 2003, the institute had been announced but not yet opened its doors: its home was going up next to the Whitehead on Main Street. Its formal name was the Eli & Edythe Broad Institute of MIT and Harvard. But it was soon known simply as the Broad.

18   ROAD TO THE BROAD

Just a few blocks north of Biogen's swank headquarters in the Philip A. Sharp Building, a drab two-story structure sits in institutional beige glory. Back in the 1980s, the block-long building served as a Budweiser distribution plant. Later, it became a warehouse for a concessions company that supplied popcorn, candy, and soda to sports arenas, airports, and the like. Today, though, the warren of rooms behind its doors holds twenty gene sequencers with the potential to churn out up to sixty terabytes of genomic data each day. That's about the equivalent of five million phone books.[1]

And that's just its "day job," or normal workload. For years, the center concentrated solely on genomic sequencing and analyzing cancer data, but after a long weekend in late March 2020, the facility was reconfigured so it could also process COVID-19 tests at scale. Initially, it churned through roughly a thousand tests per day. By that fall, the largely automated center was averaging around seventy thousand tests per day, with capacity to handle a hundred thousand tests daily—and it planned to double capacity by spring 2021.

This is the Genomics Platform for the Broad Institute of MIT and Harvard. The Broad (rhymes with *road*) is the largest nonprofit research institute in Kendall Square—edging out Draper Laboratory—with roughly four thousand affiliated scientists and more than 1,800 employees (called *Broadies*), making it one of Cambridge's top employers. Besides the Genomics Platform at 320 Charles Street, the Broad has three other facilities in the square. These include its glitzy flagship on Main Street and a connecting fifteen-story laboratory building on Ames Street, where most of its bench science is done. The institute also rents two floors from Biogen on Broadway almost around the corner from the genomic sequencing facility. That's home to its 250-strong software engineering team that, among other things, develops open-source genomics analysis tools made freely available to scientists worldwide. The institute's annual budget has crossed $500 million, and since its 2004 inception its scientists have identified more than one hundred genes causing cancer and shed light on fields as diverse as gene editing, mental illness, the microbiome, diabetes, arthritis, heart disease, neurology, and immunology.

The path has not always been smooth. The Broad rankled some with its very existence, and there's been at least one high-profile controversy that added to the angst. Still, the Broad's launch was part of the early 2000s wave that tipped the makeup of

Kendall Square from historically tech and computing heavy to the life sciences. Of all the nonprofit institutes in the square, including the Whitehead and the Koch Institutes, both immediate neighbors, the Broad is easily the largest. And, says Phil Sharp, the Nobel Laureate cofounder of Biogen and Alnylam and a member of the Koch Institute, "it's by far the most significant institute in Boston."[2]

* * *

Eric Lander is the seemingly tireless, whirling dervish–like force behind the Broad—a superstar in an area teeming with hypertalented individuals. He was the mathematician who, without a degree in biology, joined the Whitehead Institute at its inception, soon thereafter won a MacArthur "genius" grant, then went on to help crack the human genome and cofound Millennium Pharmaceuticals and Infinity Pharmaceuticals, two publicly traded biotech powers. In early 2021, he was named then president-elect Joe Biden's choice as presidential science advisor and director of the Office of Science and Technology Policy. For the first time in US history, this was to be a cabinet position.

In the small world category, as related in chapter 1, Lander had brought home top honors in the 1974 Westinghouse science competition, the same year future Akamai CEO Tom Leighton placed second. Having the top two finishers from a single year heading major organizations within a long block of each other—one in the exploding internet field, the other in the equally hot biomedical and genomics arena—serves as a testament to the magnetic attraction of places like MIT and Kendall Square.

The Broad Institute's roots trace to the former beer distribution center and concessions warehouse on Charles Street. In 1990, Lander and research scientist Nicholas Dracopoli started a project called the Whitehead Institute/MIT Center for Genome

Figure 28 Eric Lander, center, after winning the 1974 Westinghouse Science Talent Search. Future Akamai CEO Tom Leighton, upper left, placed second. Third place Linda Bockenstedt, upper right, became a professor at Yale School of Medicine.
Source: Society for Science.

Research. It began in a spare lab on the first floor of MIT's Cancer Center, then shifted to rented space in One Kendall Square. Then, in the mid-1990s, the center expanded to Charles Street, ultimately taking over the entire building. Their landlord was Harvard alum and Boston Concessions CEO Joseph O'Donnell, whose first child had been born with cystic fibrosis. "He rented a portion, and eventually all, of the building to the Genome Center because he believed in the importance of genetic medicine," says Lander.[3]

The basic idea behind the project was to bring robotics automation and computing power to bear on the burgeoning field of genomics—and share results openly. The Whitehead/MIT center soon became a major force in the government-funded, international effort known as the Human Genome Project that was racing to sequence the human genome. Ultimately, it proved to be the single largest contributor to the project—which involved twenty universities, government research centers, and nonprofit organizations worldwide.[4]

One participant in those early efforts was David Altshuler, a physician-scientist who would later become a founding core member of the Broad. Altshuler had been coming to Kendall Square since the late 1960s, when he waddled in to nursery school at MIT, where his father was a professor. He attended MIT as an undergraduate, went on to get a joint MD-PhD degree from Harvard and MIT, and would come back again to work with Lander. "For me, this is a lifetime story," he quips.[5]

Altshuler joined Lander's lab as a postdoctoral fellow. He came to study the emerging field of genomics, which risked derailing his advancement because it was so unproven. "Every mentor I had told me it was career suicide," Altshuler says. But he saw the collaborative studies the center was pursuing as opening a world of possibility that would otherwise not be remotely within his grasp. He had already banked eleven years of training after getting his bachelor degree. That included eight years earning his MD-PhD, then three more years between an internship, his residency, and a clinical fellowship. Plus, he still had three years to go on his postdoctoral studies before he could have a shot at a faculty position and his own lab. "So the problem of that, and it's a huge problem for anyone who's really looking at academic life science, is to have someone spend from age twenty-two to thirty-six in a training position is some of the best years of their life. So my options were in 1997, at thirty-three, I could be on a bench with my own pipettes for the next ten years, limited by what I could do with my own two hands, or I could work with Eric. I had a vision that had to do with categorizing and cataloging human genetic variation and working on type-2 diabetes. And I could marshal remarkable resources and collaborations because of the environment and the fact [you didn't have] to go through this pinhole of every career move."

Around the time Altshuler joined the genome center, "two things happened which in my mind are key to the Broad story." One was Lander's announcement the very day Altshuler started of a five-year collaboration around functional genomics involving the Whitehead/MIT center, Bristol Myers Squibb, California-based Affymetrix, and Millennium. Functional genomics was a just-emerging branch of science that attempted to mine the growing trove of genomic data to understand the complex interactions between genes and the proteins they coded for and how it all played out to influence biological processes and disease. "Back then, people thought genomics was mindless and boring," says Lander. "I made up the term *functional genomics* for this collaboration to distinguish it from simply mapping and sequencing the genome, which I called *structural genomics*."[6]

Lander had raised $40 million to support the effort, and was busy recruiting a group of rising stars to join him. Among them were Altshuler; Todd Golub, then a young faculty member at Dana Farber; Pamela Sklar, a postdoc at Massachusetts General Hospital interested in psychiatric genetics; Whitehead Institute fellow George Daley; and Mark Daly, who as an MIT freshman had been the first person to join Lander's lab and had grown into a leader in human genetics.

All those mentioned were in their thirties, except Mark Daly, who was still in his twenties. Most importantly, relates Altshuler, "In 1997, the genome's not supposed to be sequenced for years, but [the collaboration] is about the postgenome world." That set the stage for the inspiration behind the Broad.

The second big occurrence came the next year, when it was announced that Celera Genomics, a spinoff from Perkin Elmer's life science division under J. Craig Venter, was jumping into the genome sequencing race. Despite his late start, Venter thought Celera could get there faster and for a tenth of the cost—$300 million versus the $3 billion budget of the Human Genome Project.[7] That upped the ante on everything, spurring even more thoughts of what might be possible after the genome was sequenced. Relates Altshuler, "All of a sudden this thing was turbocharged."

* * *

That was putting it mildly. The race turned feverish, drawing headlines worldwide. Well before 2000, the writing was on the wall—humanity's "code" would soon be cracked. As things transpired, the international Human Genome Project and Celera both completed rough draft sequences—containing about 90 percent of the code. Their achievements were announced in joint ceremonies by US president Bill Clinton at the White House and British prime minister Tony Blair at 10 Downing Street on June 26, 2000. This was followed by publication of key scientific papers the following February. Celera then dropped out of the race, but the international project continued and produced a near-complete sequence in April 2003.[8]

With ultimate success in sight, a debate arose around what would happen with the Whitehead Genome Center when the project was completed. "Most people's assumption was we should just go away," says Altshuler. But for those already part of collaborations looking farther into the future, that made no sense. "We said we want to keep doing what we are doing. We have a group of people from across this remarkable ecosystem who in the morning can go to clinic or their own lab at the hospital, and then come together and do things at a scale and scope with robotics and computers and technology that we could never do on our own."

"Sequencing the genome was a wonderful ground-floor foundation, but the buildings would need to be built," Lander adds. Around the year 2000, he counted sixty-five collaborations the Whitehead Genome Center was part of—all outside normal academic boundaries. "Something magical had happened," is how he once described it.[9] "No inter-institutional agreements had been signed saying that people from *here* could come over and be working *there*. We just kind of did it without asking the permission, which is often a very good way to get started," he says. "But it was very clear that as the genome project came to an end, we would need to somehow become respectable—turn into whatever version 2.0 would be."[10]

By fall 2001, the idea had developed into an overarching vision for what was called the X Institute. This new organization was proposed as a collaboration among MIT, Harvard, and Harvard's affiliated hospitals—a plan Lander says was ultimately supported strongly by both Chuck Vest, MIT's president, and his Harvard counterpart, Larry Summers. An open question remained, though: how to fund the effort.

Fate soon provided the answer. That October, as Lander was considering potential donors to approach, philanthropists Eli and Edythe Broad asked to visit the genome center while in town for Eli's induction into the American Academy of Arts and Sciences. Broad had earned two fortunes—first through home-building company KB Home, then by founding the retirement savings concern SunAmerica. The couple's foundation supported scientists conducting innovative research into inflammatory bowel disease (IBD), which afflicted one of their family members. Lander had received a $100,000 grant to study the genetics of IBD.

The couple arrived on a Saturday morning for what was supposed to be a short visit—but the Broads were so captivated by the young scientists and warehouse full of robotic sequencers, they ended up staying hours. A few months later, Eli Broad called to say he had heard about Lander's dream to create an institute and asked if he would visit the couple in Los Angeles to discuss the idea. Lander flew out and described the scientific vision for an interdisciplinary institute to bring genomic advances to patients, as well as a financial plan that included an $800 million endowment. "Eric had a vision for a new way of conducting science, breaking down the silos that usually keep medical researchers, biologists, and engineers from collaborating on common projects," Broad once recalled. "I was intrigued, but when Eric told us he needed $800 million to start it, I just wished him luck."[11]

But Broad remained intrigued. After vetting the idea with other scientists, the philanthropist made a counteroffer: a $100 million gift to be spent over ten years. Broad also tried to convince Lander to create the institute in the Los Angeles area, perhaps as a collaboration between USC, UCLA, and Caltech, where he served on the board and whose president, David Baltimore, had spearheaded the Whitehead Institute's creation. He offered to double the bequest if the institute formed out west.

Lander countered that it would be impossible to recreate the scientific environment in Boston and the core team already in place in Kendall Square. He then spent

Box 7
Don't Let What Happened to Computing Happen to Genomics and Biotech

> The Broad Institute of MIT and Harvard marked a collaboration of two great universities—and was announced in June 2003 with strong support from their respective presidents, Chuck Vest and Larry Summers. But according to Eli Broad, while Vest almost instantly embraced the concept, his counterpart's reaction when they met to discuss the idea, probably in early 2002, was to plead Harvard did not have the money. According to Broad, "After the meeting, Eric and I agreed that, despite whatever Larry said, there was no way Harvard would want to be left out of something so big, an institute that would draw talent and attention from around the world. That appeal to Harvard's self-interest was our leverage—and we were right."*
>
> Help in tipping the scales, though, might have come from Vest. At least, that's the suspicion of Broad founding core member David Altshuler. Not long after the meeting, in August 2002, the MIT president penned an op-ed for the *Boston Globe* entitled "Genome Research Presents Opportunity For Hub." Recounts Altshuler, "I can't prove it moved anyone, but I remember at the time thinking Chuck was talking to Larry and the faculty through the *Globe*."**
>
> Vest's article never mentioned hopes for a new institute. However, it did reference the tremendous accomplishments in sequencing the human genome by Lander and the Whitehead Institute/MIT Center for Genome Research, as well as advances at Harvard and other Boston area institutions. Perhaps most importantly, the article hearkened to Boston's losing the digital computing race to Silicon Valley and urged the region not to make the same mistake by squandering its strength in the blossoming field of genomics.
>
> "We are at the center of action in the emerging world of systems biology. The future is ours," Vest wrote.***
>
> "But wait—we've seen this movie before. We were once the center of the computer industry. MIT and its Lincoln Laboratory spawned the minicomputer. Digital Equipment Corp. became the epicenter of technological entrepreneurship. Companies and jobs grew around Route 128, and the Massachusetts Miracle occurred.
>
> "And then, we largely missed the fundamental transformation brought about by the 'silicon revolution' of the microprocessor. The Massachusetts computer industry struggled, sputtered, and spiraled downward while Silicon Valley was born and the industry, jobs, and hot economy moved to California."
>
> Vest urged that history not be repeated. "Boston must seize this opportunity and live up to its name as the Hub—not a spoke—of the next-generation biotechnology industry."
>
> In conclusion, the MIT president wrote: "Our world-class universities and hospitals can and will lead the scientific revolution. But the companies, jobs, and economy can grow here, or they can flow elsewhere. This time, let's be nimble and committed to grow them here. Now is the time to start."

*"Broad, *The Art*, 96.
**Altshuler email, March 18, 2021.
***Vest, "Genome Research." All Vest quotes in this box are from this article.

months in parallel negotiations—with Broad on the one hand and MIT and Harvard on the other. Ultimately, the two paths were united.

As that was playing out, the institute concept also faced significant resistance from the local academic and medical communities, Altshuler relates. "The reaction of most people in Boston to the Broad was why would we do it with them? We should do it ourselves. No one saw why this vision of Harvard, MIT, and the hospitals working together made sense."

At the time, Altshuler was finishing his postdoctoral work and looking at possible faculty jobs. He told one potential employer he was seeking a position that supported collaborations with the institute Lander was planning. "Why would I give you money to go work with Eric Lander?" the man asked. "We're competing with Eric Lander."

Altshuler fired back. "I have no interest in competing with Eric Lander for two reasons. One, I've been working with him for three years and I'm part of that place. And second, have you ever met Eric Lander? Why would you want to compete with Eric Lander? You're going to lose."

*   *   *

The Eli & Edythe Broad Institute of MIT and Harvard was announced to the world on June 19, 2003. It formally began operations the following May, based temporarily in the Whitehead Institute/MIT Genome Center's quarters at One Kendall Square and on Charles Street.

Soon, a permanent headquarters was going up at 415 Main Street, next to the Whitehead Institute and across the street from MIT's biology department and Center for Cancer Research. A Harvard official had suggested the old Watertown Arsenal campus, where they could get a good deal, Lander says.[12] "I went out there, but I said, not a chance, because it's not in Kendall Square. It's not going to work." He also nixed taking over the still under construction building Vertex had leased opposite the Genzyme Center. Even that, he says, was too far from MIT—about six blocks from the core biology buildings. "Even though the place we are now was more expensive, we made the decision to put it cheek to jowl with MIT, catty-corner to the Cancer Center, across from Legal's. Because, I think I said at the time: interaction falls off like one over R to the sixth ['$1/R^6$,'], the binding force. It was just a passing physics joke, but the idea of proximity was important because there was a flow of students, it was next to the T, and it was across the bridge from MGH."

The seven-story building opened in spring 2006. That same year, the Broads pledged another $100 million. Three years after that, they contributed an additional $400 million for a permanent endowment. And in 2013, the philanthropists chipped in $100 million more to open investigations into new areas—bringing their total commitment at the time to $700 million. "Of all we have done over the past six decades, the effort I am most proud of is the creation of the Broad Institute," Eli Broad wrote.[13]

When it opened in 2006, its lobby boasting floor-to-ceiling windows that allowed pedestrians to see in, the Broad's new home added zest to an already surging Kendall

Box 8
Broad Institute Founding Mission and Organization

> The Broad's founding principles and organization were detailed in Lander's report to MIT's president to close the 2003–2004 academic year, when the institute had operated for just two months.
>
> The Broad Institute's scientific mission is
>
> - to create tools for genomic medicine and make them broadly available to the scientific community and
> - to apply these tools to propel the understanding and treatment of disease.
>
> Its organizational mission is
>
> - to enable collaborative projects that cannot be accomplished solely within the traditional setting of individual laboratories and
> - to empower scientists through access to cutting-edge tools.
>
> The institute had four founding core members: *Eric Lander*, its director; *David Altshuler*, an associate professor of genetics and medicine at Harvard Medical School and Massachusetts General Hospital; Harvard professor of chemistry and chemical biology *Stuart Schreiber*; and *Todd Golub*, then at Dana-Farber Cancer Institute and Harvard Medical School. Althsuler and Golub had been affiliated with the genome center, and Schreiber, a leading chemist, was founding director of Harvard's Institute of Chemistry and Cell Biology. All worked with Lander to shape the Broad's vision, joined by Altshuler's wife, Jill, a former management consultant who helped write the plan.
>
> The Broad also named fifty-seven associate members from a host of area institutions.* Initial research focused on eight areas—four core programs and four initiatives described as pilot areas that might turn into programs.
>
> *Programs:* Cancer; cell components, states, and circuitry; medical and population genetics; chemical biology
>
> *Initiatives:* Inflammatory disease; infectious disease; psychiatric disease; metabolic disease

*Lander, "Broad Institute." Affiliated institutes included MIT, the Whitehead Institute, Harvard Medical School, Harvard Faculty of Arts and Sciences, Harvard School of Public Health, Dana-Farber Cancer Institute, Brigham and Women's Hospital, Massachusetts General Hospital, Beth Israel Deaconess Medical Center, and Boston Children's Hospital.

Square. "There is a lot more foot traffic on Main Street in Cambridge these days," began the *MIT News* article about the opening.[14]

Traffic soon got heavier. The Broad cracked the top twenty list of Cambridge employers in 2010, coming in at fourteenth, with 870 employees. By 2020, it had more than doubled in size, ranking eighth, with 1,880 employees. One early hire was Jill Mesirov, who had joined the Genome Center after leaving Thinking Machines. She would serve as the Broad's chief informatics officer until 2015, when she became an associate vice chancellor at UC San Diego's medical school. The ranks of nonstaff trainees and affiliates, who were generally faculty at Harvard, MIT, and the Harvard teaching hospitals, also swelled. To accommodate the growth, the institute erected a fifteen-story building that housed eight hundred: it was connected to the headquarters via walkways across several different floors. Among other things, that became home to the Stanley

Center for Psychiatric Research, formed around a staggering $650 million donation from another philanthropic couple, Ted and Vada Stanley. With other gifts, the Stanleys contributed more than $825 million to the Broad before Ted died in 2016—even more than the Broads.[15]

The institute expanded well beyond its initial areas of focus. In addition to the psychiatric center, studies at which included autism, bipolar disorder, and schizophrenia, it began to explore areas such as epigenomics and even drug discovery, as well as programs to identify genetic factors influencing a range of areas: obesity, infectious disease, rheumatoid arthritis, multiple sclerosis, diabetes, inflammatory bowel disease, kidney ailments, rare diseases, and more. In 2017, it also became coleader, with the UK's Wellcome Sanger Institute, of the Human Cell Atlas project, an international collaboration to map every cell in the human body to provide a foundation for a variety of scientific studies and efforts to improve human health.

"Half the place is devoted to finding the basis of disease and half is devoted to trying to transform and accelerate the development of therapeutics," Lander once told the *New York Times*. "It's different from what you find in many university settings where you have many labs, each of whom does its own thing."[16]

As it ramped up, the Broad almost immediately made its mark on science. "The Broad rose from nonexistence in 2003 to the pinnacle of molecular biology," wrote STAT's Sharon Begley in 2016. "By 2008 three Broad scientists, including Lander, ranked in the top 10 most-cited authors of recent papers in molecular biology and genetics. In 2011, Lander had more 'hot papers' (meaning those cited most by other scientists) in any field, not just biology, than anyone else over the previous two years, according to ThomsonReuters' ScienceWatch. By 2014, eight out of what ScienceWatch called 'the 17 hottest-of-the-hot researchers' in genomics were at the Broad."[17] As of late 2020, the main faculty had grown to sixty-seven, of whom twenty-two were women. The sixteen *core institute members*, including Lander, maintain their primary laboratories at the Broad itself. The fifty-one *institute members* operate their main labs at their home institutions, but many also conduct research at the Broad. The institute puts out more than a thousand research papers a year.

Lander continued piling up personal achievements as well. In 2008, he was tapped by Barack Obama to cochair the President's Council of Advisors on Science and Technology. Five years later, he was one of eleven winners of the inaugural Breakthrough Prize in Life Sciences founded by Yuri Milner, Sergey Brin, Anne Wojcicki, Priscilla Chan, and Mark Zuckerberg. The $3 million prize, awarded for Lander's pioneering work in genomics, is more than double the size of the Nobel.[18] Then, in January 2021, he was named to the cabinet position mentioned earlier. Lander took an unpaid academic leave of absence from the institute, and Todd Golub, a Broad founder and its chief scientific officer, was named director.[19] Lander didn't last long in Washington, resigning in February 2022 after an internal investigation found evidence he had demeaned staff, and that women in particular had complained about his behavior.

* * *

Figure 29 Eric Lander in 2019, celebrating the Broad's fifteenth anniversary. *Source:* Broad Institute.

When the grim reality of the COVID-19 pandemic hit home in March 2020, it spurred a massive pivot at the Broad. Even as the Charles Street Genomics Platform center was revamped to allow it to process COVID tests, much of the laboratory science the Broad's researchers had been pursuing was put on indefinite hold. "That meant an absolute scramble to shut down all of our science that wasn't COVID, which was tough, because it meant suspending key cancer research, for example," says Lee McGuire, the Broad's chief communications officer, who was named chair of the Kendall Square Association that June. "This is research that takes months to develop. You can't just put it on pause without careful planning. So we had to spend a good week in March figuring out how do you safely shut down most lab operations with a minimal loss of research."[20]

While most teams were able to do analysis and other work remotely, for a period the only active scientific work in institute labs involved coronavirus research. Broad scientists joined a number of global partnerships aimed at understanding the body's response to COVID in hopes of helping answer a series of bewildering questions. How does it target specific cells in the lung? Why was it hitting men more often than women? Why adults far more children?

The institute's scientists had little to do with developing therapies or vaccines; that work was better suited to others, including Moderna, the biotech company just a block and a half away that was advancing a vaccine candidate at a dizzying rate. "We see our position here as helping to explain the underlying biology to inform those decisions," McGuire explains.

Things took a step toward normality in May 2020, when the Broad brought more of its scientists and researchers back to their labs, coupled to an elaborate system set up to screen every employee for the virus twice a week.[21]

The COVID testing program continued to dominate the Broad's on-site activities, however. Partnerships were being formed with the Commonwealth of Massachusetts and a variety of groups around surveillance testing. Such programs hinged on regular testing of large groups—in nursing homes, schools and colleges, even housing blocks. The idea was to quickly identify newly infected people, many without symptoms, get them isolated, and trace who they had been in contact with before that became too difficult—thus minimizing the risk of spreading the virus.

Among its efforts, the Broad was working with the State of Massachusetts to bring surveillance testing to every nursing home and long-term care facility in the state. Another effort focused on mobile testing in economically disadvantaged areas. And by fall 2020, the Broad was also working with 108 colleges and universities throughout New England to regularly test students and staff. "It's a huge logistical challenge for a school," says McGuire. "So we're helping them to do that. We can handle all the tests."[22]

* * *

Although the Broad rapidly proved itself a heavyweight power in genomics science—attracting hundreds of millions of dollars in philanthropic contributions and even more in research grants and support from industry—its success did not come without some bad blood and controversy.

At least some hard feelings could be traced to the Broad's very existence. But one incident in particular—an article Lander wrote about the gene-editing technology CRISPR in 2016—sparked social media firestorms and a barrage of personal attacks. The paper, *The Heroes of CRISPR*, was published in the journal *Cell*. STAT's Begley wrote a comprehensive article about the backlash, headlined "Why Eric Lander Morphed from Science God to Punching Bag."[23]

Among other things, a number of people felt Lander had downplayed the critical research done by Jennifer Doudna at UC Berkeley and key collaborator Emmanuelle Charpentier of the Max Planck Institute of Infection Biology in Berlin, and overstated the work done by the Broad's Feng Zhang. As Stephen Hall wrote in *Scientific American*: "Zhang's discovery narrative is long, detailed and colorful; Doudna's . . . work doesn't get nearly the same star treatment."[24] To some, that was sexist—minimizing the role of two women important to the story. Others saw it as an attempt to manipulate perception of who deserved the most credit for the technology in order to influence a major patent fight about key CRISPR inventions ongoing between the Broad and the University of California system.[25]

In short, it got ugly—and it went viral. "Those missteps triggered a bitter online war, including the Twitter hashtag #landergate," wrote Begley.[26] Stephen Hall labeled the controversy "the most entertaining food fight in science in years . . . The spat is like an escalating and increasingly ugly domestic dispute: no one wants outsiders to get involved but the screaming has gotten so loud that somebody has to call the cops."[27]

Lander quickly issued clarifications to explain certain points, and insisted he was only trying to showcase the complicated history of CRISPR. In a note to Broad staff, he said: "Needless to say, 'Perspective' articles are personal opinions. Not everyone will fully agree with anyone else's point of view. In the end, we come to understand science only by integrating a diverse range of thoughtfully expressed perspectives. And, when scientific discovery is also the subject of patent disputes (as is the case with U.C. Berkeley and Broad–M.I.T.), intellectual disagreements can, as here, give rise to vigorous online discussion."[28]

Nearly five years later, in October 2020, Doudna and Charpentier were awarded the Nobel Prize in chemistry for their contributions to genome editing. Lander tweeted his praise to the winners at 6:45 that morning, before most people awoke: "Huge congratulations to Drs. Charpentier and Doudna on the @NobelPrize for their contributions to the amazing science of CRISPR! It's exciting to see the endless frontiers of science continue to expand, with big impacts for patients."[29] Doudna showed her appreciation in a Zoom call with reporters later that same day. "I'm deeply grateful for the acknowledgement from Eric. It is an honor to receive his words."[30]

*   *   *

David Altshuler left the Broad in 2015 to join homegrown biotech Vertex as executive vice president for global research and chief scientific officer. "There are critics of the Broad. There are people who criticized it when it opened. There are people who criticize it today," Altshuler says. "Their criticism usually stemmed from some mix of the scale of activity, the concern that people working together in teams would squelch the creativity of individual scientists, and that there was too much influence concentrated in too few people. I always believed, and with the benefit of hindsight still believe, it has had a positive impact on the city, the region, and the world of science, but it's not without complexity. Any experiment in the organization has upsides and downsides—and time will tell when the whole story is written. But I believed then, and I believe now, that the field of science at that moment in time required bringing technology and disciplines and even institutions together in a way that could not be done in the traditional model, and we clearly had a significant impact on the collaborative culture of science in Boston.

"What we see now is a lot of young stellar scientists and physician-scientists, just like computational people, are not going through the traditional academic route because the options are very limited in the resources you have available in your late twenties and thirties. Whether it's at a place like the Broad or a VC-backed company, you can have many more resources if your vision can attract the support. This has opened up additional opportunities, and that's great for helping society. We now have Harvard, MIT, BU, the hospitals, the biotech companies, the pharma companies, [and the Broad] all so close together. And that is the magic of Boston."

At the time the Broad was conceived, Lander adds, collaborative efforts were still not a normal thing for universities and other organizations. "The knee-jerk reaction was not to say, 'How do we join forces?' Because each of those institutions felt, rightly, that they were amazing, world-leading institutions and why did they need anybody else? But to take on genomics the way it needed to be taken on to build on the human genome project was actually a tall enough order that it would require all the institutions."

In Lander's view, such collaborations are vital to deciphering the hugely complex puzzles that remain to be solved. What are those challenges? "I think increasingly the next frontier is to read out all the programs in human cells. Human cells only have a finite number of *programs* or *tricks* that they can do—that's a finite program library," he says. "The only way we're going to be able to do that is by a combination of sophisticated biology and medicine, and very large-scale data collection and machine learning. But to the extent that we know all the circuits, maybe we can think about real, programmable therapeutics. So that sometime in the twenty-first century, young students in Kendall Square will look back with some combination of bemusement and horror at how ancient humans in the 2020s tried to make drugs by finding molecules rather than sitting down and writing the code."

19 THE CORPORATIZATION OF KENDALL SQUARE

George Scangos was hunkered down in a second-floor office at 10 Cambridge Center, just off Broadway. It was June 2011, and he had carved out temporary quarters—not happy about the circumstances that had put him in the makeshift space. Scangos had previously led South San Francisco–based oncology company Exelixis. He had taken over as Biogen Idec's CEO the previous summer—moving into the company's recently opened new headquarters in Weston, Massachusetts, about a fifteen-mile drive from its longtime base in Kendall Square, where R&D and production had remained. His predecessor, James Mullen, had decided a few years earlier to move business operations to the suburbs in a cost-cutting measure—and the newly erected Weston campus had begun opening just months before Scangos took the helm. The new boss almost immediately took steps to move everything back.

While this boomerang maneuver was being sorted out, Scangos had established an interim office back in Kendall Square, where he tried to spend at least one day a week. "What I don't like is the split where we have sales and marketing sitting in Weston and R&D sitting here," Scangos said at the time. "Today, I'm the only member of senior management who's here. Everyone else is in Weston. That's not optimal. If we could get everybody together, I would very much like that. It's more sensible to have everyone in one location. You can talk, go down the hall, have lunch together, just meet people randomly, and get really good discussions going between departments of R&D and commercial."[1]

It took a while to nail down the details. About a month later, though, Biogen announced plans to bring some 530 employees back to the square, many of whom would eventually be housed in two new buildings that would be constructed along Binney Street.

At the time, Biogen had not brought a drug to market since its 2004 introduction of Tysabri for multiple sclerosis. "We have a lot on our plate, and execution will be absolutely critical," Scangos said. "Being geographically separated, there's a barrier."[2]

* * *

Biogen's about-face didn't have the bombshell character of Novartis's announcement nearly a decade earlier that it was moving R&D to Cambridge. Still, it was symbolic of another major phase of Kendall Square's evolution—a period of explosive growth that had begun soon after the turn of the century but largely characterized the second decade of the 2000s.

During this time, outside of a few special cases like Vertex that escaped the square's orbit, an almost gravitational power seemed to be pulling companies and people into the ever-denser hub. Accompanying these changes, Kendall Square's streets finally showed real signs of life. A wave of restaurants, bars, and cafes opened up and down Main Street, over near One Kendall Square, and back on the other side of the Broad Canal along Third. Several new "luxury" apartment buildings also opened their doors, and alongside them appeared a smattering of additional manifestations of city neighborhoods—live music and a farmer's market in the summers, a dry cleaner, an ice cream parlor, a liquor store, a couple of bank branches, and, at end of the decade, the square's first supermarket.

Perhaps the most visible aspect of this period was a building boom that harkened back to the early 1900s, when the area teemed with factories—soap plants, rubbermakers, iron works, tanneries, confectioners, printing houses. Only this was the modern-day equivalent—dominated by large corporations that needed offices and labs, not manufacturing plants that spewed pollution and filled the air with pungent odors. These often sported ground-floor retail space, chiefly for the new restaurants and cafes. In the face of this activity, most of Kendall Square's remaining undeveloped or underdeveloped blots were expunged.

Although the growth was driven by companies across a range of industries, from software to drug-making, the culture and character was shaped by life sciences. Indeed, many felt that the Boston area, led by Kendall Square, had separated from San Francisco to clearly become the world's densest and strongest life science hub. "It is to life science what Hollywood is to movies, what Wall Street is to finance," proclaims Broad Institute founding member David Altshuler. "That's what it is."[3]

* * *

If there was an overriding theme to this period, it was the corporatization of the square. Only a decade or so earlier, Kendall Square had still been largely funky, grungy, and startup-rich. Now, by most yardsticks, it was pretty much just for the rich.

The area's homespun biotechs, several of them having become corporate powerhouses in their own right, were part of the latest surge—as Biogen's reconsolidation and plans to erect two new buildings exhibited. In 2014, Vertex finally pulled off its long-delayed move to an $800 million facility in Boston's Seaport: its Kendall Square space was the last foothold it moved out of in Cambridge.

When announced in 2008, Vertex's relocation to Boston was seen as a big blow to Cambridge's biotech stature—and the outlook had worsened when Biogen revealed its plans for its suburban migration. However, just a few years later, not only was Biogen coming back and expanding but all the nonretail space in the six-story Kendall Square structure Vertex had partially occupied—a nursery, dry cleaner, restaurant, café, and smoothie shop occupied its ground floor—was gobbled up by Alnylam. The upstart biotech was overflowing its headquarters building just across Linskey Way, and the expansion created an urban minicampus that housed some 1,400 employees.[4] Meanwhile, in 2017, well-heeled gene therapy company Bluebird Bio, coming off an IPO four years earlier, moved into a newly erected office and lab complex down Binney Street from Alnylam toward the river, on the site of a former parking lot serving the historic Athenaeum Building. For its part, Moderna established itself next to Polaroid's former headquarters in Tech Square: launched at the start of the decade, the messenger RNA company, later to become world-famous for its coronavirus vaccine efforts, would celebrate raising $604 million in the biggest IPO in biotech history to date in December 2018.

All of those locally grown biotechs except Moderna—Biogen, Alnylam, and Bluebird—would rank among Cambridge's top twenty-five employers in 2020. But an even bigger force of change was big pharma. The wave that Novartis had kicked off back in 2002 turned into a tsunami, aided by a series of state moves to cut red tape, increase tax credits, and invest in infrastructure—capped by the $1 billion Massachusetts Life Sciences Initiative enacted by Gov. Deval Patrick in 2008 (see chapter 27). One after the other, the world's drug giants began taking up residence in Kendall Square. By 2020, at least thirteen of the world's top twenty drugmakers had established a presence—and that didn't include two, Shire and Baxalta, whose operations had been absorbed after their acquisitions by Takeda. In total, pharma companies employed double the numbers of the big four Cambridge biotechs, with operations that ran from those with more than a thousand employees, many doing R&D in key clinical areas, to much smaller "innovation outposts" designed to give companies a toehold for building scientific and business collaborations.

"It really was a fascinating period," recalls John Maraganore. "All the pharma companies wanted to be biotech. I think they just suddenly realized that biotechnology had become the source of their innovation, so they therefore wanted to be that type of company themselves." At one point, the Alnylam CEO undertook an informal survey and saw that a slew of the drug giants had even changed their taglines to reflect this shift. "They were no longer calling themselves pharmaceutical companies. They either called themselves biopharmaceutical companies or had the audacity to call themselves biotechnology companies," he relates. "And of course, if you want to be a biotechnology company, where do you have to be? You can't possibly be in New Jersey. You can't possibly be in Switzerland. You gotta be in Kendall Square."[5]

Big pharma also brought big cash, and drugmakers used their financial clout to make acquisitions that boosted their local presence and profile. Sanofi's $20.1 billion purchase of Genzyme in 2011, the largest acquisition of a Boston biotech to date, served as a bellwether—but Takeda took the top prize. The Japanese drugmaker added to its 2008 Millennium acquisition by purchasing Kendall Square-based cancer drug developer Ariad Pharmaceuticals in 2017, then buying Shire the next year for a whopping $62 billion. Shire itself had recently acquired Baxalta—and the combined operations were based at 650 East Kendall Street, across the small park from the former Vertex

Box 9
Big Pharma's Kendall Square Gold Rush

The arrival of "Nibber," the Novartis Institutes for BioMedical Research, proved a turning point for Kendall Square—helping spark Big Pharma's stampede to the area. Takeda, Novartis, and Pfizer all maintain huge presences on the square's westerly end, extending from Central Square up to Tech Square. Most of the others, including another large Takeda presence, are grouped around Binney Street and Third Street on the east. The table ahead is roughly arranged by order of appearance in the area, with 2020 locations and employees.*

	First moved to or near Kendall	Location(s)	No. employees 2020 (est.)
Novartis (NIBR)	Opened in 2003.	Multiple locations, Central and Kendall squares	2,330
Merck	2004	Merck Exploratory Science Center, 320 Bent St.	65
Shire	Came to Massachusetts in 2005 through TKT acquisition. In 2016, expanded through Baxalta acquisition. Acquired by Takeda in 2018.	700 Main (TKT), 650 E Kendall St. (Baxalta). Expanded to 50 Binney St.	833 in 2018 pre-Takeda acquisition
Takeda	Acquired Millennium Pharmaceuticals in 2008; Ariad 2017; Shire 2018.	Eleven locations in Cambridge. Three in Kendall Square: 650 E. Kendall St, 500 Kendall, 125 Binney.	3,484
Sanofi Genzyme	Sanofi acquired Genzyme in 2011.	First 270 Albany St., then 50 Binney St.	1605
AstraZeneca	2012, office in KS closed in 2016.	141 Portland St.	
Johnson & Johnson	2013	One Cambridge Center (245–255 Main St.). 700 Main St. (LabCentral).	200
Pfizer	Moved into current location in 2014, consolidating spaces from around the area.	1 Portland St., 610 & 700 Main Street South.	941
Ipsen	2014. Moved headquarters to square in 2018.	650 E Kendall St.	200+
Lilly	2015	450 Kendall St.	40
Baxalta	2015. Acquired by Shire, now part of Takeda.	650 E. Kendall St. (previously)	Planned for 500
AbbVie	2016	200 Sidney St.	75+
Bayer	Presence since 2016, via Casebia Therapeutics deal. Innovation Center also created in 2016.	245 First Street, Cambridge Science Center	150
Servier	Presence began in 2018 when Servier purchased Shire's cancer drug business. Official US HQ opened in 2019.	One Broadway, Kendall Square; 200 Pier Four Blvd, Boston, MA 02210.	300, most in Boston
Bristol Myers Squibb	2018	100 Binney St.	Plan to move 2,700 MA employees into Cambridge Crossing

Source: Table put together from press accounts, plus correspondence from most companies involved. Also interviews with Kane, Emmens, Vasella. For overviews, see Weisman and Newsham, "Bristol-Myers Squibb."

*Exceptions are Shire and Baxalta. Both were acquired by Takeda. However, both had a significant Kendall Square presence at the time of their acquisitions—hence their inclusion.

building Alnylam was taking over. That made Takeda the largest corporate employer in Cambridge, with some three thousand workers. Takeda and Sanofi also vaulted to the top of wider Massachusetts rankings: they were, respectively, the state's number one and number two life science employers in 2020.[6]

Big pharma's onrush, especially after 2010, filled most of the remaining holes around the old brownfield site that Lyme Properties had originally developed, as well as down Binney Street towards the Charles. Nearly seven years after its Genzyme purchase, Sanofi moved what was then called Sanofi Genzyme into a newly constructed Alexandria building a few blocks away at 50 Binney Street that also offered lab space. Save for some ground-floor retail areas, the move left the award-winning Genzyme Center Henri Termeer had seen as a beacon for attracting and retaining employees completely vacant. Shire ultimately signed a long-term lease—which was then inherited by Takeda when it bought Shire.[7]

"It came down to availability of deeper R&D talent," sums up former Shire CEO Matt Emmens, explaining Pharma's rush to Kendall Square.[8] In 2005, Emmens engineered the then UK-based drugmaker's first foray into the square via its acquisition of Transkaryotic Therapies (TKT). "The attractiveness of the Boston area, despite the premium real estate prices, is it's where you can get talent literally to walk down the street and join you without moving a lot of people and upsetting their world." The need to establish operations closer to large talent pools—in industry and academe—became even more glaring given the changing nature of drug-making and its increasing dependence on relatively new fields like molecular biology and genomics, he says. "The emphasis on the high science and the very deep stuff associated with genetics has amplified that need to be near a mecca of science as dictated by academia and associated research science. It's difficult to get the very best people to go anywhere else."

Emmens had several vantage points on the Kendall Square story—including the downsides of price and space. In 2009, he became CEO of Vertex (he had joined the board while at Shire) and engineered the move out of Cambridge to Boston's Seaport District after its failed attempt to find enough room in Kendall Square. Later, after his retirement from Vertex, he sat on the board of Bristol Myers Squibb when it opened a lab on Binney Street in 2018.

In Shire's case, the acquisition of TKT for $1.6 billion served a double purpose that spoke to the changes many drugmakers were facing. Shire at the time had one major product—Adderall, a mixture of amphetamine salts used to treat attention deficit hyperactivity disorder (ADHD). With a shrinking period of exclusivity—a so-called patent cliff—Emmens believed the company needed to replace Adderall (which it did with Vyvance) and simultaneously move into specialized drugs for rare diseases. TKT was based at 700 Main Street, the historic building that had served as Charles Davenport's carriage works and Edwin Land's private lab. The company had two approved drugs—one for Fabry disease, the other for kidney disease–related anemia. Moreover, its pipeline held several other candidates with orphan drug potential. So the acquisition gave Shire both a leg up on its pivot to specialty drugs and access to researchers versed in the latest science. "ADHD was relatively easy scientifically—Adderall was mixed amphetamine salts," says Emmens. "And now we needed to think how to prepare the company for much more complex products. We had to get to a center that had research capability and talent in genetic diseases. We had to get to a place where we could find many more different R&D employees that we had not needed previously."

The TKT acquisition turned out extremely well, Emmens says, and presaged Shire's later move of its US headquarters to Kendall Square. But things didn't exactly start on a high note. The purchase was vilified by many analysts and in the press—particularly in the UK, where Shire was then headquartered. It didn't help that soon after the acquisition was announced, TKT's manufacturing plant outside of Boston caught fire. As Emmens recalls, one British paper ran a big photo of the fire, carrying the headline: "Shire's Hot New Deal."[9]

Takeda, which ultimately purchased Shire, told a similar story—without the fire. It had purchased Millennium in 2008. Seven years later, as mentioned in chapter 1, it moved its global research and development headquarters from Japan to Cambridge. The new head of R&D, Andy Plump, had attended MIT as an undergraduate, then gotten his MD and PhD degrees at UC San Francisco before becoming a senior executive at Merck and then Sanofi. "I never thought I would come back, [but] it just became a magnet," he says. "It's truly the center of healthcare innovation."[10]

The changing world, plus some changes at Takeda, shaped the drugmaker's thinking. Former GlaxoSmithKline executive Christophe Weber had taken Takeda's helm just before Plump joined. "We embarked on this assessment of what the external landscape was, where we were relative to all of that. And, you know, we just realized that we needed to make a huge strategic change," recalls Plump.

That change involved shifting from a broad-based approach to drug-making to a more targeted focus on four core therapeutic areas: oncology, neuroscience, rare diseases, and gastroenterology. To excel in those, Plump concluded, Takeda needed to reorganize its research groups and develop new expertise. And because no single company could do everything itself, it also needed to pursue external partnerships and collaborations. "So those are the core elements of our strategy," Plump relates. "Then, from a structural standpoint, we were spread too thin. We were in Japan, multiple spots in the US, multiple spots in Europe. And we made the decision that Boston would be our hub."

In 2018, after its Ariad and Shire acquisitions, Takeda announced plans to move its US headquarters and commercial operations from Chicago to the Boston area, complementing its R&D shift. As of fall 2020, four of its eight global business units, as well as biologics manufacturing, were centered in Massachusetts—along with seven of eighteen executive team members. To make room for its growing Cambridge operations, Takeda undertook a big renovation of the long-vacant Genzyme Center. The work was slowed by the COVID-19 pandemic, but the building opened on June 1, 2021.[11]

The fact that so many major pharmaceutical companies have come to a similar conclusion about the attractiveness of the Boston market creates a huge competitive challenge, Plump notes. "But those are the kind of challenges I like, because you need to be competitive. You need to be able to attract and retain talent in a strong marketplace—because if you're in a weak marketplace, there's not turnover, so you don't really know how attractive you are as an organization."

*   *   *

For the vast majority of the post–World War II era, Kendall Square's high-tech ethos had been largely characterized by computing and software companies—the shadow of digital computing Noubar Afeyan had felt hanging over the area as he entered the life science arena in the late 1980s. Now that had been usurped by the gene shadow. But that didn't mean digital had gone away. In fact, in concert with pharma's incursion, many of the country's computing and technology giants established or significantly expanded Kendall Square operations in the second decade of the 2000s, providing another dimension to the ecosystem's corporatization. Google, Microsoft, and IBM had the biggest footprints, but they were joined by Apple, Facebook, Amazon, Mitsubishi, Toyota, and Twitter, among others.

Among the big tech players, Big Blue had been in the square the longest. Its presence dated back at least to its 1964 establishment of the IBM Cambridge Scientific Center, an early inhabitant of the Tech Square complex that two years later would become Polaroid's world headquarters. The scientific center, in the same building as Project MAC, had hosted work on a variety of computer and software issues until its closure in 1992. Three years later, IBM purchased Lotus, giving it a much bigger presence on the other side of the square. Reluctant to fiddle with Lotus's culture, it largely left the software company's researchers alone for five years. Around 2000, though, it absorbed Lotus Research under Irene Greif into a new Cambridge branch of IBM Research set up in the former Lotus space at One Rogers Street.[12]

In 2016, while pharma was making its big push into Kendall Square, IBM started moving into a new building at 75 Binney Street—subletting space from Ariad Pharmaceuticals. This became the headquarters of IBM Watson Health. Big Blue soon shifted its research group to the building, and also brought in IBM Resilient, the cybersecurity operation it formed after acquiring privately held incident response company Resilient Systems in 2016. All told, IBM had roughly 750 employees in Cambridge in 2020, according to city figures. The company would not confirm employment details, but the bulk of those, if not all, would almost surely be in Kendall Square.

Microsoft made a splash in 2008 when it opened Microsoft Research New England, near the Longfellow Bridge at One Memorial Drive, a seventeen-story luxury office tower abutting MIT's campus.[13] The lab was Microsoft's sixth research outpost worldwide, and the first in the United States outside the West Coast. To lead it, the software maker tapped Jennifer Chayes, who had headed Microsoft Research's theory group at its flagship lab in Redmond, Washington. She was the first woman to manage one of the company's research centers. Her husband, Christian Borgs, also from the theory group, was named associate director. Both were renowned mathematicians and computer scientists, and Microsoft supported them with some heavyweight talent. Among them: computing legend Butler Lampson, winner of both the Draper Prize in engineering and computing's Turing Award; and economist Susan Athey, on sabbatical from Harvard University, the first woman to win the Clark Medal, a prize given to rising-star economists under forty.

Just as its counterparts in the life sciences had done, Microsoft leveraged acquisitions to build its Kendall Square stronghold. A series of recent buyouts—Groove Networks, Softricity, Fast Search & Transfer—had seen its Boston-area workforce surge from around two hundred to more than six hundred. Microsoft had rented almost half the building, which already played host to some one hundred employees from the company's SoftGrid unit, its new name for Softricity, a virtualization and streaming technology company. It also harbored a new product development and innovation center run by Reed Sturtevant, the former Lotus employee and two-time MIT dropout mentioned in chapter 11, who had most recently served as chief technology officer (CTO) of Eons, the briefly high-flying social networking portal for the over-forty crowd.

Chayes explained at the time that she and Borgs had proposed the research lab the previous November in part to tap into Microsoft's growing regional workforce, but also because the Boston area was chockablock with "phenomenal people Microsoft would just love to attract who are not going to move to the West Coast."[14]

The software giant's collection of groups at One Memorial Drive soon became known as Microsoft New England Research and Development (NERD).[15] The research lab's initial focus areas included computer science theory and mathematics, economics, social science, and design. Over the years, work extended into security, privacy, and cryptography, as well as artificial intelligence and machine learning. In fall 2020, in addition to the research arm, One Memorial Drive housed half a dozen or so Microsoft development teams working in areas that included Intune, the company's mobile device management service; security work for Office 365; and various machine learning and artificial intelligence efforts, according to site manager Eric Jewart. The company would not reveal its current local headcount, but given its footprint was likely around five hundred in 2021.

Google, as briefly covered in chapter 1, had dipped a toe in the Kendall Square waters back in 2005 with its acquisition of mobile operating system developer Android. It wasn't until 2008 that it established a formal Google office in Cambridge Center. Things had then exploded from there, as the company created an urban campus connecting large parts of three adjacent buildings in the heart of the square. In 2019, it announced plans to expand to a sixteen-story tower Boston Properties was building next door to the Marriott Hotel and grow to some three thousand workers—rivaling Takeda.

The Android acquisition, and hence Google's presence in Kendall Square, had not been announced for several years (Google had opened a Boston sales office in 2002). Android maintained teams in the Boston and Palo Alto areas. Shortly after the acquisition, Android cofounder Rich Miner and his small Boston-area contingent took up residence in the Cambridge Innovation Center—without revealing they were part of Google.[16]

Google held a big opening bash at its new headquarters at Five Cambridge Center in May 2008, just a few months ahead of Microsoft Research's own grand opening.[17] Deval Patrick was on hand for the ceremonial festivities and, after crowing in the elevator about his table tennis skills, he was challenged to a match by the site's new director, Steve Vinter.

Vinter had first worked in the square around 1990, in a small Siemens Nixdorf operation. "I thought it was amazing. It just felt so prestigious," he recalls thinking.[18] The company moved out to Burlington, Massachusetts, shortly after he arrived. "I was really disappointed," Vinter says. But now, some eighteen years later, he was back.

Ahead of the opening festivities, Vinter had spent some fourteen months scouting for talent out of the company's CIC quarters. When he joined, Google only had about forty employees in Boston—roughly twenty-five in the sales office and fifteen engineers, about half of whom hailed from the Android team. But Google was out to quadruple the number of its worldwide sites, or campuses, from about ten to around forty. Vinter's charge was to develop Cambridge/Kendall Square. "It was just this huge expansion," he says of the global push. "There was a lot of joking by the executives that it was completely out of control, partly because it was a little bit." As if in evidence of that, Vinter soon learned Google had no firm plan about how big to grow the site or how fast. When he asked his boss about a budget, the reply came: "We don't do budgets at Google." The company preferred to gauge what was needed and spend accordingly.

Less than a year after Vinter's arrival, it became clear he needed to find bigger, dedicated space. The hunt had led to Cambridge Center, on the other side of the Marriott from the CIC. "I made the decision that there ought to be a compelling reason to leave Kendall Square, because we were just in an amazing location and it was going to grow," says Vinter. "I deeply believed that. We were going to grow Google, but Kendall Square was going to grow. Microsoft was across the street. It was only a matter of time before other tech companies were coming in, and biotech was already prominent. So it was pretty clear it was going to be a growth area."

Google leased several floors in Five Cambridge Center, which made up its main footprint. But it also rented space in Three Cambridge Center next door, which shared the same entrance atrium. The search giant also took out an option for additional room in adjacent Four Cambridge Center, where Vinter had worked with Siemens Nixdorf.

By the time of the grand opening, the workforce totaled 175, roughly split between engineering and sales. The early engineering focus was on a small group of projects—among them, Android for mobile phones; YouTube; content delivery; search; the Chrome web browser; Friend Connect, a then-new platform designed to let website owners easily add social networking functions (it was retired in 2012); and Google Books, then known as Book Search, a controversial effort to scan public-domain books and other material and make the full text searchable online.[19]

The Great Recession had officially begun in December 2007, a few months before the opening, and wasn't declared over until June 2009. Google continued hiring during this period, though at a slower pace than it would have otherwise, says Vinter. In 2009, the company also launched Google Ventures, later known as GV. The investment arm was based in Mountain View, California, under Bill Maris, but Rich Miner was given charge of Kendall Square operations. Then, in mid-2010, Google acquired ITA Software for $700 million. ITA had been founded by a group from MIT's AI Lab back in 1996, the tail end of the AI Alley days. It developed behind-the-scenes software that organized and tracked flight prices and itineraries and served as the engine behind a slew of major airlines and travel sites, like Orbitz and Kayak.

ITA was based only a few blocks away—in five floors of 201 Broadway, the same building that housed Mitsubishi's research lab. When its lease ran out a few years later, Google brought everyone over to Cambridge Center. The few blocks made a difference, explains Vinter. "When you're not colocated, you just don't have the same level of engagement and interaction." The ITA technology became the basis for Google Flights.[20]

The ITA acquisition added several hundred people to Google's local payroll. By the time Vinter stepped down as site lead in fall 2016, Google had grown its Cambridge presence to around 1,200 people and added connecting corridors to link the three adjacent but separate buildings on several levels. By then, the big focus areas included search, Chrome, and YouTube, as well as efforts in image search, networking, Google Play, and Google Ads. In addition, a number of smaller engineering project offshoots had reportedly started up in image processing, Google Compute Engine, Privacy Sandbox, and the Go programming language, among others.

For 2020, Google reported 1,800 employees in Cambridge, ranking as the city's tenth biggest employer, and second among tech companies to homegrown, inbound marketing company HubSpot, headquartered in East Cambridge, just outside the usually recognized Kendall Square boundary. That year, it also announced plans to occupy virtually all of the sixteen-story office tower Boston Properties was building in the same complex. The space would allow Google to almost double its local workforce. The COVID pandemic slowed construction, but the building was on track to open in early 2022, although Google gave no timeline for adding employees.[21]

A contingent of other tech heavyweights soon followed in the footsteps of IBM, Microsoft, and Google, mirroring what was happening with pharma (see box 10).

*   *   *

At the Google grand opening, Vinter made a comment that spoke to issues Kendall Square would still be facing in 2020—to a magnified degree. Not everyone would agree with his answer, but it provided some perspective on the square's corporatization. He was responding to a premise put forth by Wade Roush, chief correspondent of my news company, Xconomy: that heavyweights like Google might be sucking up talent that might otherwise go to startups—thereby depriving Boston of entrepreneurs and possibly curtailing innovation.

Box 10

Big Tech Rides the Kendall Wave

IBM's presence in Kendall Square dates back to 1964. A few other tech and computing companies started small labs or outposts in the 1980s and 1990s. But those operations were typically small, and several soon closed down—among them Atari, Nissan, and Digital Equipment Corporation (see chapter 14). It wasn't until the 2000s, and especially after 2010, that big operations took root—led by IBM, Microsoft, and Google, with Amazon and Facebook on the rise. This table shows those operating in 2020, along with a few that opened after 2000 and later shuttered.

	First moved to Kendall	Location(s)	No. employees 2020 (est.)
IBM	1964. Opened at 545 Technology Square. Acquired Lotus in 1995. Moved most operations to 75 Binney Street in 2016.	1 Rogers St., 75 Binney St.	750 in 2020
Apple	1989 when it acquired Coral Software. Established Advanced Research Lab in 1989.	1 Broadway, 314 Main St.	In 2018, around 200 members working in the Boston metro area. Planning to add several hundred more by 2026, many likely in Kendall Square
Mitsubishi Electric Research Lab	Founded in Cambridge in 1991.	201 Broadway	80 regular employees and 60 summer interns in 2020
Orange Labs[a]	2002–2009.	175 Second St.	Peak of about 60 hardware and software engineers
Google	2005. Recently acquired Android unit moves to Cambridge Innovation Center. Opens big office in Five Cambridge Center in 2008.	Three buildings in Cambridge Center. Plans to occupy most of new 16-story tower at 325 Main St. when it opens in 2022.	1,800
Microsoft	New England Research and Development center (NERD) opened 2007. Microsoft Research opened 2008.	One Memorial Drive	450
Dell EMC[b]	EMC opened research arm in 2009, but had a presence for nearly a decade before that.	11 Cambridge Center (now closed)	Around 50 at height
Amazon	First temporary location opened in or around 2012.	101 Main St., 9th Floor.	Leased space for more than 600 people.
Facebook	2013	100 Binney St., expanding to 50 Binney	Room for 650+
Twitter	2014.	141 Portland St.	
Toyota Research Institute	2016	One Kendall Square	

[a] For a fabulous feature on Orange Labs Cambridge, see Roush, "R.I.P. Orange Labs Cambridge."
[b] See Buderi, "EMC Opens Research Arm."

The question was posed during a time when many felt that a high-tech brain drain was taking place in Massachusetts. The Boston area had a higher concentration of colleges and universities than probably anywhere in the world. But stories were rampant—in news articles, op-ed columns, talks at events, and chatter at bars and cafes—of young graduates leaving for Silicon Valley. People couldn't seem to get over the fact that just a few years earlier, Mark Zuckerberg had started Facebook in Silicon Valley after dropping out of Harvard and failing to find local investors.[22]

"I think that kind of misses the bigger point, which is that too many smart people are leaving this area," Vinter retorted. "We can't do enough to create more opportunities for them. The more we can do to build a mixed ecosystem of small, medium, and large-sized companies, the more it will be self-sustaining and self-expanding, which leads to more competition, which leads to more opportunities."[23]

20 VENTURE MIGRATION AND THE TECH STARTUP SQUEEZE

Gordon Baty considered himself part of the country's second wave of venture capitalists when he and two partners founded Zero Stage Capital in 1981. "It was still sort of the Neolithic period of venture capital," he says. "It wasn't a particularly respected branch of the finance world. It was barely a branch of the finance world at all." But Zero Stage was in the first wave of venture firms in Kendall Square—in fact, it *was* the wave. "There weren't any others that I was aware of," he recalls.

Baty's cofounders were businessman Paul Kelley and Ed Roberts, a rising star professor at MIT Sloan. When Zero Stage launched with $5 million in its coffers to invest, the partners set up shop at Sixth and Binney Streets, renting desk space from Icon Corp., a firm Baty had cofounded in 1964. Not long after Zero Stage got going, Biogen had started its Cambridge operation directly across Binney. But it was mostly a sketchy, downtrodden area. A few blocks toward the river, at Third and Binney, sat a vacant lot that became a winter dumping ground for the city's plowed snow. Glaciers of the dirty stuff would rise up forty or fifty feet, and kids would sled down the slopes. "It would take practically until summer to melt, and when it finally melted there was all this trash and broken baby carriages and junk that it had picked up," Baty recalls.

After a few years, the firm moved to the Badger Building at Third and Broadway, in the midst of what was then AI Alley. Later, Zero Stage redeployed to more plush billets in the newly erected 101 Main Street closer to the Charles; Roberts's Sloan office was almost directly across the street.

The premise behind Zero Stage inspired its name. "We were going to provide the first round of institutional money to startups of various types and try to help them along," Baty relates. In short, they wanted to provide seed funding before the first official stage of financing, typically called the series A round. "We thought that Kendall Square would be a great place to establish ourselves, because there was so much activity even then. The backside of Kendall Square [toward Lechmere] was full of one-story garage buildings that were very cheap. So three graduates could go over and rent some office and lab space for very short dollars. Kendall Square was at the opposite end of the economic spectrum than it is now."

* * *

It would take more than twenty-five years, and by then Baty had retired, but eventually other venture capitalists descended on Kendall Square. The rush started around 2009, almost in concert with big companies upping their presence. It was as if investors had suddenly heard the same clarion call of students, emerging science and technology, and deals as the major corporations—and they no doubt had.

The venture field had been largely pioneered by General Doriot and American Research and Development in 1946. One of ARD's earliest investments had been in Ionics, a company with novel water purification technology that was based at 146 Sixth Street, which now makes up one flank of Biogen's headquarters building.[1] ARD itself operated out of Boston's financial district, and several of the area's early venture firms took root in Boston as well. Later, during the heyday of Route 128 and beyond, a large number moved to be near the tech scene there. Well into the 2000s, a warren of leading venture firms congregated along Winter Street in Waltham, many in a soul-starved modern office park called the Bay Colony Corporate Center. There was probably no greater concentration of venture investors anywhere in the world, outside of Sand Hill Road in Menlo Park.[2]

The sands of Boston's venture community began to shift just a few years later. As Baty puts it, "They realized they were in the middle of nowhere. Everybody discovered what we'd figured out earlier, which is you want to be where the startups are."

One of the first salvos came in 2009, when the wildly popular Techstars "entrepreneurship boot camp" expanded to Cambridge. Based in Boulder, Colorado, and run by a cadre of renowned entrepreneurs and investors, Techstars not only trained hopeful entrepreneurs in exchange for equity in their startups, but also invested in those companies and helped attract other backers. The program was highly competitive: hundreds of young companies applied for ten spots a year at the time. Techstars tapped a colorful local entrepreneur, Shawn "Doody" Broderick, to manage the Cambridge operation, its first expansion outside its home base. (In 2020, Techstars oversaw more than thirty accelerator programs in fifteen countries and had helped launch more than 2,100 startups.)

One of the four Techstars founders, Brad Feld, was an MIT graduate who had launched his first company in Massachusetts—and had invested in several area startups. Feld had also famously tried and failed to endow a bathroom at MIT (he was successful with a University of Colorado men's room). But he and his partners saw a different type of void they could help fill. "We always thought Boston would be a great next logical step just because there's so much going on there, and so many great schools and entrepreneurs around," summed up cofounder David Cohen.[3]

Techstars initially set up operations on the top floor of a run-down building at 727 Massachusetts Avenue in Central Square. After two years, under new managing director Katie Rae, the accelerator would move to One Cambridge Center in Kendall Square, initially on an unused floor that Microsoft loaned out. It later crossed the street to the eleventh floor of One Broadway, where it stayed until uprooting again to Boston's Leather District in 2014.

On the heels of Techstars' arrival, leading venture firm Polaris, based in Waltham's Bay Colony center, opened the second branch of its Dog Patch Labs startup incubator in the American Twine Building in Kendall Square. The original "kennel" was in San Francisco's Dogpatch District. "What we're trying to do is key off some of the things that are working in San Francisco, but recognize that this is a different market," explained Polaris partner Dave Barrett.[4]

Polaris's move, and Barrett's remarks, spoke to several forces that were converging at the time. One was particular to the area's entrepreneurial climate, but others related to trends in technology and the country more broadly.

Locally, a strong feeling pervaded the Boston tech community that the region needed to do more to match Silicon Valley's startup prowess. The angst was almost palpable, with constant comparisons in the press and debates and diatribes at entrepreneurial events. As part of it, a major push was launched to eliminate noncompete clauses, which were unenforceable in California but commonplace in Massachusetts—something many felt curtailed innovation and entrepreneurship by deterring people from taking a job with a competitor or launching their own company without fear of being sued by their previous employer. (After years of failed efforts, noncompetes were largely rendered toothless in the Bay State in 2018.)[5] Another chief lament was that stodgy, old-guard VCs in Boston didn't think big enough, didn't understand the consumer internet, and were too often loathe to take risks on unproven, first-time entrepreneurs—as witnessed by their passing on Zuckerberg. Nor was criticism spared of local entrepreneurs themselves, who many felt all too often built their companies to flip rather than grow into the new Google or Microsoft and become anchors for the ecosystem.[6]

On a broader front, the tech scene was also changing. As the world went increasingly online and mobile, many entrepreneurs were inspired to launch web and software ventures—not the kind of "big box" electronics, telecom, and computer-aided design companies that had populated Route 128 and farther out along the I-495 corridor. At the same time, the country itself was witnessing a major trend among talented, often technology-oriented young people to eschew the suburbs in favor of city life. This migration has been chronicled by urban studies theorist Richard Florida in his 2002 book *The Rise of the Creative Class*, along with more recent works. It began in the early 2000s but became more evident after the 2008 recession, he says. "It's really at that point that we saw this sweeping shift of technology-based business back to two kinds of places. One is to urban centers like lower Manhattan, like San Francisco's downtown and Mission District. And the second one, and we documented this in a series of studies looking at zip code location of venture capital investments, was to places like Kendall Square, and the area probably from Central Square to Kendall Square."[7]

Such forces manifested themselves in different ways in Boston's innovation ecosystem, and notably in Kendall Square—but they all contributed to the pull of startups and investors to the city. For startups, the new phone- and app-centered world helped counterbalance the effects of rising rents driven by corporatization. An internet or mobile company could launch and grow in one of the incubators or the CIC; it didn't need a lot of square footage. A large percentage of tech entrepreneurs were also young and single. They preferred to be in the city, not out in the suburbs. Many didn't even have cars: the subway didn't run to Waltham, and they often had no easy way to get to Winter Street to make their pitches.

Increasingly, VCs decided they needed to get closer to the startups, both to fit the more casual, approachable climate entrepreneurs preferred and to decrease the odds of missing something big. Proximity also made it easier to take part in events, bump into people on the street, and meet entrepreneurs for coffee or beers. One after the other, beginning almost three decades after Zero Stage blazed the trail, more firms began to come in from the burbs. In 2010, two top-tier funds took the plunge. Bessemer Venture Partners moved from high-brow Wellesley to Broadway in Kendall Square. Atlas Venture closed its Waltham headquarters and uprooted to First Street in East Cambridge, by the CambridgeSide Galleria. CommonAngels, the area's top angel investor group (it would evolve into the early-stage venture firm Converge), opened an office in the Cambridge Innovation Center. The next year, two more leading firms announced they would also move into Kendall Square—Highland Capital Partners and Charles River Ventures.[8]

For Charles River Ventures, rebranded in 2014 as CRV, it was coming full circle. The venerable firm, which has raised more than $4.3 billion since its 1970 founding, opened its first office at 300 Tech Square in Kendall Square. About five years later, it moved to Boston, and from there to Waltham. In 2011, though, it returned to the square—renting quarters in the CIC, just blocks from where it had all begun. "When we moved, Kendall Square was going through its revival, but real estate was still inexpensive, and many tech startups moved there too; we wanted to be near them," relates Izhar Armony, a CRV general partner who joined the firm in 1997. At the CIC, he says, "we thought we would meet a lot of entrepreneurs more casually and naturally 'in the elevator' versus asking them to make a special trip to Waltham (in days when young folks stopped owning cars)."[9]

The allure of Kendall Square wouldn't hold in CRV's case. Armony watched as young tech entrepreneurs and startups were steadily driven back out of the square. "After a few years, the biotech industry took over, building huge office buildings and driving rents way up. Around 2015, many of the tech startups could not afford Kendall Square anymore." Many, he says, found far cheaper digs in downtown Boston. "So in 2018, we followed them again." Just as it had done some forty-five years earlier, the firm moved across the Charles—this time to Boston's Back Bay.

It's tough to gauge the extent of the startup migration Armony perceived, but it was real. Not just a migration, either. An increasing number of fledgling companies never came to the square in the first place: they launched in Boston. "Boston suddenly finds itself the state's tech startup capital," proclaimed a *Boston Globe* headline in April 2014. The article looked at the city's surge of startups, noting that much of it had come at the expense of Kendall Square. Well-known local angel investor Jean Hammond weighed in on the shift. "If you went back even a couple years, it would have been unthinkable," she assessed.[10]

Part of the change could be credited to the debut of the MassChallenge accelerator and startup competition cofounded by John Harthorne and Akhil Nigam, both strategy consultants at Bain & Company when they hatched the idea. The nonprofit's first cohort arrived in 2010: 111 startups took part in the 2.5-month program, which included mentoring and concluded with a final competition. By 2020, MassChallenge had become one of the world's most prestigious startup competitions, offering nine accelerator programs in seven locations globally. It counted nearly 2,500 alumni companies.

MassChallenge had started in Kendall Square in 2009, spending its first six months inside the CIC, but was wooed away by the city of Boston—becoming a linchpin of Mayor Thomas Menino's 2010 campaign to rebrand the largely isolated Seaport District as the Innovation District. Once replete with loading docks, warehouses, artist's lofts, and open-air parking lots, the district was transformed into a trendy, tech-savvy scene. Even before Vertex moved into its new buildings in 2014, more than eighty new organizations and businesses had moved there, including a number of startups.[11]

Price was a core part of Boston's appeal. Rents for top-tier commercial space ran 15–20 percent higher in Kendall Square than downtown Boston. But that was for prime space, and few startups could afford that out of the block. That was a big problem with the square: pretty much all it offered anymore was premium space. Boston, meanwhile, held a variety of neighborhoods—the Leather District, Back Bay, and Downtown Crossing, among them—that sported grittier space more conducive to startup budgets and ethos. Affordability also extended to living arrangements. There was hardly any place in Kendall Square to live, unless you were a well-heeled biotech or tech executive and could pony up for the luxury apartments or smattering of condos. Boston harbored a much wider array of housing options for all levels of employees—often close enough to walk to work.

The city also had more room for coworking spaces, and a number soon set up shop, including national chains like WeWork and Industrious. The Cambridge Innovation Center was not to be left out. In spring 2014, it opened a branch in Boston's financial district, barely 1.5 miles from its original hub, but with lower prices. In Boston, though, coworking was just one option for startups. In Kendall Square, it was increasingly the only option as corporatization spurred developers to raze or refit older spaces and replace them with premium lab and office buildings. To help ease the squeeze on startups, the CIC expanded its square operations, while others arose specifically to cater to life sciences startups (see box 11).[12]

Price wasn't the only factor discouraging startups. As big corporations moved into Kendall Square, many with hundreds of employees, commuting became an ever-bigger nightmare. Car traffic was brutal at rush hour. The poorly maintained, delay-prone Red Line offered the only subway option—unless you counted the rickety and sketchy Lechmere Station in East Cambridge, a good fifteen- to twenty-minute walk from meeting spots like the Marriott and Legal Sea Foods. Boston held its own traffic horrors, but

Box 11
Coworking in Kendall Square: Increasingly, a Startup's Only Option

Startup culture had been part of Kendall Square almost since its earliest days. After MIT's arrival in 1916, the pace had slowly picked up, and it had kept on accelerating into the twenty-first century as a multitude of new ventures found cheap space in run-down factories, warehouses, and various other neglected nooks and crannies.

As late as 2010, perhaps even a few years after that, scrappy entrepreneurs could find affordable space in the square—in places like the former American Twine building on Third Street, the warren of structures in One Kendall Square, and a slew of low-slung, almost garage-like structures along Rogers Street between Second and Third Streets. But as the big players moved in, run-down buildings were overhauled or torn down to create room for higher-priced structures. Rents that were affordable for young companies became harder and harder to find. Increasingly, if a startup wanted to be in the square, its only options were professionally managed spaces like the Cambridge Innovation Center. Prices were still sky-high when gauged by the square foot. But you could get bite-sized chunks—for a few desks—and short-term leases, and take advantage of shared conference rooms, kitchens, and common areas. That made it work for many.

Accordingly, the CIC increased its presence in the square—expanding into a second building in 2013 and a third in 2019. Pre-COVID, its Kendall Square space totaled some three hundred thousand square feet. In 2020, Cambridge ranked it the city's fifteenth-biggest employer, with 1,490 workers spread across 698 client firms.

In a sense, the CIC represented the corporatization of startup space. A cadre of additional professionally managed startup spaces also took root in Kendall Square, mostly focused on providing lab spaces for biotechs, something CIC did not offer. The nonprofit LabCentral launched in 2013 and gradually took over 700 Main Street, the site of Charles Davenport's carriage works. By 2020, it had grown to offer space for seventy-odd companies and some two hundred workers and had partnered with Pfizer to add a second facility next door for another half-dozen midsize biotechs—and roughly 175 more people. LabCentral, as previously noted, also signed on to expand to a third Kendall Square location—in the clock tower building at 238 Main Street. It was scheduled to open in late 2021.

At least two other places for biotech startups operated in the square as well. One was BioLabs, a for-profit company founded by Johannes Fruehauf, who also cofounded LabCentral. It had been hatched a few years ahead of LabCentral, and in 2020 it operated in Kendall Square as Ipsen Innovation Center BioLabs, named after its corporate sponsor, whose premises it occupied. The other was Alexandria LaunchLabs, based at One Kendall Square and run by Alexandria Real Estate Equities. It opened in December 2018.

CIC, BioLabs, and Alexandria LaunchLabs all belonged to larger networks of startup spaces with sister facilities around the United States. CIC operated internationally as well, and BioLabs had plans to expand overseas. LaunchLabs was part of a publicly traded real estate trust. They were all part of the square's corporatization.

However you looked at it, the days of the scrappy, garage-style startup in Kendall Square were pretty much long gone.

was served by a network of subway stations. "There are more subway lines, and it's a little hipper in the downtown in many ways because there's better food," noted Raj Aggarwal, CEO of software startup Localytics, which had launched in Kendall Square but uprooted to Boston in late 2012. "Even if the prices were the same in Kendall, we would stay downtown."[13]

When it came to venture firms, the picture was mixed. CRV's move to Boston in 2018 was in a sense late to the party. A variety of fellow VCs had beat the company to it. A new seed stage fund, Nextview Ventures, had opened in Boston in 2010. Battery Ventures, another well-known veteran of the Bay State's venture scene, moved from Waltham to Boston's Seaport District in late 2013—bypassing Kendall Square entirely. A few months later, Polaris announced it would follow suit. It soon moved its headquarters to the Seaport, settling into the same building as Battery, one floor below it.[14] Techstars itself moved to Boston in 2014. And that is just a sampling.

A variety of other VCs, though, found the latest iteration of Kendall Square worth any added wallet and commuting pain. More moved into town—and stayed. In 2020, at least twenty-one venture firms maintained offices in Kendall Square. A few, such as Highland and Google's venture arm, GV, continued to invest across many fields, including tech. But most firms with a presence in the square in 2020 were focused chiefly on biotech and the life sciences.

* * *

Noticeable changes in deal flow patterns accompanied the movement of startups and venture firms. At first, all signs pointed to Cambridge. A report from market intelligence firm CB Insights released in late 2012 looked at the past five years of US venture capital investments. It found that Massachusetts had firmly established itself as second only to California in terms of the amount of venture dollars companies in the state raised. However, inside Bay State borders, the momentum had shifted from suburbs like Waltham and Woburn to Boston and, especially, Cambridge. Companies in "the Bridge," as some locals called Cambridge, had roughly doubled the amount of venture capital they attracted during that period—and they had absolutely dominated in the past year. Over those last four quarters, Cambridge firms had taken in $991 million in 122 deals. In the same period, Boston companies had raised $541 million in sixty-three deals, while Waltham, the old-guard leader, edged out Boston with about $575 million raised, but did that in just forty-one deals. "The clear winner in Massachusetts has been Cambridge with the lead in both VC deals and dollars," CB summed up.[15]

Just two years later, though, the latest CB Insights report found Boston had surged to the deal-making top. All told, Boston-based startups had closed ninety-seven venture deals, sharply up from sixty-six the previous year. Cambridge companies, by contrast, had seen their deal numbers tumble from ninety-nine to seventy-nine over the same period. The Bridge still held a solid lead in dollars raised: $820 million to $593 million. But that gap had also narrowed.[16] All this gave some statistical evidence to what people

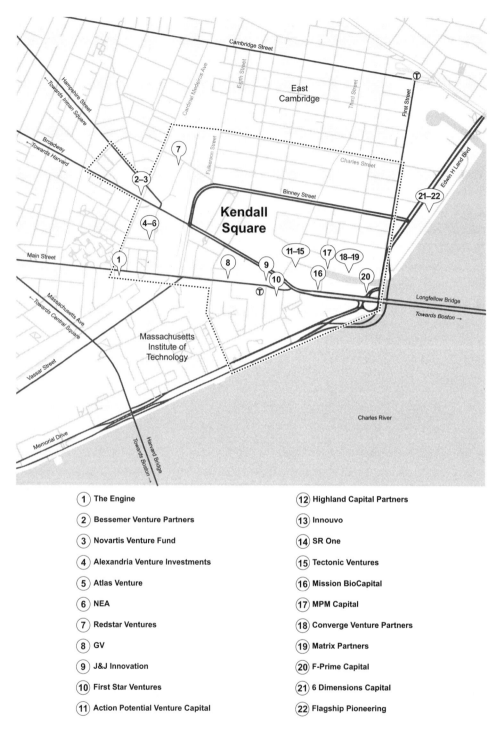

Figure 30 Most venture offices in Kendall Square are grouped along Main Street. All are walkable from the Red Line.

had been noticing—and pointed to the growing strength of Boston's startup ecosystem, at least to some degree at the expense of Cambridge and Kendall Square.

Since that time, despite the trends in deal-making, Cambridge companies appear to have continued to lead the way in dollars raised. A variety of metrics show companies in Kendall Square accounting for the lion's share of these deals. As might be expected, it's especially dominant in the life sciences. That's partly due to the field's preeminence in the ecosystem. But it's also because the high costs of research and time needed to bring drugs to market—often more than a decade—typically require biotech companies to raise substantially more investment capital than tech startups.

Table 1
Top Massachusetts biotech venture financings 2019–2020[a]

2019	Financing ($ million)	Investment round	Cambridge?	Kendall?
Anthos Therapeutics	$250	Series undisclosed	Y	Y
ElevateBio	$150	A	N	N
Beam Therapeutics	$135	B	Y	N
FORMA Therapeutics	$100	D	N	N
eGenesis	$100	B	Y	Y
Black Diamond Therapeutics	$85	C	Y	Y
OncologiE	$80	B	N	N
Oncorus	$80	B	Y	Y
Arrakis Therapeutics	$75	B	N	N
Inozyme Pharma	$67	A	N	N
2020				
Atea Pharmaceuticals	$215	D	N	N
EQRx	$200	A	Y	Y
ElevateBio	$170	B	Y	Y
C4 Therapeutics	$150	B	N	N
Cullinan Oncology	$131.2	C	Y	Y
Affinivax	$120	B	Y	Y
Dyne Therapeutics	$115	B	N	N
Generation Bio	$110	C	Y	Y
Vor Biopharma	$110	B	Y	N
Praxis Precision Medicines	$110	C	Y	Y

Source: MassBio/Evaluate.
[a]Half the top ten venture deals for Massachusetts biotechs in 2019 and seven of the top ten for 2020 were based in Cambridge. All but two of those twelve were in Kendall Square.

Table 2
Top Massachusetts Biotech IPOs 2019–2020[a]

2019	Amount raised—Millions	Cambridge?	Kendall?
Stoke Therapeutics	$142	Y	Y
Akero[b]	$105	N	N
Morphic Therapeutic	$103	N	N
Karuna Therapeutics	$102	N	N
Frequency Therapeutics	$84	N	N
Stealth Biotherapeutics	$78	N	N
Kaleido Biosciences	$75	N	N
TCR2 Therapeutics	$75	Y	Y
Fulcrum Therapeutics	$72	Y	N
Axcella Health	$71	Y	N
2020			
Relay Therapeutics	$460	Y	Y
Forma Therapeutics	$319	N	N
Atea Pharmaceuticals	$300	N	N
Dyne Therapeutics	$268	N	N
Akouos	$244	N	N
Black Diamond Therapeutics	$231	Y	Y
Generation Bio	$230	Y	Y
Praxis Precision Medicines	$219	Y	Y
C4 Therapeutics	$217	N	N
Beam Therapeutics	$207	Y	N
Kymera	$200	Y	Y
Sigilon Therapeutics	$145	Y	Y
Pandion Therapeutics	$135	Y	Y
Inozyme Pharma	$129	N	N
Foghorn Therapeutics	$120	Y	Y
Keros Therapeutics	$110	N	N
Oncorus	$87	Y	Y
Codiak	$83	Y	N
Imara	$75	N	N
Checkmate Pharmaceuticals	$75	Y	Y
SQZ Biotech	$71	N	N

Source: MassBio.
[a]Ten Bay State–based biotech companies went public in 2019, followed by twenty-one in 2020, a 110 percent increase. Sixteen of those thirty-one companies were based in Cambridge—and twelve of the sixteen were in Kendall Square.
[b]Moved to South San Francisco, CA.

* * *

If there was one venture capital firm that exemplified the emerging era, it was Atlas. Another celebrated Boston fund, Atlas was founded in 1980. For decades, it invested in both technology and life sciences companies, ultimately raising over $3 billion serving that diversified mission while expanding to several European offices and one in Seattle.

In 2004, Atlas had moved its headquarters from Boston to Waltham. When it moved to Cambridge in 2010, ahead of nearly every other venture outfit then out by Route 128, it was responding to the latest shift in the entrepreneurial winds that, among other things, indicated investors needed to be more accessible. "The decision to come back into Cambridge was just this sense that out in Waltham, up on the hill, people had to come pay homage to the big venture capital firms that were out there," says Atlas partner Bruce Booth. "And, you know, we really wanted to break with that and become much closer to the entrepreneur."[17]

Jeff Fagnan, another Atlas partner at the time, adds that the rise of the internet and web services made it much easier to start a company without a lot of experience. "We saw a trend where founders were younger and younger. Previously, the 128 and I-495 [entrepreneur] was your two-tour-of-duty VP of engineering who was going to start a company," he says. "They were early forties and seasoned." The new breed was often in their twenties—and they wanted to be in the city, near each other. "And we're like, 'If we're going to be the early-stage venture firm for New England, let's put our money where our mouth is. Let's pick up and go there.'"[18]

Atlas moved to 25 First Street in East Cambridge, a downtrodden building on the edge of Kendall Square that had been previously occupied by Art Technology Group and was then also home to HubSpot. In 2014, the partners concluded another major change was in order. That October, Atlas announced that going forward, it would be a life sciences–only fund. Its tech side—Fagnan, Ryan Moore, Chris Lynch, and Jon Karlen—would split off and form a separate fund. "One of the things that we realized as a firm was that our tech franchise was very strong, and our biotech franchise was very strong, but there really was no synergy between the two," relates Booth. "Frankly, we had to dilute our messaging to say, 'We back early-stage innovation,' instead of saying, 'We're trying to make drugs to cure people,' or 'We're trying to be the leading-edge software and technology innovator.' And so by splitting, we were able to crystallize our missions much more clearly and communicate the strategic aspirations of our respective firms."

The Atlas partners liked to say they had an amicable divorce: "We got the name, they got the house," quips Booth, referring to the fact the biotech firm retained the name Atlas Venture while the tech side stayed in the First Street offices. The tech group then launched a campaign to crowdsource a name for their upstart firm—in the interim doing business as Formerly Known as Atlas (FKA). They offered $100,000 in "carry" to whoever submitted the chosen name—meaning they would treat that person as if she or he had invested $100,000 in the new fund, which ended up raising $205 million.[19]

More than 1,600 ideas poured in. Top contenders included Rig Assembly, Motive Labs, Fuseyard, and Accomplice. Not so popular with the team: "4 Bald Guys."

The following year, the verdict was in, and the new firm was christened Accomplice. In 2018, after its Cambridge lease expired, the partnership moved to Boston's South End. The life sciences–only Atlas Venture, meanwhile, established its headquarters in Tech Square. Its offices were just over a mile from the First Street quarters. "Even just that difference was enormous," says Booth. "Twenty-five First is out there on the periphery of the innovation centers of Cambridge. And so by coming right into the heart of Kendall Square, you can't help but meet people, bump into people, share ideas. The impromptu connections, the sort of talent flows, the engagement in the community—it's just an incredible driver."

Not just their split, but the physical location of the two firms—the tech arm in Boston, biotech in Kendall Square—seemed to symbolize the changes affecting venture firms and startups alike. But it was all part of a bigger transition that, especially after 2010, brought a larger share of deal-making and company creation back to the city.

As Booth says, "The gravity of Cambridge just kept getting stronger and pulled a lot of us back in—and in doing so created even more energy in the space."

SPOTLIGHT: FLAGSHIP PIONEERING—KENDALL SQUARE COMPANY CREATOR

Where venture firms hung their hats wasn't the only thing changing about the industry. Especially in the life sciences, several were doing at least some business in a different way—conceiving of new companies, undertaking proof of concept, and then launching startups themselves, rather than investing in outside entrepreneurs who pitched ideas to them. Venture firms had long dreamed up startups in-house: TA Associates reaching out to scientists to form Biogen offered an early Boston example. What seemed most different in the emerging era was the magnitude and formalization of the practice. It became known as *venture creation* rather than *venture capital*.

Boston-based Third Rock Ventures (box 6) was a posterchild for this approach. Mark Levin had envisioned it as "a venture firm that's not really a venture firm." To one degree or another, a variety of venture groups engaged increasingly in the practice. In the life sciences, three other leading Bay State names included Kendall Square–based Atlas, and Polaris Partners and 5AM Ventures over in Boston. Stromedix, founded in 2007 to pursue treatments for organ failure and fibrosis and later sold to Biogen, was one of the first companies Atlas hatched in-house. Today, venture creation is close to "maybe 80, 90 percent of what we do," says partner Bruce Booth. "Most of the time we found, cofound, seed, and incubate companies in our office."[1]

One Kendall Square outfit has its own unique take on the idea, though—to the point that it doesn't even call what it does venture creation. Instead, it describes itself more as a science and invention factory, offering investors a "return on innovation" rather than a return on their investments.

The firm is Flagship Pioneering. It's based in a Land Boulevard office building overlooking the river that once housed Lotus. Scattered around Kendall Square and other parts of Cambridge, though, the firm maintains some 1.3 million square feet of lab space—making it one of the city's biggest leaseholders. In 2020, it employed five hundred people—including some seventy full-time inventors. Flagship maintains its own patent portfolio. It engages in scientific studies and forms prototype companies around its advances that must prove their potential before launching as standalone entities. In recent years, it has founded or cofounded some of the biggest biotechs in the country—chief among them high-flying coronavirus vaccine maker Moderna. By a couple of different measures, Flagship has provided one of the best investor returns of any life sciences fund. And to fuel its company creation engine, it closed three

Figure 31 Noubar Afeyan at the Nasdaq for Moderna's initial public offering in 2018.

successive funding efforts of over $1 billion in less than two years—the last one, totaling $2.2 billion, coming in June 2021.

Flagship's cofounder and longtime CEO is Noubar Afeyan, who upon leaving grad school in the late 1980s had little inkling of the biotech ascendency that lay ahead. By 2020, of course, everything had changed. In fact, arguably more than any other venture—or venture-like—firm, Flagship had something to do with that.

*   *   *

What became Flagship began as NewcoGen, based in West Cambridge near the Alewife T Station, next to Genetics Institute. The name was shorthand for *new company generation*. It wasn't exactly a venture capital firm—but it was created to form and finance startups. Afeyan founded it in 1999 and officially launched the next year after raising $60 million from investors for a fund called Applied Genomic Technology Capital (AGTC).[2]

He was able to attract that capital because of his past success. In 1988, a year after earning his doctorate through MIT's new biotechnology process engineering program, Afeyan had founded PerSeptive Biosystems to build instruments for the still-nascent biotechnology industry. One of his MIT professors had been Ed Roberts at the Sloan School. Roberts helped him hone the business plan, then provided seed funding through Zero Stage Capital. PerSeptive's first offices were in the American Twine Building on Third Street in Kendall Square. But it grew quickly, moving to bigger quarters near Central Square and eventually sprawling into eleven buildings scattered around Cambridge—before consolidating everything at a suburban campus in Framingham, MA, about twenty miles west of Boston. In 1992, before the move, the company went public. Then, in 1997, it merged with leading instrument maker Perkin Elmer.

The merger brought Afeyan into a large corporation. He was made chief business officer and in that role wrote the business plan for a genome-sequencing venture Perkin Elmer wanted to spin out. The spinoff was called Celera; it was headed by a Perkin Elmer scientist named Craig Venter. Until Celera launched in May 1998, both it and Venter reported to Afeyan.[3]

The Celera experience proved key to setting Afeyan on the path to Flagship. At PerSeptive, he had also taken bits and pieces of technology or expertise and spun out a few ventures—in diagnostics, vaccines, and genomics—but with nothing like the resources Perkin Elmer had put behind Celera. "So I had seen the hard-fought way with PerSeptive of figuring out how to get some money and building something from scratch and starting a spinout. I had also gotten involved with additional biotech start-ups as part of a founding team," he says. "And then with Celera, I was deeply involved in a corporate intrapreneurship way to cofound an internal venture and have it resourced through the parent company, with lots of people and money."

That had inspired NewcoGen. "I basically decided to form an entity that would be an experiment on whether you could institutionalize the act of innovation and company creation. By *institutional*, I mean organized, purposeful, mindful, corporate—as opposed to kind of winging it, which is what the startup community has long cherished as an enabler, but which I doubted."

But it quickly became clear to Afeyan that to form, nurture, and grow several companies at once—parallel entrepreneurship—would take far more than $60 million. That led him to partner with veteran venture capitalist Ed Kania, managing director of Boston-based OneLiberty Ventures. Together they raised $150 million for a life science fund to be managed by newly formed Flagship Ventures, which also absorbed Afeyan's NewcoGen and oversaw its fund.

Afeyan and Kania were listed as Flagship's cofounders when it formed in 2003, with Afeyan as CEO—a position he has held ever since. Flagship started in the NewcoGen space. But in 2005, the firm moved to One Memorial Drive in Kendall Square—the building later dominated by Microsoft. Coming out of the dot-com crash, the square offered real bargains. "There was basically nobody else there," says Afeyan. "So it was a lot cheaper for us to end up in Kendall Square."

Flagship's original strategy was to create companies both from its own ideas and around academic innovations—in partnership with other venture firms. But after the Great Recession of 2008–2009, when a few VCs began bailing on various co-investments, Flagship decided to pursue only internally generated ventures and finance them itself through the early growth stage—before launching them as standalone companies and inviting other VCs to invest. "The notion that somebody could just say, 'Oh, well this is too risky, I'm outta here' a year or two into an investment just didn't look like it could be sustainable," Afeyan says. Under its new model, because Flagship had absorbed all the early-stage risk, other VCs got less of a stake in the company for their investment. (These days, even after other investors join in, Flagship typically retains 40–60 percent of a company, Afeyan says.) That rendered Flagship less vulnerable to partners dropping out. Moreover, because it had a bigger share of the startup than in the typical venture syndication model, it garnered higher profits if the startup went public or was acquired—or both.

To pursue its revised strategy, Flagship developed a multistage process of company formation—running from "explorations" to forming prototype companies to launching new companies and then to the growth stage. "Our model is not primarily an investment model," Afeyan sums up. "Our model is an innovation model. We are in the business of return on innovation; we are not in the business of return on investment. So we don't buy shares in a company by putting money into it. We own the company outright, and we capitalize our own company. Then other people come to put in money, and if you're successful, you create a ton of value. So that's different." Such changes took Flagship farther from the traditional venture capital model. In December 2016, to drive home the idea it didn't consider itself a venture firm, Flagship Ventures rebranded as Flagship Pioneering.

To fuel this innovation engine, Flagship has grown into a small corporation in its own right. In 2020, it counted about 150 full-time staffers. This included its core operating team and roughly seventy PhD scientists who invent things for a living. "We file hundreds of patents that Flagship owns—just like MIT owns and Harvard owns," says Afeyan. These inventions are then licensed to its startup companies as needed. He has great respect for other firms that do venture creation—being "purposeful" about company formation, as he calls it. But the kind of intellectual property engine Flagship has created, he adds, "I don't believe anybody else has."

In a given year, Flagship conducts between fifty and a hundred scientific "explorations" designed to determine if something might merit forming a company. "We launch a bunch of different trial balloons, if you will, in terms of experiments. And whichever ones show progress, we pursue further," Afeyan says. That step means forming a protocompany. These are pods of about five to seven people who try to reduce various ideas to practice. Somewhere between ten and fifteen protocompanies are typically operating at a given time, scattered around various spaces that Flagship leases, almost all in Cambridge. In 2019, Flagship spent roughly $200 million on this internal research and development machine.

Everything up to this stage falls under what's called Flagship Labs. "Those are all wholly owned and operating entities, so they're not yet spun out," Afeyan explains. "The purpose of the enterprise is to spin out thriving, growing companies. That's what Flagship's business model is. We think of producing companies who in turn produce solutions, hopefully to important problems of the world. We're not making cookie-cutter companies. We're not making, you know, e-sunglasses, e-diapers, e-whatever. These are all bespoke works of science."

The firm invests between $1 million and $2 million in a pod and gives each team nine to twelve months to demonstrate it is worth launching a company. "If during that period of time something does not come out that portends a big revolution, we will not pursue it," Afeyan says. Flagship's monetary rewards policy encourages pod teams to be objective about the potential of their projects, he adds. That's because team members often have a stake in other pods—and if theirs doesn't work out, they go on to a new group; they aren't out of a job. "Unlike the traditional startup world, where an entrepreneur has every reason to make whatever slides they have look as good as they can to attract capital, we just let the projects decide. You eliminate the bullshit, as opposed to trying to sell your bullshit to somebody else so that you can continue to work on the problem."

Flagship estimates that about seventy-five companies graduated from the prototype stage between 2008 and 2020—roughly five to seven a year. These new companies are almost all based in Cambridge. Between them and the protocompanies, Flagship counts another 350 workers and leases about two hundred thousand square feet of lab space.

It's only when a new company has established a platform and product-pipeline, and gets ready for its next round of growth financing, that it officially spins out of Flagship and other investors are invited in. In fall 2020, the firm counted twenty-seven of these growth companies. Together they accounted for another roughly 1.3 million square feet of lab space and some four thousand employees. Moderna is the most famous of these ventures—almost in a class by itself, although it came together in a somewhat different way (see "Spotlight: Mapping the Moderna Network"). Others in recent years include Rubius Therapeutics, Kaleido Biosciences, Tessera Therapeutics, Inari Agriculture, Sigilon Therapeutics, Repertoire Immune Systems, and Foghorn Therapeutics, which raised $120 million in its October 2020 IPO.

There are rare exceptions to Flagship's homegrown model. The firm invested in Denali Therapeutics, a South San Francisco–based biotech, because of long-standing ties to Denali cofounder Marc Tessier-Lavigne, a former biotech executive who's now president of Stanford University. Denali raised $250 million in the largest biotech IPO of 2017.

Almost everything else originated internally, and the results have been phenomenal by some key financial yardsticks. Rubius, which bioengineers red blood cells to treat cancer, rare diseases, and autoimmune disorders, raised $241 million in its July 2018 public offering, the largest biotech IPO of 2018 at the time and the largest in Massachusetts history. Both records were shattered by Moderna, which raised a staggering $604 million that December. In its tally of VC biotech exits in 2018 and the first half of 2019, consulting firm Bay Bridge Bio ranked Flagship number one by a large margin, putting the market value of its portfolio companies that went public during that time frame at $2.9 billion. That was versus $1.1 billion for number two Orbimed. In a subsequent version of this list, tallying the market value of companies that had gone public as of September 30, 2020, Flagship ranked number two, at $4.7 billion, behind only the Strüngmann family investment arm that had been a major backer of German biotech BioNTech, which partnered with Pfizer on its COVID vaccine.

Meanwhile, life sciences news site STAT's deep dive into the performance of top biotech venture funds pegged Flagship's $270 million 2012 fund as the single best

Table 3

Total market value of investments in companies that went public[a]

Firm name	At June 30, 2020, price	At IPO price	No. of portfolio IPOs
AT Impf GmbH (Strüngmann family)	$7,902,017,430	$1,712,122,800	1
Flagship	$4,672,382,920	$2,933,735,489	5
Medicine	$2,886,265,853	$625,364,550	1
Viking Global	$2,299,621,806	$902,846,877	3
AstraZeneca	$2,207,424,246	$885,439,521	3
Orbimed	$1,578,226,390	$1,444,059,891	19
T. A. Springer	$1,427,266,761	$524,241,926	3
KKR	$1,403,187,171	$641,465,189	2
Pfizer Venture Investments	$1,306,272,567	$611,182,966	6
Baker Brothers	$1,257,933,073	$826,412,824	7

[a]Only includes data for companies where investors own >5% of common stock.

performer—providing a nine times return on its investment. Across all funds from 2007 to 2019, Flagship ranked third in investment return.[4] As a private company, Flagship does not discuss return specifics. However, notes Afeyan, "I believe there are public reports about our net IRR [internal rate of return] for the past 12 years across all funds being over 30%."[5]

Besides the fact that companies are well-vetted before they formally launch, Afeyan says one reason for Flagship's success is that even though its ventures go off on their own at the new company stage, they are not really on their own. Throughout their lifetime, they draw on Flagship's fundraising abilities, its ties to pharma companies for potential partnerships, and much more. "We're not just a shareholder in the company," he says. "We're basically an institution that attracts the capital, the people, and helps these things grow."

21   FORTY MISSING COMPANIES

Sangeeta Bhatia grew up outside Boston, raised by Indian immigrants. Her father was a serial entrepreneur—and encouraged her to follow in his footsteps. When Bhatia decided to accept a job in academe, she shared the news with some trepidation. "He really thought I was going to be a captain of industry," she says. "I told him, 'Dad, I'm going to be a professor. But don't worry, I'm not going to be locked in an ivory tower.'"[1]

"Okay," he said, absorbing the decision. "But when will you start your first company?"

Bhatia laughs, recalling the exchange. "It wasn't like, 'When will you start *a* company?' It was a very clear expectation that I would be a serial entrepreneur."

She didn't disappoint. Bhatia, a bioengineer whose research has mostly focused on treating diseases of the liver, cofounded her first company in 2004. As of late 2020, she had founded or cofounded four others. But she was acutely aware that few other women followed her example. In fact, from MIT alone, where Bhatia is a professor, there are likely at least forty fewer biotech and health-tech companies than there ought to be. That's the estimated number of companies that would've been spun out of the institute if women faculty founded companies at the same rate as their male counterparts. "It's a tragedy," says Bhatia. "There are inventions and discoveries happening with taxpayer dollars and the hard work of students that simply are not seeing the light of day by virtue of where they're occurring—in a woman's lab."

Bhatia, joined by MIT president emerita Susan Hockfield and Nancy Hopkins, a professor emerita, is spearheading a unique program to do something about it. Their effort is called the Future Founders Initiative, a collaboration of about thirty Boston-area faculty members and businesspeople. (*Disclosure:* I am a member.)[2] As of early 2021, the group had taken several specific steps designed to significantly reduce the "failure to launch" numbers by helping women faculty become more familiar with the process of starting companies and build the connections to make it happen. These included an entrepreneurship boot camp, led by Bhatia and Genzyme cofounder Harvey Lodish; a fellowship program that will enable women faculty to take a paid semester to learn about company formation at venture capital firms; deeper dives into the data to illuminate the situation more clearly and inspire new ideas to make things better; and a $250K prize competition for women faculty with startup ideas. Perhaps even more importantly, the group was exploring ways—and funding—to expand

Figure 32 From left, Sangeeta Bhatia (*photo credit:* Scott Eisen/AP Images for Howard Hughes Medical Institute), Susan Hockfield (*photo credit:* David Sella), and Nancy Hopkins (*photo credit:* Donna Coveney).

the program nationally and at least provide a template for those in other regions to follow.

The "missing companies" finding stems from the group's initial analysis of MIT data about faculty and entrepreneurship, completed in early 2020. "I mean, that just kind of takes my breath away," says Hockfield. "Forty more companies, forty more drugs, forty more diagnostic devices, forty more cures. We're just missing the opportunities that are really sitting in our hands."[3]

*   *   *

The working group's founders are taking a systems approach to addressing this challenge, in large part because they know the system as well as the issue.

Hockfield came to MIT as president in 2004—the first woman and first biologist to lead the institute—and served until mid-2012. She had previously been provost of Yale, where she had been on the neurobiology faculty since 1985.

Hopkins arrived in 1973—joining Salvador Luria's cancer center—after earning her doctorate from Harvard and doing postdoc work with James Watson at Cold Spring Harbor. She rose to national prominence in the late 1990s after famously measuring the lab space given MIT women faculty and showing that men of equivalent standing had labs up to four times larger than their female counterparts. An important committee was formed, with Hopkins as its chair, that gathered data to document the problem. Its report, released publicly in 1999, led to systematic reforms at MIT and far beyond—in academe and business—around hiring and promotion, resource availability, day care, family leave, and more. Hopkins and her pioneering efforts toward equality for women scientists were featured as part of the documentary *Picture a Scientist* that premiered on PBS Nova in April 2021.

For her part, Bhatia earned a joint MD-PhD through the Harvard-MIT Division of Health Sciences and Technology, known as HST. After postdoctoral training at Massachusetts General Hospital, she became a professor at UC San Diego, then joined the

> **Box 12**
> Companies Sangeeta Bhatia Founded while at MIT
>
> > 2007: Hepregen (acquired by BioIVT)—drug metabolism and toxicity testing
> >
> > 2015: Glympse Bio—noninvasive, in vivo diagnostics
> >
> > 2019: Impilo Therapeutics (acquired by Cend Therapeutics)—RNA delivery
> >
> > 2019: Satellite Bio—implantable "satellite" organs that don't require removal of diseased organs

MIT faculty in 2005. As noted in chapter 1, she became only the twenty-fifth person and second woman elected to the National Academies of Sciences, Engineering, and Medicine. In addition, she belongs to an even smaller subset of those who have also been inducted into the National Academy of Inventors. Her first high-tech spinoff, a nanobiotech company called Zymera, was formed in 2004, when she was still in California. The rest have come while she has been at MIT.

All three women have offices in the Koch Institute for Integrative Cancer Research at MIT.[4] Although Bhatia had been at MIT for several years when the institute opened in 2007, she hadn't really known Hopkins. She knew her colleague's work measuring lab space, though, which had inspired Bhatia. "It led to nine institutions banding together to really structurally move the needle on gender equity in academia. And so I have been trying to just take that simple message to every institution that I've gone to. And when I came back to MIT and actually moved into the Koch Institute, I got placed next door to Nancy, who is like my hero. We became friends."[5]

The inspiration that led to the Future Founders Initiative came more than a decade later, in September 2018. Hopkins, who had retired, was given a lifetime achievement award by the Xconomy news site.[6] When approached about the award, Hopkins had one reservation: although she was proud of the influence she'd had on women in academia, she was not so happy with the situation of women faculty when it came to company formation. "I warned Xconomy that I was going to be critical of the industry in this regard," she relates.

Hopkins says she had no thoughts of her acceptance speech kicking off some sort of movement. "I didn't realize it was a call to arms at all. I just thought, 'Oh God, here I go again.'"

As it turned out, Bhatia and Hockfield were in the audience. To their ears, Hopkins's short speech was a very loud call to arms.

* * *

Nancy Hopkins did not set out to be an activist. A tipping point came in the mid-1990s. "I was a tenured full professor," she relates. "I was trying to convince the local administrators that I deserved as much lab space as an assistant professor. A guy kept telling me, 'Why do you say things that aren't true?' I said, 'But a five-year-old child can tell you it's true. And, by the way, a woman who washes glassware for the labs asked me the other day, "Why do you have so much less than these men have?'" But just to convince him, I literally got my tape measure and measured every space in the building. I mean, it was nuts. And it still didn't convince him."

Hopkins felt certain she was experiencing a systemic marginalization of women at MIT, but she didn't know what to do next. "If you said you were discriminated against, people thought you were not good enough, you know." Hopkins shared her experience with two other women professors—and was surprised to hear they had come to the same realization. "I was just astonished," she laughs. "I knew all women were having a problem, but I didn't think *they* knew."

The trio soon discovered that all the fifteen tenured women in MIT's School of Science saw similar disparities in their access to resources when compared to their 197 male counterparts.[7] They raised the issue with Dean Robert Birgeneau, who quickly formed a committee to tackle the problem. After two years of study, the committee issued a 150-page confidential report. Yes, there were clear and widespread inequities between men and women, not just in lab space, but also in pay and other benefits.

Birgeneau took it extremely seriously, and made comprehensive reforms, says Hopkins. The report remained confidential until two years later, however, when a Sloan professor, Lotte Bailyn, became chair of the MIT faculty and pressed for a public version. Hopkins drafted something for a special edition of the *MIT Faculty Newsletter* in March 1999. She didn't see it as especially noteworthy. "It was just a summary of the work we had done and the reason the group had come to exist, and what we found," she relates. To accompany the article, Bailyn asked MIT president Chuck Vest to write a comment about it. Vest readily agreed.

"He wrote this comment which really put the thing on the map," says Hopkins. "He said, 'I have always believed that contemporary gender discrimination within universities is part reality and part perception. True, but I now understand that reality is by far the greater part of the balance.'[8] And when I saw that sentence come over my computer, I have to say that was probably the most emotionally meaningful moment of my professional life. I thought I'd go to my grave with not one person ever caring about or knowing there was a problem. And here was the president of MIT saying, 'There's a problem. And now I understand what it is.' It was life changing for me."

Newspapers from the *Boston Globe* to the *New York Times* picked it up—as did wire services, magazines, radio, and television. "It just went totally around the world. Literally. I mean, nobody'd ever seen anything like it. The president, the dean—they were overwhelmed," Hopkins relates. One day she arrived at her office to find the phone ringing. She picked it up and a voice said, "Hello, you're on the air in Australia."

* * *

In her lifetime achievement award speech, Hopkins described the gratifying response at MIT. She showed a slide charting a big increase in women faculty in the School of Science since the mid-1990s—with the number more than doubling.

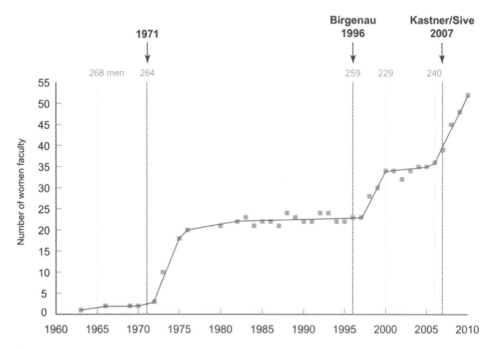

Figure 33 Slide courtesy of Nancy Hopkins. The number of male faculty is across the top. Robert Birgenau became dean of the School of Science in 1996. Marc Kastner became dean in 2007, and Hazel Sive was named associate dean.

At the same time, Hopkins recounted, MIT swept away the disparities in lab space and salary, and reviewed and overhauled hiring standards. She even held up the actual tape measure she had used to make her original point—to thunderous applause.

But then Hopkins switched gears, from celebrating the successes to sounding an alarm. A major disparity, she warned, remained between men and women scientists. She shared a story of meeting a woman from Harvard Business School who told Hopkins about a list of some one hundred scientists in the Boston area who had received venture funding to start companies: just one woman founder was on the list. That had dovetailed with Hopkins's observation of how few female biology faculty seemed to be starting companies or were even connected to the business world. In fact, she had done her own informal study of the founders, directors, and scientific advisory board members of eighteen biotech companies—and counted 223 men and just eight women. Among those 231 individuals, she identified eighty-four faculty members from various universities: just four were women, about 5 percent.

This huge gap, she told the awards crowd, was not right in a field where 50 percent of PhDs have gone to women for many years and women comprise a quarter or more of university faculty. One of her slides stated: "Time alone does not change things! People do."

When Hopkins finished her talk, the three hundred–plus attendees leapt to their feet for a standing ovation. The speech had special resonance for two of the women in the audience. One of them was Sangeeta Bhatia. "She really pulled together this kind of anecdotal observation about the work that was left yet to do," says Bhatia. "The next proximal thing was this really huge gap in entrepreneurship—and she saw it very clearly in the biology department, because the biology department at MIT has launched much of the biotech in the region. So she was sitting there watching many of her male colleagues do it, and almost none of her female colleagues."

Bhatia had been asked to introduce Hopkins to the audience. She had then gone to sit at the Koch Institute's table. Susan Hockfield was at the same table. "She'd heard the data before, but something about that moment really moved her," Bhatia noticed. Bhatia suspected that Hockfield might be a powerful ally. When she shared her hunch with Hopkins after the dinner, Hopkins was skeptical, but agreed to give it a try. "So I reached out to Susan and asked her if she would like to do it, and she said 'I'd love to,'" Hopkins relates. "Needless to say, that changed everything. She stepped up the way Vest did, and it is only through such leaders that we were able to make progress."

Sangeeta Bhatia picks up the story. "The three of us got together to figure out okay, what are we going to do? And that ultimately became the Future Founders Initiative."

* * *

The group, originally called the Boston Biotech Working Group, formally kicked off on December 20, 2018, with a dinner hosted by the American Academy of Arts and Sciences. The turnout was impressive—a strong representation of the biotech ecosystem: venture capitalists, deans, professors, executives, entrepreneurs, nonprofit leaders, and other influential figures. Women and men.

From that dinner and two others, five subgroups or *workstreams* were created to address various aspects of the job ahead: data, VCs, academic deans, innovation ecosystem, and founder development. By the third dinner in May 2019, movement was taking place on several fronts. Most notably, a grant had been awarded by the Alfred P. Sloan Foundation to study academic founder activities in seven faculty departments at MIT, as well as gather statistics on faculty who served on corporate boards or as scientific advisory board members. The work was led by Teresa Nelson, a data scientist at Simmons College—with Bhatia and Sloan School professor Fiona Murray serving as principal investigators.

Meanwhile, the working group as a whole set two broad goals based on early data:

1. Boosting the share of faculty who serve on boards of directors and scientific advisory boards in the biotech community from 14 percent women to 25 percent within two years
2. Increasing the share of faculty who had founded companies from fewer than 10 percent women to 25 percent within five years

By the fall of 2019, Nelson's team had some harder data in hand. They had surveyed current MIT faculty members in seven engineering and science departments: biology, chemistry, brain and cognitive sciences, electrical engineering and computer science, biological engineering, chemical engineering, and materials science. The first three were in MIT's School of Science; the rest were from the School of Engineering.[9]

Those seven departments had 337 faculty members, of whom seventy-three were women, and those faculty had helped launch 252 companies. Women faculty, however, had only founded twenty-five of these startups, just under 10 percent of the total—despite comprising 22 percent of the faculties overall. Women were similarly underrepresented relative to their male counterparts as board directors or members of scientific advisory boards. Male faculty served on 235 boards, compared to just ten for women faculty—and 464 scientific advisory boards versus fifty for women. Another stunning fact: in the three departments most related to biotech company formation—biology, biological engineering, and chemical engineering—twenty-six faculty members had teamed with other MIT faculty members to form twenty-seven companies. All were men. They had never once included a female MIT colleague when founding a company. What's more, earlier inquiries by Hopkins and Bhatia had determined that women had not declined to participate in company formation—they had not been asked. One woman member of the working group described her experience as wearing a cloak of invisibility, like in Harry Potter: her male colleagues couldn't see her when it came to founding companies.

The disparities were most stark in biology. Only two of fifty-six companies started by current biology faculty had been founded or cofounded by women, despite the fact that women constituted 24 percent of the department. In comparison, the numbers were more equal in biological and chemical engineering (see tables 6 and 7). One possible explanation was that engineering by its nature is focused on solving real-world problems and that founding companies have long been embedded in engineering departments. Biotech, by contrast, is a much newer field, without a longstanding tradition of commercialization, particularly by biology faculty.

Hopkins looked at the data and "did a very simple calculation," she recalls. "Take the number of companies founded by the current faculty—the number that were founded by men, the number founded by women. If the women had founded companies at the same rate as the men, how many companies would there be instead? And for those seven departments, it was forty companies were missing."

The idea of "forty missing companies" became a rallying cry for the group. The big challenge, though, was what to do about it. Members felt broadly that for a variety of reasons, such as gender bias in society and education, female faculty were generally not as familiar or comfortable with entrepreneurship as their male counterparts. This led naturally to smaller numbers of women starting companies or engaging with businesses, even if they wanted to. At the same time, companies looking for board members tended to gravitate to people they already knew—and those were usually men. Some investors also noted that it was often difficult to find qualified women for their boards, creating a self-reinforcing cycle that shut women out.

Table 4
Rates of founding companies by MIT faculty as of July 2019 in three departments of science and four departments of engineering

		School of Science				School of Engineering			
	Total	Biology	Brain and Cognitive Sciences	Chemistry		Biological Engineering	Chemical Engineering	Electrical Engineering and Computer Science	Materials Science and Engineering
Total # of current faculty (as of July 2019)	337	58	35	30		26	33	125	30
# of faculty who are female	73	14	12	6		5	5	22	9
% of faculty who are female	22%	24%	34%	20%		19%	15%	18%	30%
% of males who are full professor	70%	77%	65%	58%		76 %	68%	72%	62%
% of females who are full professor	62%	71%	58%	50%		60%	60%	64%	56%
% of males who founded at least one company	40%	(19/44) 43%	(6/23) 26%	(6/24) 25%		(11/21) 52%	(10/28) 36%	(41/125) 40%	(12/21) 57%
% of females who founded at least one company	22%	(2/14) 14%	(1/12) 8%	(1/6) 17%		(3/5) 60%	(4/5) 80%	(4/22) 18%	(1/9) 11%

Source: Nancy Hopkins. Project supported by an Alfred P. Sloan Foundation grant.

Table 5

Commercialization activity by MIT faculty in biology, biological engineering, and chemical engineering

	Science	Engineering	
	Biology	Bio Eng	Chem Eng[b]
Total # of current faculty (as of June 2019)	58	26	33
% of faculty who are female	24%	19%	15%
# of SABs (scientific advisory boards) served on by male faculty	121	63	110 (43)
# of SABs served on by female faculty	9	5	9
# of BODs (boards of directors) served on by male faculty	42	17	81 (23)
# of BODs served on by female faculty	1	2	4
# of company-founding events by male faculty[a]	63	25	71 (25)
# of company-founding events by female faculty	2	5	5
# of companies founded by male faculty	55	24	64 (18)
# of companies founded by female faculty	2	5	5

Source: Nancy Hopkins.
[a]*Company-founding event* refers to an individual participating in founding a company. Due to cofoundings, these exceed the number of companies founded.
[b]This department includes one male faculty outlier. Data are shown both including and, in parentheses, excluding the outlier.

Participants had several ideas for correcting this situation. One was building a directory of women faculty and their fields of expertise, to make it easier for corporate boards to identify people qualified for scientific advisory boards and directorships. A second idea, suggested by a venture capitalist, was for venture firms to host women faculty as entrepreneurs in residence or to offer fellowships as a way to demystify the company formation process and build networks. Yet another idea was to create a special course or bootcamp to help women learn the basics of entrepreneurship.

"We hear from women and men that one reason for women not to [found companies] is that they haven't been part of the conversations that introduce them to the routes and the people that would make navigating that path possible," Hockfield later told STAT. "The projects we propose have dual purpose: to introduce the VCs and board and SAB members to women who are ready to engage in entrepreneurship; and to give women the vocabulary and network necessary for their success."[10] By late fall 2019, for example, the venture capital workstream, led by Amy Schulman and Terry McGuire at Polaris Partners and Steve Knight at F-Prime Capital, held its own dinner looking at getting more venture firms involved.

By the fourth dinner of the entire group that December—the last in-person meeting pre-COVID—members decided to go public. In late January 2020, they shared the initiative with two reporters who had gotten wind of the working group and made inquiries: Sharon Begley of STAT, and the *Washington Post*'s Carolyn Johnson.

The plan had been to put several core ideas into action in spring 2020. That included launching the bootcamp and writing one or more articles about the data findings for the *MIT Faculty Newsletter*—the same place the 1999 public report drafted by Hopkins had been published.

Not surprisingly, the pandemic slowed the timetable significantly. But things continued to evolve, and the action picked up that fall. Here is where core pieces of the plan stood in early 2021.

The Bootcamp

Because of COVID, the bootcamp idea, originally envisioned as in-person experiences starting in spring 2020, evolved into a series of virtual events. The first six sessions featured discussions with mostly women founders and role models about starting companies and hard-won lessons around entrepreneurship.

The first took place on September 3, 2020. Susan Hockfield welcomed the crowd and explained the series. Then Bhatia was introduced by Theresia Gouw, a founder of Aspect Ventures in Palo Alto, California, one of the few women-run venture firms. She and Bhatia had been undergraduate roommates at Brown University, and Aspect was the lead seed investor in Glympse Bio, which Bhatia founded in 2015. Then Gouw interviewed her former roommate about her experiences founding companies—a virtual fireside chat that hoped to inspire women in the audience to try entrepreneurship.

The second event featured Lodish, interviewed by Hockfield. Additional chats that year featured Daphne Koller, a former Stanford professor who cofounded online education company Coursera and more recently started a machine-learning-driven life science company called insitro; Stanford chemist Carolyn Bertozzi, a founder of at least seven companies; Katherine High, a professor emeritus at the University of Pennsylvania who cofounded Spark Therapeutics; and Jodi Cook, a former executive of PTC Therapeutics.

The bootcamp's second phase, beginning in January 2021, was called Startup 101. It aimed to dive deeper into the specifics of founding companies and included sessions on patenting and licensing intellectual property, raising money and selling companies, and such issues as whether academic founders should leave with their companies or stay in universities. "Anecdotally, what we observe is the women engineers who have started companies, like myself, had exposure—when they were training—to the concept, to the value, and to the practice of the profession," explains Bhatia. "I called in a patent filing the night of my PhD defense, because my advisor told me to. That meant that when I became a professor, I knew you were supposed to file your IP [intellectual property] before public disclosure—whether or not you were starting a company. If no one ever teaches you that, then how would you know?"

While implementation of the series was delayed, the planners found a silver lining in COVID. "We're actually using the pandemic in our favor because we realized, 'Oh, well, if we could be on Zoom, we can ask amazing women founders from across the world,'" says Bhatia. Another positive was that a virtual event could be attended by a wider group than initially planned. A survey of MIT faculty held pre-COVID identified forty women who expressed interest in attending the in-person event. For the virtual events, they targeted faculty at MIT and two other area institutions—Partners Health-Care and Boston Children's Hospital, where Hockfield and Lodish, respectively, are board members. More than five hundred people, not all women, registered for at least one of the sessions.

The Accelerator Fellowships

One of the working group's early ideas was to create a sabbatical program for women faculty to work at venture firms. This evolved into what were called *accelerator fellowships*, and the idea won buy-in from department deans all the way up to MIT vice president for research Maria Zuber and president Rafael Reif.[11]

The plan was to select up to five fellows a year, who would spend at least the spring semester and possibly also the summer at venture capital firms. The first VC firms to sign up as hosts were Polaris Partners, F-Prime Capital, Pillar, Novartis Venture Fund, and The Engine. The fellowships were announced in fall 2020, with the thought that the first fellows would start in spring 2021.

The program will focus initially in biotech and related fields, but it potentially could be expanded to others, says Bhatia. "It's a scalable model."

The Venture Capital Pledge

The venture capital workstream created a nonbinding pledge that it hoped a large group of venture firms would agree to. This was separate from the accelerator fellows program. The pledge read as follows:

> As venture capital investors in biotechnology we are proud of a lot that we have done to support patients, entrepreneurs, and our investors. We aren't as proud of our record of diversity in senior management, boards and our partnership ranks. We are committed to working on this. Women comprise approximately half of graduating medical students, 45% of graduating M.B.A. students, and over 50% of students receiving both Ph.D. and Master's degrees in the biological sciences. Women comprise approximately 50% of the overall workforce in biotechnology companies. However, women fill only 25% of C-suite positions and only 10% of positions on the boards of directors of biotechnology companies.
>
> That's simply wrong and unproductive. So as a start, we, the undersigned, are committed to increasing the number of women in our ranks. To begin, we the undersigned have made a pledge to do all in our power to ensure the boards of directors for companies where we hold positions of power are twenty-five percent female by the end of 2022. Many

of us already are there. This commitment and our ongoing vigilance with respect to portfolio decisions and our own hiring will help instigate further progress towards properly bringing the advantages of full gender diversity into the senior ranks of U.S. biotechnology companies.

Increasing board participation was seen as a key way to help women develop the business contacts and experience that might make them more comfortable starting their own companies one day. "Service on a board creates access to a network of investors, leading academic scientists, and other key leaders," Polaris' Schulman related. "It's a door opener and a credentialer."[12]

In addition to Polaris and F-Prime Capital, seven other Boston-area venture firms had endorsed the pledge: Omega Venture Partners, Canaan Partners, SV Health Investors, 5AM Ventures, Venrock, GV, and Atlas Venture.

Institutionalizing the Data

The initial study looked at seven MIT departments—and only at current faculty. Many others had made notable contributions to starting companies and other commercialization efforts who were outside those departments—or who had retired or left MIT for others reasons, including some who were deceased. The aim was to gather data on their activities as well.

Even more important going forward, the three founders also hoped that the methodology developed would help make gathering data on faculty founders more accessible and routine. That way, results could be published in an annual MIT report, perhaps along with statistics about research dollars attracted and the like. Says Bhatia, "Instead of us having to do this again, it would just be in the system and it would get reported out as an outcome that's very transparent every year. Like how many women founders were there?"

Dolphin Tank and Prize Competition

One new idea gained momentum in early 2021 and was set to launch in 2022: a $250,000 prize competition along the lines of the MIT 100K, but for companies founded or cofounded by women faculty. In a group Zoom call that February, Harvey Lodish explained this as a natural outgrowth of the bootcamps—allowing women faculty to gain experience developing business plans and then pitching to venture capitalists. It would be like pitching "in a dolphin tank—not a shark tank," he told members. (The term dolphin tank, suggested by Aoife Brennan, CEO of Kendall Square-based Synologic, referred to a much friendlier version of the *Shark Tank* TV show, where judges can sometimes be brutal in their assessments of hopeful entrepreneurs.) Bhatia added that women entering the contest would have access to one-on-one mentoring to help them prepare their pitches, and described the competition as "a friendly, easy audience. The goal is not to have to attack the pitch, but actually to be constructive."[13]

* * *

How this would all play out would take years to fully assess. As of early 2021, Nancy Hopkins was optimistic overall. "Last year was the first time I've really felt this encouraged it was going to get fixed—could get fixed," she says. The new *MIT Faculty Newsletter* special edition finally came together and was published that spring.[14] The three founders, with strategy and execution help from group member and Kendall Square Association president C. A. Webb, were also pursuing some promising new ideas for significantly expanding the program nationally—though discussions were at too early a stage for them to share.

Some things, though, had not started out as well as hoped. One case in point was the venture pledge. While there was quick uptake from a few firms, it proved surprisingly hard to get a wider group of VCs to sign on. A few declined outright. Others said they had to check into legal ramifications. This proved frustrating to several working group members.

"I felt totally underwhelmed by the venture response," says one non-faculty member of the group, who thought the pledge itself fell short of the mark, never mind the fact that many seemed to have trouble signing it. "I mean, I know those guys well, and it was just on script for them—'We'll do a pledge.' Like, fuck you. Our research proved that there is so much value going unrealized, so many new therapeutics and technologies not being brought to market, because these guys aren't taking women seriously. I mean, these guys, they don't get it. They need to retire. We need the men who are under forty to become the general partners, because most of those men are married to women who are working. Most of those men are actively parenting. Most of the men, while they certainly still are benefiting from all the white privilege, they tend to be more self-aware and more genuine in their interest in working side-by-side with strong women."

The three organizers also had a different concern—that their focus on women neglected to tackle the similar issues around opportunity and company formation for faculty of color. This was even more on their minds as the Black Lives Matter movement gained power in spring 2020.

"It's something that I struggle with," says Bhatia, herself a person of color. "At the top line, our sense is that both of these stories are, sadly, old stories. So we shouldn't be deterred on trying to advance this one. But I think we are in a moment where we may want to shift the way we think about it and talk about it. As a starting point, we'll make sure the women-focused activities have diverse women, but to me it feels a little bit disingenuous to all of a sudden broaden it, because I don't think we've actually sourced the data and the solutions around race with enough focus. The truth of the matter is the problems are not exactly the same, and the pipeline breaks in different places, and the proven interventions are overlapping, but not identical. So, that is something that we have to decide about."

However the effort evolves, there is also a growing awareness among the founders that data alone can miss a key part of the solution—the human element. "You know, in the first draft, it was the data. We're scientists—we wanted everyone to look at the data and let the data speak for itself," says Bhatia. "But we've been reflecting on the power of narrative and storytelling. The three of us feel more pressed to be brave about squarely putting our finger on the challenges and what we've experienced and how we think about them."

So perhaps what's really needed goes beyond data, bootcamps, and pledges to a whole new narrative or culture that simply takes as a given that women can—and will—be as entrepreneurial as men. "Fifty-six companies were started in the department of biology, and two of them were by women," says Bhatia. "Yeah, that's a really powerful statistic, but you know, Nancy lived that. She saw Phil [Sharp] start Biogen and Harvey [Lodish] start Genzyme. Each one of those fifty-six is a real, living, breathing company that's sitting there in Kendall Square." And some of those companies—plus others that don't exist—could have been started by women if the underlying narrative had been different.

So imagine what will happen if the Future Founders Initiative succeeds and the culture finally does change. Those forty missing companies will be missing no longer.

22 700 MAIN: THE STORY OF KENDALL SQUARE—IN ONE BUILDING

Kendall Square has a sweeping history, from the area's marshland beginnings to becoming the world's densest center of research and innovation, especially in the life sciences.

A microcosm of its journey lies in the history of just one building: 700 Main Street.[1] The oldest industrial structure in Cambridge, a three-story red brick structure with a bright green door and green trim around the windows adorned by a quartet of historic plaques, 700 Main has witnessed almost every key change in the evolution of Kendall Square. From within its walls came the first modern railroad passenger car, the first pipe wrench, the first two-way long-distance telephone call, instant photography, spy cameras, and much more. Today, it's home to LabCentral, a startup space containing what is arguably the biggest concentration of fledgling biotech companies on Earth.

This is its story.

*   *   *

700 Main was "born" in 1814 or 1815. It started life as a combination house and manufacturing company—called a "store" at the time. But as described in chapter 4, it first leapt to the center of the Kendall Square story in 1842, when Charles Davenport moved in with his fledgling railroad car business and erected six single-story wooden workshops on the east side of the property. Within five years, the company added two large wings, housing the machine shop and the foundry and blacksmith shop, creating the largest factory complex in all of Cambridge.

After Davenport sold his business and retired in 1854, the warren of buildings was acquired by Caleb Allen and Henry Endicott, who operated an iron foundry suitably called Allen & Endicott. In an early version of the major role the building would play later for startups in biotech, the partners rented out some of their space to other manufacturers. One of their tenants was J. J. Walworth & Company, a pioneering maker of steam heating systems.

In the late 1860s, one of the mechanics at Walworth's Main Street plant was Daniel Stillson, who had served as a machinist under David Glasgow Farragut, the US Navy's first admiral, in the Civil War. In 1869, Stillson invented a wrench that was much better at gripping pipes and other rounded surfaces than existing monkey wrenches. His patent for what became known as the pipe wrench (and Stillson wrench) was issued the following year, and he ultimately received about $80,000 in royalties, a considerable amount at the time—worth more than $1.5 million today.²

Unlike for Davenport, there is no plaque on 700 Main Street commemorating Stillson or his wrench.

* * *

The next historic role for 700 Main Street was as a bit player in a larger story of innovation. But it was truly a world-changing innovation—the invention of the telephone. Most of the story took place in Boston, where Alexander Graham Bell made the legendary first telephone call in March 1876 to his right-hand man, Thomas Watson: "Mr. Watson, come here, I want you."

Two months after that, the telephone proved a big hit at the Centennial Exposition in Philadelphia. But would it work over longer distances than just a few rooms away? On October 9, 1876, as Watson put it, "we were ready to take the baby outdoors for the first time."³ They got permission from the Walworth company to use the telegraph line that ran from its Boston headquarters to the factory at 700 Main Street—a distance of about two miles.

Watson traveled to Cambridge that evening "with one of our best telephones." He waited at 700 Main until Bell signaled through the telegraph "sounder" that things were ready on the Boston end. At first, nothing happened. But Watson did some quick troubleshooting, and before long, Bell came through loud and clear: "Ahoy, ahoy." As Watson relates, "I ahoyed back, and the first long distance telephone conversation began." The next day, the *Boston Advertiser* hailed the event as "the first conversation ever carried on by word of mouth over a telegraph wire."⁴

Bell and Watson would eventually join Charles Davenport in getting a plaque at 700 Main Street.

* * *

In 1882, about six years after the historic phone call between Thomas Watson and Alexander Graham Bell, Allen & Endicott tore down the front building of 700 Main Street. They replaced it with the three-story brick "headhouse" that stands today. Later, they extended one wing so that it bordered Albany Street.

The building was sold again in 1927, to the Kaplan Furniture Company, a maker of Federal-style reproduction furniture that was started around 1905 by Russian immigrant Isaac Kaplan.⁵ While Kaplan was in a totally different line of business than Allen & Endicott, he continued the subletting tradition. One of his tenants began the next major chapter of trailblazing innovation at 700 Main. His name was Edwin Land.

Land moved into the building around 1942, when he was in his early thirties. In the private lab he created at 700 Main (the official address of the space was 2 Osborn Street, first door on the left as you walked down Osborn from Main Street), Land would make many of his famous inventions—including the original work on instant photography

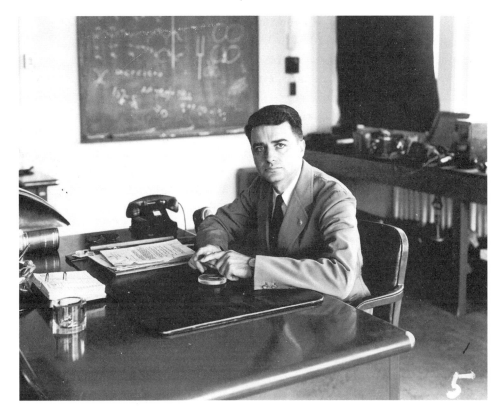

Figure 34 Edwin Land in his private laboratory, the Mole's Hole, in 1943.

announced in 1947, as well as key defense work in photographic reconnaissance. He would ultimately amass 535 patents.[6]

Even after Polaroid moved its headquarters in 1966 to Tech Square, less than two blocks away, Land spent as little time at the corporate home as possible. The inventor figured he added much more value at his lab, which he sometimes called the Mole's Hole after the underground lair of Mole in the children's novel *The Wind in the Willows*.[7] Notes Victor McElheny, "Like 'Doc' Draper and many others in the MIT ecosystem, Land was content with ad hoc arrangements, in basements and rented buildings, and never really operated in some of the fancier buildings that Polaroid built or acquired in later years."[8] He stayed put at Osborn and Main.

Land implemented tight security measures at the lab, which he needed because of the many secret government projects he worked on, from the U-2 spy plane to the Corona reconnaissance satellites. A private security car—said to be manned by an off-duty policeman—was long stationed in the driveway of his house on Brattle Street in West Cambridge.

For decades, the building at 700 Main was an innovation dynamo. The Mole's Hole itself nestled behind a thick wooden door on the Main Street side of the building. It was equipped with a free-standing, two-sided blackboard for brainstorming, a

Barcalounger, and an array of desks and tables. "Dotted around tables and shelves were numerous telephones. It was said that Land knew who would call on which phone," remembers McElheny. The bookshelves, he adds, "were filled with an eclectic collection of frequently consulted books, everything from volumes on dye chemistry to psychology."[9]

From Land's office in Mole's Hole, a door led to a lab stretching further along Main Street back toward Kendall Square. That lab, in turn, connected to a seemingly endless warren of corridors, labs, and offices that made up the rest of the building. This was the domain of Land's "elves," his laboratory workers, "many of them strong-minded women who knew their stuff," recounts McElheny. They worked on their own projects and in their own groups, but were there to be summoned to assist their boss directly when a problem arose.

One floor above Land's labs and offices was the black-and-white laboratory long led by Meroë Morse, a highly talented, energetic, and cheerful woman who joined Polaroid directly upon graduating from Smith College with a degree in art history and not a single course in physics, chemistry, or even business administration. She quickly grew to be one of Land's most trusted lieutenants and, perhaps because of her art history background, managed Polaroid's relationship with photographer Ansel Adams, who long worked as a consultant for the company and helped to promote its film.[10] Polaroid's original instant film only produced sepia-colored prints. Morse's black-and-white lab led the charge to figure out how to make true black-and-white film, which it did successfully. Her lab also dealt with the crisis that arose in 1950 when it turned out the black-and-white print material faded too quickly. Land liked to come out of the Mole's Hole and hop up the stairs two at a time to meet with Morse, who died of cancer in 1969, at just forty-six.[11]

Next door, in a part of the building apparently shared initially with the Kaplan Furniture Company—Polaroid would take over the entire complex by 1960 and purchased it outright in 1998—Howard Rogers led a team of "color elves."[12] His group worked for more than two decades on color films. A short, wiry man with spiky hair who, along with his wife, traveled the world to view solar eclipses, "Howie" Rogers quickly became Polaroid's most important inventor in color work. He had started off at 2 Osborn sharing a desk with Land himself. Then he had moved up to the second floor, near Monroe, and finally expanded next door. His work—and in particular the patents it produced—proved critical to successfully fending off Kodak's later bid to muscle into aspects of Polaroid's instant photography business.[13]

Land got two plaques at 700 Main. The first, to commemorate him, was put up by the Polaroid Retirees Association in 2014. The following year, the American Chemical Society designated the building a National Historic Chemical Landmark. Land's role in Kendall Square was not over either—the Rowland Institute for Science that he founded in 1980 is still a presence in the square and has since merged with Harvard University. But even without that, he left a legacy that few can match, much of it tied to 700 Main Street. At a 1944 conference called The Future of Industrial Research, held in

Figure 35 Modern-day face of 700 Main: entrance to LabCentral in the courtyard, just off Main Street.

New York even before World War II ended, Land laid out a "doctrine" for the science-based companies he saw as key to leading the nation in the postwar, post-Depression era. "The small company of the future will be as much a research organization as it is a manufacturing company," he told attendees. He called such companies "the frontier for the next generation."[14]

*   *   *

Perhaps nowhere in Kendall Square is Land's hoped-for frontier of science-based startups more evident (and densely concentrated) than at 700 Main Street today. Inside the same red brick walls that witnessed the creation of luxurious railroad cars and the invention of instant photography, some seventy fledgling biotech companies are now being nurtured in a venture called LabCentral.[15]

In addition to its commitment to the idea of science-based startups, LabCentral epitomizes the evolution of Kendall Square, from manufacturing to physical sciences and computing and then to biotech. If just one of the ventures inside its walls turns out to cure a form of cancer or make a huge advance in treating or avoiding Alzheimer's, say, LabCentral might get its own plaque at 700 Main, or at least share one with that tenant.

LabCentral, launched in late 2013, rapidly become a powerful force in the life sciences ecosystem. In its first six-plus years, it was home to 126 startups, according to its *2019 Impact Report* released in 2020. These companies had raised a cumulative $5.9 billion—through angel and venture investments, grants, partnerships, and more—nearly $4 billion of that in 2018 and 2019. LabCentral said its current and alumni companies took in $582 million in series A funding in 2019—representing 20 percent of total US series A investment and enough to rank as the country's fourth-largest "state" in terms of attracting early-stage investment. Current and former LabCentral companies had also added 2,395 jobs to the state's economy, it said.[16]

Johannes Fruehauf, the main visionary behind LabCentral, comes from a medieval German town east of Frankfurt called Steinheim am Main. He had no idea he was going to be in the United States running a startup space. If fact, he went to medical school in Germany and France and trained as an obstetrician—planning on being a doctor, like three of his four grandparents and many others in his family. He only moved to Boston when his wife, also a doctor, got into a postdoctoral program at Tufts University in 2002. Fruehauf continued thinking he would work as a physician even when he found his own postdoc position, a one-year program at Beth Israel Hospital. "My sense was always, 'I'm a physician,'" he sums up.[17]

But Fruehauf's research at Beth Israel turned into a long-term project—and then into a startup. It still blows his mind a bit. He was working in the lab of Chiang Li, a renowned academic and serial entrepreneur, studying an inherited form of colon cancer. After publishing some interesting data from mouse experiments that indicated progress might soon be made against the disease, they suddenly started getting calls from venture capitalists. Li listened. "So we spun out a company, and he asked me to go find lab space," says Fruehauf.

This was now late 2005 and into 2006. Fruehauf got steeped in the world of brokers, commercial leases, and other real estate details—things he had no previous experience with. "It took us a long time to even know what we were looking for," he recalls.

Eventually, they found space at One Kendall Square, a renovated complex of buildings that plays a central role in the Kendall Square evolution story (see chapter 4). Their startup was called Cequent Pharmaceuticals; at its peak, the venture had about thirty employees, with Fruehauf serving as head of research and development. Cequent was acquired in 2010 in an all-stock deal worth about $46 million—merging with MDRNA, a company in Washington State specializing in RNAi-based drug discovery and development.[18]

The experience launching Cequent, however, convinced Fruehauf that there must be a better—and quicker—way to start science-based companies. With Cequent, it took eight or nine months of hard work just to find space and set up labs, he recalls. During that time, the company wasn't doing science and generating the data it needed to move ahead with drug discovery. It also ate up a third of its first venture round (roughly $2 million) on equipment and improvements to the space, Fruehauf estimates. "The investors get no upside from that, nor do we," he says. "Instead, we as founders have to raise more money to pay for this stuff and suffer more dilution. It's not a good model."

That experience, coupled with what Fruehauf learned while working on his next company, Vithera Pharmaceuticals, led to what he's convinced is a far better model for launching startups. Instead of raising venture money for Vithera, which was focused on genetically engineered probiotics for inflammatory bowel disease, he decided to cover the bills for Vithera's own science by conducting paid research for other people on the side, using space he leased at One Kendall Square.

It didn't take long for Fruehauf to realize that the side contract research operation, dubbed BioLabs, was a much better business than Vithera. Not only was there great demand for research, but the projects that panned out often turned into companies that wanted to continue to do their science right at BioLabs. "They needed space, and we said, 'Yeah you can use this bench,'" says Fruehauf. "We just split the cost of the rent by the number of benches and we came up with the pricing model. And then, the biggest breakthrough—we saw that it's actually real fun to work with other smart people around you. Here's two or three other groups, now five, now a dozen, that work around us in our own lab doing their own thing, and it's a wonderful community."

All this unfolded between 2010 and 2013 in the BioLabs space at One Kendall Square. And things kept growing from there. "People started calling us even if they didn't need our services—they just wanted space," Fruehauf says. He opened another BioLabs-type operation in Durham, North Carolina, and then created seven more around the United States.[19]

Meanwhile, Fruehauf moved ahead with an even bigger and more ambitious version of BioLabs: LabCentral. It was structured as a nonprofit in order to apply for a grant from the Massachusetts Life Sciences Center (MLSC), a state effort to boost its life sciences sector then run by Susan Windham-Bannister that had already awarded what it called an accelerator loan to a startup called Moderna. "We had all this data from these wonderful companies that we had built in the first three years of BioLabs," Fruehauf recalls. "Hundreds of jobs created and hundreds of millions of dollars that went into companies that came out of this admittedly B-quality space that we had at One Kendall. We showed them this big vision of what this could be."

The core message was that startups could focus on the science, and not on operations and finding equipment. Fruehauf soon developed this into a buzz phrase: "We're going to change the way people think about building biotech companies." And he had his eye on the perfect location for LabCentral—the historic building at 700 Main Street. The property had been purchased from Polaroid by MIT in the late 1990s and converted to a biotech research facility by late 2002. But in 2013, it was in a kind of "dead zone" between Kendall Square and Central Square, making rents reasonable, Fruehauf explains.

The team that applied for the grant—Fruehauf and two other cofounders of LabCentral—was handpicked. One was Peter Parker, an experienced venture capitalist who had been an investor in Cequent and then stepped in as its CEO. The other cofounder was Tim Rowe, founder and CEO of the Cambridge Innovation Center. In between Cequent and Vithera, Fruehauf had rented space in the CIC and had grown to love the model, adopting parts of it for both BioLabs and LabCentral.

Windham-Bannister and the MLSC were convinced. They awarded LabCentral a $5 million grant that was used to build out twenty-eight thousand square feet on the

Figure 36 Johannes Fruehauf (left) with Peter Parker in LabCentral.

ground floor of 700 Main Street. In addition to labs, the facility contains community spaces, shared conference rooms, and telephone booth–like cabins for private calls. It also offers free lattes, espressos, and other beverages (sometimes including beer and wine), fruit, and a variety of other snacks, and it hosts educational and fun events. "A company comes in here on Monday. They go through safety training; they can be doing experiments on Tuesday. Think about that," proclaims Fruehauf. Plus, they get access to millions of dollars' worth of electron microscopes, mass spectrometry instruments, and other equipment. There's even a sort of vending machine for lab supplies.

LabCentral quickly proved to be a kind of "field of dreams" for biotech: *build it and they will come*. "The place filled up much faster than I had expected, even after I had been running the only place in town that offered this. It was amazing," says Fruehauf. "Now everybody is doing it this way. We have brought a new model to the industry. This is the no-brainer standard for knowledgeable VCs. They will make their CEOs do it here: it is so much more efficient, so much faster, so much cheaper. And it's more fun."

A typical LabCentral denizen starts with two or three people; the facility can accommodate a company's growth up to about fifteen personnel and kicks everybody out after two years whether or not they reach that figure. The price tag can seem hefty, Fruehauf acknowledges. In early 2020, LabCentral was charging $4,600 per month for one bench in a row of lab benches (you get a little sign with your company name, too). But to many, it offers great value, and the demand for LabCentral grew so fast that Fruehauf snatched up the upper floors of 700 Main—forty-two thousand additional square feet—when their occupant, Pfizer, moved out in late 2016 into its own buildings nearby. That was made possible by another $5 million grant from the MLSC, plus $4.5 million from corporate sponsors.[20]

By late 2017, however, that space was completely full as well. So LabCentral partnered with Pfizer to open a sister facility a stone's throw away at 610 Main Street. LabCentral 610 offers room for slightly bigger companies than the original operation. Fruehauf calls it the *graduation lab*. "It's built specifically for companies growing from fifteen to forty employees," he says. "We have eight suites around a common marketplace. So the company is still enjoying the community, but they have more privacy. They have their own offices. They have their own rooms. We still share a café, break rooms, conference rooms, and a common reception area." Pfizer, as the main supporter, took two of the suites for its own external innovation efforts—embedding itself in the community.

Fruehauf stresses that LabCentral is a break-even business. "It basically distributes and divides the cost of running the thing by the number of vendors," he says. "I think that's really what sets us apart from other places. [Users] get a ton of value for the rent that they pay because we don't have to pay investors and we don't have to pay any loans."

Another thing that sets LabCentral apart, he says, is that not just any startup can get in. "We are very selective. We reject four out of five applicants because we want to have the best companies here. You cannot do that if you're a real estate entity because then you have to have the real estate filled. For us this is strategic."

That selectivity makes it easier for LabCentral to attract corporate support to augment the rents it collects and any grants it wins. A sponsor wall showcases the logos of all the corporate benefactors. They include a dozen leading pharmaceutical and biotech companies and a host of smaller concerns, from drugmakers to equipment manufacturers, who each pony up a minimum of $100,000 per year. All told, LabCentral took in about $40 million in sponsorship support over its first five years, Fruehauf says. The total was close to $65 million by early 2021.

Yet another way it's strategic is that Fruehauf and Parker joined with two other biotech veterans to form a life science venture capital fund, called BioInnovation Capital. The $135 million fund first closed in 2017 to make seed and series A investments specifically in companies that come through LabCentral and BioLabs doors. It later merged with San Francisco–based Mission Bay Capital, and the joint firm was rebranded Mission BioCapital. Says Fruehauf, "This is how we leverage what we have built with these labs. This way we can help entrepreneurs through space, connections with pharma, and now also through capital."

Next to the sponsor board is another display bearing the logos of all the companies that have been part of LabCentral. Some went on to large venture rounds, and a few even went public, such as Rubius Therapeutics, Fulcrum Therapeutics, Surface Oncology, and Unum Therapeutics. Others didn't survive. But even if a company goes bust, Fruehauf says, "It's great for us, because we get to know these founders and executives—and then they come back and say, 'Hey, my company didn't work out. Can you connect me with another group?' And so we became a very active hub of exchange."

In fact, much of LabCentral's value is the connections it offers to interesting people and ideas, Fruehauf argues. That's why there are common spaces offering coffee and other beverages, snacks, and comfortable chairs and couches—complemented by events, book clubs, and rotating art in the hallways. "We want to allow for this human collision that happens in the hallway and the café, at our events," says Fruehauf. "The real value is not in the flow cytometry—the real value is in all these other people around me."

A last thing he wants to talk about is the LabCentral Learning Lab, a collaboration with a group called BioBuilder that offers middle and high school students especially, as well as teachers, access to state-of-the-art lab equipment and training in areas such as DNA transfer, synthetic biology, and even cloning in a dedicated space. It's sponsored by New England Biolabs, a leading enzyme producer for the biology field, which makes a sizable annual donation to support the program.

Most tenants at LabCentral are part of small company teams, but a few are lone wolf inventors or scientists working on a concept or idea. The problems these fledgling companies are tackling can be seen as windows into the future as well; maybe someone will find a biological breakthrough that will change the world as much as the telephone. The vast majority of the startups are therapeutics companies developing drugs for many types of cancer, Alzheimer's, autism, a variety of rare diseases, and more. Others work in medical devices, diagnostics, imaging, dermatology, and audiology.

One is an MIT spinout dubbed Asimov, which is combining synthetic biology with computer science to fashion microprocessor-like biological circuits it thinks have a wide range of applications in therapeutics, food, and beyond. It raised $4.7 million in a seed round in late 2017, led by high-profile venture firm Andreessen Horowitz. Another is Affinivax, which is developing next-generation vaccines from technology licensed from Boston Children's Hospital. It follows a different startup strategy, eschewing venture funding to remain more independent and instead relying on an initial investment from the Bill & Melinda Gates Foundation and strategic pharmaceutical company partnerships. It quickly pivoted during the COVID pandemic to focus on a coronavirus vaccine.

A third LabCentral inhabitant is Dyno Therapeutics, which emerged from the lab of Harvard University geneticist George Church. It employs machine learning to engineer super-efficient viruses that can evade the immune system and carry therapeutic genes with more precision to key targets inside the body, including organs such as the kidney and lungs that have historically been very hard to treat.

And one of the highest-flying members of the LabCentral crew is Kronos Bio, which was formed around a drug discovery platform called a *small molecule microarray*, invented by Angela Koehler of the Koch Institute for Integrative Cancer Research. The array employs advanced printing technology to attach small molecules to its surface so that thousands of molecules can be evaluated almost at once, holding the potential of not only more efficient and less costly drug development but also unlocking drug targets long thought practically "undruggable." Launched in 2017, Kronos took in a whopping $105 million in a series A funding round in July 2019. By then, it had attracted Norbert Bischofberger, former chief scientific officer of Gilead Sciences, to be its CEO and had moved its headquarters to San Mateo, California. However, it keeps a key research arm at LabCentral, just down Main Street from Koehler's lab.

Will one of those, or some other early-stage effort inside 700 Main, find breakthroughs that cure a disease, or rise to be the next powerhouse public company, ultimately building its own labs? Almost certainly. But which one? John Armstrong, former worldwide head of IBM Research, once told me that only half the projects and other efforts going on inside Big Blue's labs would ever prove successful. "The problem is, I don't know which half."

It's all about making smart, calculated bets on the future—and finding the people and environments that provide the greatest odds of success. It's hard to imagine that some of LabCentral's tenants or former tenants won't blossom and even exceed expectations. That, after all, has been the story of 700 Main in all of its incarnations for more than a century and a half.

23 NEXUS OF COLLABORATION

A nice crowd should have been gathered in the tiny park opposite the Genzyme Center, which featured an ice-skating rink in the winter and live concerts in the summer. Instead, with everyone still hunkered down due to COVID, attendees tuned in to the festivities via a webcast.

It was December 3, 2020, and the digital guests were on hand to celebrate Henri Termeer and witness the unveiling of a life-sized sculpture of the longtime Genzyme CEO placed in what was now called Henri A. Termeer Square (I described this scene in chapter 1's walking tour). The Dutch-born Termeer had died unexpectedly on May 12, 2017, at age seventy-one. Within a few months of his death, a movement in the biotech community arose to honor him. A tribute committee raised more than $3 million to redesign and rename the park outside his beloved Genzyme Center—and commission the sculpture. The park had been renamed in 2018, with the landscape work largely finished later that year. The ceremony and placing of the artwork was the final piece. It had been originally scheduled for the third anniversary of Termeer's death in May 2020, but the pandemic had forced the delay.

The outpouring of support to honor Termeer spoke directly to another aspect of Kendall Square: its spirt of mentorship and collaboration, in which Termeer had played a leading part. After his retirement following Genzyme's 2011 sale to Sanofi, Termeer expanded his role of counselor and convenor—often welcoming hopeful entrepreneurs to his "yellow house" in Marblehead, investing in startups, and hosting events for up-and-comers and established players alike. In addition to the park, a foundation had been formed by his wife Belinda and daughter Adriana. Under its auspices, the Henri A. Termeer Legacy Program had been created to provide fellowships for young biotech entrepreneurs. The inaugural crop of five fellows had been welcomed in May 2018, the first anniversary of his death, with nine more named over the next two years. The fellowship didn't come with a monetary award. Instead, it offered something of potentially much more value: access. A large, star-studded group of biotech leaders committed to being available for advice and consultation as the fellows pursued their careers. "We are the number one life science cluster because of people who can work together and still be competitive, and I think Henri was all about making that happen," says Belinda Termeer. "He inspired so many people to dream bigger."

Figure 37 The Termeer sculpture shortly after it was unveiled in December 2020, before the fountain to his right was flowing.
Source: Author.

His statute was meant to evoke this directly. The artist was Bolivian sculptor Pablo Eduardo, who fashioned the statue of former Boston mayor Kevin White that stands near city hall—and one of Charles Darwin at Cold Spring Harbor. Termeer is seated on a bench inside the park, near a curved granite waterfall that feeds into a reflection pool. His legs are crossed, a hand open in welcome, head tilted as if listening or conversing. Robert "Bob" Coughlin, former CEO of MassBio and cochair of the committee honoring Termeer, put it this way to close the livestream ceremony: "I'm going to enjoy going there to sit alongside Henri and be able to finish conversations that we weren't able to finish, as well as just ponder what would Henri do in a situation like this."[1]

*   *   *

Alnylam CEO John Maraganore served as the tribute committee's other cochair. In his view, his friend and mentor Termeer was just one face—albeit a key one—of an extraordinary community. "The collaborative spirit across the leadership of biotech companies is amazing," he says. "It is so incredible to see how at the level of the CEO, the level of the CMOs, the heads of R&D, CFOs, finance leaders, et cetera, how well connected and outwardly engaged in gathering that community is. I have had discussions with our counterparts in the Bay Area, who lament that they did not have that type of spirit in South San Francisco or up and down the peninsula. Even when two companies are competing in a given therapeutic category or clinical indication, you'll find those two CEOs sitting down together and talking about the broader environment and working together to make sure the environment is sound."

"Boston is a small town. Biotech is a small town. You put them together, and it's a really small town," says Katrine Bosley, board chair of Arrakis Therapeeutics and former CEO of Editas Medicine and Avila Therapeutics, who in 2021 was heading her third biotech company, a startup that was still in stealth mode. The much more sprawling physical distribution of the Bay Area precludes at least some of that close-knit culture, Bosley believes. "You have this density of interactions here across all of the different parts of the ecosystem. I always found people to be incredibly helpful and welcoming—there's free advice everywhere you turn. Lots of informal mentoring that people do just to try to pay it forward."[2]

No one has quantified this phenomenon, if that is even possible—but it is real. And especially in the life sciences, these networks make up an essential part of the square and the greater Boston innovation cluster. They take a variety of forms. Some are formal, for-profit ventures such as BioPharma Hub, where peers share best practices in a confidential setting, or a variety of Boston-based executive conferences. They also include nonprofit mentoring programs such as the Termeer Foundation Fellows, and ad hoc initiatives like the Future Founders Initiative that seeks to inspire and support women faculty members as company founders. Various incubators and startup programs also provide mentorship and connections.

Just as often, though, the networks function informally—via dinners, cocktail parties, or simply a willingness to get together and talk over coffee. Despite forming nearly forty companies himself and winning just about every prize possible, Bob Langer is legendary for responding to almost any email within minutes. A communique from him can invigorate young entrepreneurs. And he's just one of a plethora of examples. To a remarkable degree, people's doors are wide open.

Deep interconnections are also shared and cultivated by the alumni of the region's pioneering companies. Time and time again, former employees of these enterprises—and to a lesser degree, pharma concerns like Novartis and Pfizer with long-standing presences in the region—have gone on to lead new startups, back them financially, or serve on advisory boards. Some of these veterans go back to the earliest days of biotech and have truly seen it all—and a surprising number of these pioneers are not yet retired. They have often crossed paths multiple times in their careers, sharing successes and failures, forming bonds in one endeavor that serve them in the next. Read deeply into the biographies of a startup's CEO and other top executives, and you will often find they cut their teeth at the likes of Biogen, Genzyme, Genetics Institute, Vertex, or Millennium—or several of them.

Interconnections and relationships are part of every ecosystem, and in Boston's case can be traced to the Brahmin families who helped finance the West Boston Bridge or supported inventors like Alexander Graham Bell. In the 2020s, in their more professional manifestation, networks form a core part of both the tech and life sciences scenes. However, based on scores of interviews and an informed observational perspective, the bonds seem unusually strong in biotech.

In this regard, the Boston biotech community of the 2000s seems to have succeeded where the tech community of the 1980s and 1990s failed. It has for the most part avoided the go-it-alone silos that dissuaded Route 128 electronics companies from collaborating and cross-fertilizing ideas and people, making it harder to stay innovative and competitive. In contrast, Boston biotech is widely seen to have separated from the Bay Area and established itself as the world's leading life sciences ecosystem. "I think it has an edge for several reasons," says MIT's Phil Sharp, who cofounded Biogen and Alnylam and, more recently, was advising the startup Dewpoint Therapeutics. "One is that there is so much interaction and sharing of people back and forth between companies. And mentorship—that's never talked about, but that happens all the time between CEOs and others. And the pressure to actually do something new, to really move the needle, is very high. Fossilization has not set in. It's still a dynamic community. Just to step out and say 'I gotta do something new or I can't justify my existence' is really quite, quite prominent in Cambridge."[3]

The dynamic of Boston versus Bay Area tech companies was detailed by UC Berkeley professor AnnaLee Saxenian in *Regional Advantage: Culture and Competition in Silicon Valley and Route 128*. Both MIT and Stanford had housed labs during World War II that advanced electronics and computing and led to large pools of technical talent that fueled postwar industrial growth. As a result, writes Saxenian, "During the 1970s Northern California's Silicon Valley and Boston's Route 128 attracted international acclaim as the world's leading centers of innovation in electronics. Both were celebrated for their technological vitality, entrepreneurship, and extraordinary economic growth."[4]

In Boston, this rise of the tech industry—led by minicomputer companies like Digital Equipment Corporation, Wang, Prime, and Data General—was hailed as the "Massachusetts miracle." Within a few years, however, things had starkly changed. Minicomputer makers were hit hard by the emergence of personal computing. As their layoffs mounted, startups failed to arise to fill the void. By the end of the 1980s, it was all over but the lamenting.

The UC Berkeley scholar cited several factors that caused Boston-area firms to fall behind their California counterparts. High on the list was Route 128's insular culture. In Silicon Valley, she observed, "the region's dense social networks and open labor markets encourage experimentation and entrepreneurship. Companies compete intensely while at the same time learning from one another about changing markets and technologies through informal communication and collaborative practices."[5] Around Boston, Saxenian found the opposite. "The regional economy remained a collection of autonomous enterprises, lacking social or commercial interdependencies . . . engineers generally went home after work rather than getting together to gossip or discuss their views of markets or technologies. The social gathering places that were common in Silicon Valley did not appear to have existed on Route 128."[6]

Location and geography influenced this divergence, she concluded. Massachusetts companies had spread their bases out along Route 128 and beyond, often miles from each other. "Unlike Silicon Valley, where firms clustered in close proximity to one another in a dense industrial concentration, the Route 128 region was so expansive that DEC began to use helicopters to link its widely dispersed facilities,"[7] she wrote.

The relative lack of social interaction among Route 128 firms made it harder to build mutual trust and a feeling of being in it together. These and other factors made Boston-area companies slow to respond to evolving markets and technologies. The bottom line, Saxenian found, was that while insularity offered some advantages of scale and employee stability, overall it made companies less competitive and innovative.

In the current iteration of Kendall Square, while the tech giants and pharma companies all pursue proprietary advances and do have their versions of silos, the sequestered Route 128 structure and experience seems to have been almost completely avoided. Large corporations take an active role in groups like the Kendall Square Association and MassBio, host outside events on their premises, enter into partnerships, and fund a multitude of joint projects with universities—another element Saxenian found missing in Route 128 companies.

Part of this reflects a near-universal shift in attitude. In the age of the internet, mobile phones, and social networking, more and more entrepreneurs and employees have sought out city life and interaction with their peers. That more community-oriented, connected, and open mindset is part of what drove entrepreneurs and venture firms back into Boston and the square. "The younger people had no interest in being out at Route 128," recalls former Atlas Venture partner Jeff Fagnan, now at Accomplice, recalling Atlas's 2010 move into Cambridge from Waltham. "They were really all about a lot more touchpoints, a lot more community building with each other, and a lot of birds of a feather kind of meetup groups."[8]

Kendall Square's extremely tight geographic confines helped break down barriers as well. Meetings or an after-work drink were a lot easier on schedules. Even as restaurants, bars, and cafes proliferated in the 2010s, there were only so many—and you couldn't help but run into people from other companies. "There's probably not a single day, if I'm walking through Kendall Square for some reason, I'm not going to run into two or three people," says Alnylam's Maraganore. "We might have a side conversation, might have an idea, might say, 'Hey, let's get together in a day or two.' You just can't recreate that if you're in your car going up 101 in San Francisco to the UCSF campus or going across the bay to Berkeley. That stuff is a grind."[9]

*   *   *

To one degree or another, such factors apply to everyone in Kendall Square. But the roots appear to run deeper and the collaborative, interactive networks seem strongest in biotech. How did this come about? It wasn't as if many biotech executives had read Saxenian and studiously sought to avoid the traps that had ensnared minicomputer makers.

Industry insiders have some answers—or at least plausible theories. One stems from the fact that biotech companies, far more than the minicomputer and computer makers, sprang from shared academic roots. The founders or major scientific figures behind trailblazing companies like Biogen, Genzyme, Millennium, and Genetics Institute had been fellow professors at MIT and Harvard—and many of them trained under Salvador Luria or James Watson. In fact, some still taught at those institutions—and over the years had spawned additional companies together. As a cohort, they had been central to advancing an entirely new field—molecular biology—and pioneering the biotech industry. That sense of shared creation forged mutual respect and strong bonds that went far beyond any single company. Those ties were further strengthened by the fact that bringing a drug to market was extraordinarily difficult and complex and often involved working with others.

"It's hard to capture. I've thought about this. Everybody has their own formal structures and owns their piece of the technology, and it's supported by IP and regulation of the whole pharmaceutical industry. But the pace at which science was advancing was understood by this community," says MIT's Sharp. "No one person, no team, could master it if they were somewhere else and isolated. To get this science to a patient, you have to have sophisticated physicians who know how to test new science in patients. You have to have chemical engineers and engineers who can make the drug. You have to have people who understand the regulatory process. You have to have financing. And to lead one of these companies, it takes a multifaceted skill set. It doesn't work if you are just a bottom-line person. You can't just crank it out like Chevrolets. And the communication between the CEOs in this community is just amazing—they're unofficial advisors of each other."[10]

The extraordinarily long time horizons involved in getting a product to market—often a dozen years or more for a new drug—compared to fields like software, electronics, and computing also made it easier to be open and collaborative, many say. "In high-tech or information technology, if they talked about what they were doing, someone could take that knowledge, go back and reproduce it, and get it to market much faster. So you couldn't share at the same level, and you didn't want to share—you were far more competitive," says Janice Bourque, MassBio's first full-time president. "Our industry was not competitive in that sense. It was not like other industries. They all wanted to go and be successful and be leaders, but it was so challenging that they could just talk about it freely and no one could go back and beat them to market. So that pressure was off of you, and that created a dynamic, in Kendall Square particularly, because of the proximity of MIT and the available real estate."

Related to this, few biotech companies directly competed. They were generally pursuing different diseases or segments of diseases, or exploring distinctive approaches. "You can have a hundred companies within several blocks of each other, and none of them are strictly competitors with each other because they're doing different things," says Navitor Pharmaceuticals CEO Tom Hughes, who had helped open the Novartis Institutes for BioMedical Research before moving into biotechnology. But biotechs did share challenges, such as how to approach clinical trials, navigating the FDA, or even just creating a self-sustaining industry that provided jobs and opportunities for advancement. All this facilitated interactions and strengthened networks, he says.

Hughes and others also point to venture capital—much more sophisticated than during Route 128's tech heyday—as further helping cement ties between people and companies. "They're placing bets across a number of different companies. They want them all to succeed, and the best way to ensure that is that they're all learning from each other and sharing best practices," says Hughes. "They have this inherent benefit that they derive from enhancing collaborative interactions between their companies."

While this also holds true for tech venture capitalists, Hughes believes it's especially pronounced in biotech. "Biotech requires building external capabilities that sometimes take you to Europe, Australia, or Asia to carry out the work. This differs a bit from tech, which is usually driven by internal work with no need for lab space and might be more focused on developing a transition to commercial activities much sooner, and at much lower cost. The VC-startup network is really important in biotech to understand which vendors are good—and for what—how much you should pay, how to problem solve, clinical strategy, and more."[11]

*   *   *

For Boston's biotech community, the tight interactions coupled to the complexity of the industry further compelled people to band together and reinforced the cluster. At the core, says Sharp, sat the science. "That is the essence. Deeply big science, bets on new science, trying to figure out how to use that science. That specifically was the Biogen culture. It's how Millennium worked. It's how Vertex worked. And that culture is what brought these pharmaceutical companies to Cambridge and allowed companies to grow here."

In more recent times, perhaps the most high-profile example of biotech networks in action lies in the formation of Moderna, shown as a map in "Spotlight: Mapping the Moderna Network." The richness of the life sciences community that evolved, especially after 2000, was almost mind-boggling. The increasing density of biotech companies, sophisticated investors, experienced managers, and the overall talent pool, coupled with the attractive power of leading universities and hospitals, provided even more fodder for growth. That was then further augmented by the growing presence of just about every big pharma company in the world. It all fed back on itself. Given the complexity of making drugs, many biotechs failed. The fact that there was likely another opening next door made the region even more attractive for employees. You could get laid off or seize an opportunity to advance without uprooting your family, changing schools, or even adjusting your commute. Similarly, if you were forming a company, there was arguably no better place to find your investors, employees, or partners. In a sense, the entire ecosystem had become a gigantic network.

But while success bred success, it also spawned new challenges. The thickening web of Kendall Square—in tech as well as biotech—put more strain on the already challenged public transportation infrastructure. Traffic grew ever worse. Real estate prices rose precipitously, increasing the squeeze on startups and putting pressures on the residential neighborhood around Kendall Square. Mark Levin also worried about a dichotomy he saw around all the area's life science talent pool. "Competing for talent in Kendall Square—there's great plusses for the people in high regard, but it does become very expensive," the Third Rock Ventures and Millennium Pharmaceuticals cofounder says. "The other extreme is that people are taking a lot of jobs earlier—more jobs earlier and earlier in their careers. And we're finding that they're taking jobs probably a little too early. So there has to be this balance about quality and making sure that we don't create all these companies with people that are smart, but not experienced enough."[12]

Some felt that corporatization even threatened the collaborative networks that had formed a key ingredient of Kendall Square's special sauce. Unlike with biotech companies, says Tom Hughes, "if you have even two major pharma companies next door to each other, they're competitive. And so you don't get the sense of cross-pollination of ideas or sharing solutions. You can sense it when you walk in; it's like the oxygen is gone. And it isn't just pharma either," he adds. "What really made Kendall Square in particular this hotbed of innovation, largely it's been crowded out by the larger groups."

Ideas for addressing these challenges were in various stages of consideration or even adoption in early 2020 when the coronavirus hit. As happened almost everywhere, many efforts around issues like transportation, real estate, and office space moved to the back burner, as the pandemic added question marks about the future. By late in the year, though, things were grinding slowly back into gear. Meanwhile, notes KSA president C. A. Webb, action on some other fronts, such as the more universal challenges of diversity and inclusion, received more attention in the wake of George Floyd's killing, and many corporate and collective initiatives were accelerated during the pandemic.

For Webb, the challenge for the next evolution of the square lies in making bigger strides in such areas and "really stewarding this place as an innovation district—not letting it just go the way of another high-rent office district that happens to be next to a major university."

SPOTLIGHT: MAPPING THE MODERNA NETWORK

One uplifting story emerging from the coronavirus pandemic was the record-shattering speed of developing highly effective vaccines. In Kendall Square, that success was embodied by Moderna. By early 2020, a decade after its formation, Moderna had forged a rich pipeline but had not yet developed an FDA-approved drug or vaccine. Yet in December 2020, it became just the second company, after Pfizer, to see its COVID-19 vaccine win federal approval. The key figures behind Moderna's creation highlight the Kendall Square–Boston biotech network that for many sets the region apart from any other.[1]

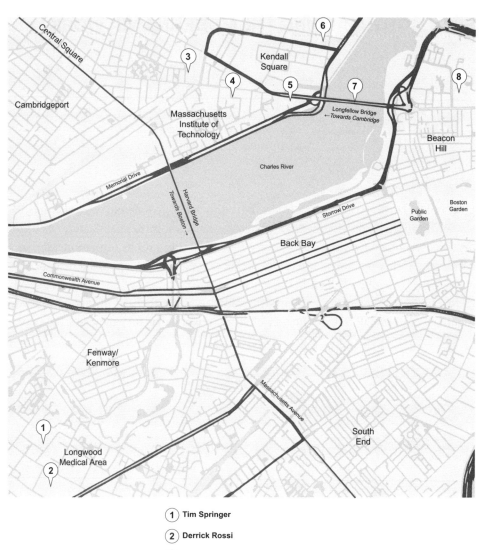

1. Tim Springer
2. Derrick Rossi
3. Moderna offices
4. Bob Langer
5. Flagship Pioneering/Noubar Afeyan
6. Moderna offices
7. Stéphane Bancel
8. Kenneth Chien

Figure 38 Moderna cofounders listed alphabetically:
- Kenneth Chien, physician and cardiology expert at Massachusetts General Hospital
- Bob Langer, MIT bioengineer and serial entrepreneur (see "Spotlight: Bob Langer—Personification of Kendall Square's Secret Sauce")
- Flagship Pioneering/Noubar Afeyan (see "Spotlight: Flagship Pioneering—Kendall Square Company Creator)
- Derrick Rossi , Harvard's Immune Disease Institute

Key non-founders and locations:
- Stéphane Bancel. Then CEO of French diagnostics company bioMérieux, whose US headquarters was based in Kendall Square, Bancel served as board chair of one of Flagship's diagnostics startups, BG Medicine. He was walking across the Longfellow Bridge when Afeyan called him to pitch joining the new company. He was named Moderna CEO in late 2011.
- Tim Springer. Harvard immune disease expert and colleague of Rossi's. Springer served on the board of, and was an investor in, Selecta Biosciences, one of Langer's startups. He helped connect Rossi with Langer and Flagship, and became the first individual investor behind Moderna. Springer served on the company's board or as a board observer from its inception until it went public in December 2018.
- Moderna's first official home was at 161 First Street, the same building that had previously housed Lotus Development. From there, it uprooted to 200 Technology Square.

24 CHALLENGES AND REGIONAL ADVANTAGE

Susan Hockfield stared out at a largely blighted area of parking lots and older, low-flung structures. She thought about how the urban eyesore sprawled before her must have been how Kendall Square had looked not too many years earlier.

Except she was in Boston. It was December 2010. Standing beside the MIT president was Boston mayor Thomas Menino. They were visiting Boston's Seaport District, which the mayor was still actively rebranding as the Innovation District. They had ridden an elevator up one of the still under construction towers destined to become Vertex's headquarters. "It was a raw building," says Hockfield. The area was raw as well. But it wouldn't be for long.[1]

Losing Vertex to Boston had been a blow to Cambridge—and the Innovation District's emergence had stirred some angst around MIT and Kendall Square. Then Menino had invited Hockfield on a tour. "I'm not sure why he invited me. Maybe it was an antidote to the fear," she supposes. "Maybe he wanted MIT to play some role. That was never very clear. But he said something that just has stayed with me. He said, 'You know, Silicon Valley isn't one place. It's a lot of separate places connected by terrible highways.' And he said, 'Our region should have many different innovation districts.'"

"I bought into it instantly," Hockfield recounts, ticking off nearby areas with their own innovation initiatives—from the Seaport to Harvard's emerging complex in Allston to a growing number of suburbs pushing their advantages of price and space. "Let's hope they all flourish," she says. "We've got a transportation challenge that we've got to solve, but if people can actually move around, it's even better than Silicon Valley."

That sentiment underscores an important point about Kendall Square in the 2020s: in the decade following the debut of Boston's Innovation District, the tapestry of science, high technology, and innovation that made the square a posterchild for innovation extended anew well beyond its tight geographic boundaries. Delegations from far and wide come to grok the secrets of this unique hub. But their quest doesn't only take them to Kendall Square. A group might look in on MIT's "tough tech" incubator, The Engine, in Central Square. Many segue out to Harvard's Innovation Labs in Allston, next to the business school. They ponder places like MassChallenge in the Innovation District, or become flies on the wall at Greentown Labs in Somerville, home to a bevy of climate-tech startups. A bit farther out, they might explore labs at Tufts University or the Entrepreneurship for All (EforAll) program at UMass Lowell. And there are still many companies and labs to visit along Route 128 and beyond. In short, Menino's vision of multiple innovation districts has been at least partially realized.

The nexus of this increasingly intertwined hub remains Kendall Square. The concentrated excellence of the people, research institutes, and companies inside its tiny borders is unsurpassed. Yet the square faces formidable challenges. Most of these have been discussed, some in depth: corporatization, rising prices that put the squeeze on startups, a lack of diversity, a dearth of ground-floor retail. Add to that list a shortage of housing, especially affordable housing, and inadequate transportation infrastructure.

Steps to address all these issues were underway in early 2020, some farther along than others. Then the coronavirus hit the Boston area with a vengeance. Virtually everything was put on hold for months. By that fall, though, most efforts were ratcheting back into gear, although ground-floor retail had taken a major hit, with several restaurants and bars already forced to shutter their doors for good.

Even setting COVID aside, none of these challenges was unique to Kendall Square. Diversity is an important issue everywhere, and strategies to address it have near universal application. Transportation is similarly a widespread problem. But here the solutions are likely more unique—such as a proposal to create a new commuter line that would open another pathway into and out of the clogged-up square.

Whether the issue involved a broad challenge like diversity or a narrower one like the envisioned train line, many sensed Kendall Square had a unique opportunity to lead in finding solutions. Nearly ten years after her tour with Menino, in late August 2020, Hockfield was the closing speaker at MassBio's State of the Possible conference—held virtually. She titled her keynote "Building Regional Advantage," harkening to AnnaLee Saxenian's book *Regional Advantage*. "Our regional advantage is extraordinary," she told attendees. "But we have the opportunity to increase that regional advantage."

*   *   *

On February 26 and 27, 2020, Biogen held an offsite leadership meeting at the Boston Marriott Long Wharf hotel. Some 175 managers from around the world attended the annual get-together. As the *New York Times* later reported: "Colleagues who hadn't seen each other in a year shook hands and vied for face time with bosses. Europeans gave customary kisses on both cheeks."[2]

Biogen soon emerged as one of the first coronavirus super-spreaders, and the story of its meeting made international news. But it wasn't until the following December, when a study led by Massachusetts General Hospital and the Broad Institute was published in *Science*, that the full impact of that event came into view. The study, which included more than fifty scientists from an array of institutions, estimated the Biogen event had likely led to between 205,000 and 300,000 infections worldwide by November 1.[3]

Whatever the exact number, and not only because of the Biogen event, Massachusetts emerged as an early US hotspot. As happened pretty much everywhere around Boston, virtually everything in Kendall Square soon shuttered. But there were also significant differences. Despite the widespread shutdown, COVID provided a call to action for many in the square. A number of efforts were either already underway or quickly geared up to bring science to bear in fighting COVID directly. By far the most publicized was Moderna's race to create a vaccine: the effort had begun early in the year and entered Phase 3 trials late that July—on its way to FDA approval before year-end. In late March, the Broad Institute converted its data center to handle COVID testing. Numerous other efforts to understand or fight the virus arose at corporate, research institute, and university labs.

Outward signs of such activity, though, were hard to see. The state allowed non-essential labs to open in the last week of June 2020. Still, relatively few people lived in the square, and only a subset of people commuted to work. When I walked the streets on July 1, the few restaurants open served only a handful of customers, and a number had already gone out of business—for lease signs or hand-scrawled notes in windows proclaiming their status.

More than three months later, in mid-October, the pace had noticeably picked up. More people were out and about. Restaurants seemed more alive, tables set up on sidewalks. But offices remained mostly empty, and overall Kendall Square still exhibited only a fraction of the hustle and bustle it should have offered.

None of that was very different from other cities. But while the general expectation was that many Boston area companies would be forced to abandon costly downtown office space in the struggle to rebound financially, the overriding sentiment was that such an exodus would not happen in Kendall Square. "What's fascinating about COVID is people have learned how to work in very different spaces. And so part of what I think is going to happen in Boston is you're going to have huge amounts of what was crazy expensive office space that they're going to be looking at desperately to figure out how to fill," says Juan Enriquez of Excel Venture Management, a life sciences venture firm based in Boston's Back Bay. "It doesn't work for biotech because you need very specific lab layouts and stuff. [Kendall Square] is so biotech-based, there aren't that many potentially empty office buildings."[4]

It isn't just about lab space, either, adds Tom Andrews, then copresident of Alexandria Real Estate Equities. "The whole premise of Kendall Square is that proximity and density breeds collaboration and entrepreneurial behavior," he says.[5]

There are also strong financial arguments for why Kendall Square will likely fare better than many other places. Biotechs are typically financed to operate in the red for years while moving a drug into development: in many cases, the pandemic meant little or no change in their development timelines. Meanwhile, pharma's profits remained strong. "Nine of the 10 biggest profit margins recorded as of July 31 belonged to drug companies," Axios reported in 2020, mid-pandemic.[6]

It was pretty much the same story for big tech firms operating in Kendall Square. Alphabet (Google's parent), Microsoft, Facebook, Apple, and Amazon remained in solid financial shape, as all saw revenues and profits grow, if not soar, in 2020. They continued strong in early to mid-2021.[7]

The outlook for startups was hazier. The Cambridge Innovation Center added flexible membership plans for individuals or companies who preferred to work remotely most of the time, but wanted in-person space a day or two a week or a place for team meetings. More clients began trickling back during fall 2020, keyed to schools reopening, says CIC founder and CEO Tim Rowe. Early in 2021, he expressed optimism about the longer-term outlook. "Clearly, with vaccines beginning to be given out, we think the notion of return to communal workspaces will be back. Slowly initially, and eventually very strong once most folks who want to be are vaccinated."[8]

Things seemed better for biotech startups. LabCentral, which Johannes Fruehauf and Peter Parker cofounded with Rowe, was proceeding full-speed ahead with plans to expand to a facility being developed as part of the Kendall Square Initiative.

"We did not see the impact that we had expected or that has been experienced in almost every other industry," reported chief operating officer Maggie O'Toole. "We lost less than five resident companies out of sixty." As of fall 2020, LabCentral was seeing no decline in new company applications. "In fact," O'Toole noted, "it seems like more than usual with all the attention to, and appreciation of, biotech research right now."[9]

* * *

As far as could be determined, not a single major Kendall Square construction plan was canceled due to COVID. Construction was halted completely in March, but resumed May 18 in the first phase of Governor Charlie Baker's reopening plan. "We were back in about two months," says Steve Marsh, who oversees real estate for MITIMCo, the MIT Investment Management Company. Because projects spanned multiple years, he notes, "clearly that is manageable."[10]

MIT presided over the square's two most ambitious projects, likely involving more than one thousand workers. Farthest along was the Kendall Square Initiative described in chapter 2 that encompassed six new or renovated buildings along Main Street. The first of these, a new residence for graduate students and their families, opened in November 2020. Less than a year behind it was Pi Tower, at 314 Main Street. Anchoring the first three floors will be the new home of the MIT Museum. Other early committed tenants included Boeing, Apple, and Capital One—and a series of additional companies signed leases during COVID. The building was on track to debut in mid-2021, although the MIT Museum was eyeing early 2022 for its public reopening. Two other Kendall Square Initiative buildings, a lab facility (the clock tower building at 238 Main) and an apartment complex on Main Street, were set to open in mid-2022, with the remaining pair likely a few years behind.

All around these buildings, greenspaces and pedestrian walkways were being created to connect the campus to Kendall Square.

The second major project involved the long-maligned Volpe Transportation Center. MIT had worked out a deal with the federal government to redevelop the 14.5-acre site in 2017. Work had begun in June 2019 on the first building, which involved consolidating Department of Transportation operations into a single, 223-foot-tall edifice on

Figure 39 Artist's vision of the new Volpe Center, which was eyeing a late 2023 opening. *Source:* Skidmore, Owings & Merrill LLP.

just over four acres in one corner of the plot, across Binney Street from Biogen's headquarters. Parking, previously sprawled across ground-level parking lots, would mostly go underground, allowing for public walkways that fused into the surrounding neighborhood. This reconstituted Volpe was set to open by late 2023.

Under its deal with the government, MIT will spend up to $750 million to develop the new Volpe Center. Anything left unspent from that amount will be given in cash to Uncle Sam. Once the reimagined Volpe is complete, the government will transfer ownership of the site's remaining ten acres to MIT. Then the institute will embark on transforming the parcel into a modern, mixed-use urban landscape blending some 1,400 apartments, lab and commercial space, ground-floor retail shops, and more. The project will probably take ten or even fifteen years to reach fruition, but the first additional elements should debut a few years after the new Volpe Center is complete. "It'll come probably in waves. We'll try to pace it so that we're doing the housing with the commercial and retail along the way—and we'd like to front-end load as much of the open space as we can," Marsh says.

*   *   *

No one could be certain when or if things would really bounce back. But to the extent that the new normal was something close to pre-COVID life, it meant that the challenges facing the square precoronavirus would still loom over it postpandemic. In 2019, the Kendall Square Association identified three areas where significant improvement was needed: place-making; transportation; and diversity and inclusivity.

Place-making and "activating the streets" were catchphrases for what had been a goal since at least the urban renewal days of the 1960s. Whatever the name, it meant more of a lot of things beyond office and lab space: restaurants, bars, cafes, groceries, pharmacies, banks, parks, community centers, and arts and cultural venues. It also meant more people living there—and not just highly paid biotech and tech execs. In short, it meant making Kendall Square less of a business district and more of a multi-dimensional neighborhood.

All of this was reflected in ongoing plans to develop the square. Somewhere around 450 apartments came online between 2018 and the end of 2020, in addition to the roughly 1,700 planned with the Volpe Center overhaul and the Kendall Square Initiative. Nearly 4.5 acres of parks, bike lanes, pedestrian walkways, and open space had also been approved.

A variety of the planned buildings also included ground-floor retail space. But retail had arguably also taken the biggest hit from COVID. Graffito CEO Jesse Baerkahn had been working to develop retail in the square since 2007. "We came from almost nothing when you think about ground-floor activation and placemaking—to something," he says. But that success, and plans for the future, hinged largely on the square's daytime population and their after-work activities. "That's gone. That's eviscerated right now," he said during the pandemic. "Kendall Square retail requires density, just like any urban neighborhood. But with Kendall, in particular, the retail is largely successful based on the strength, the spending power, and the numbers of the daytime working population. So, you know, it's going to be intense. There are a lot of TBDs."

Additional housing should provide a stronger footing for restaurants, cafes, and other retail, rendering them less dependent on the daytime population. Arts and entertainment offered another area for huge improvement—both to keep employees in the square after work and to attract outsiders to evening and weekend events. A long-standing dream was for Kendall Square to host a world-class performing arts center. A one-acre block had been zoned for this purpose as part of the ten-acre brownfield redevelopment that had enabled the Genzyme Center. The parcel still sat vacant in late 2020, little more than a gravel lot. Developer David Clem had sold the plot to businessman and philanthropist Glenn KnicKrehm, who envisioned a cultural and arts complex he called the Constellation Center. But the challenge of realizing that dream appears to have been too tough. The result, according to BioMed Realty's Bill Kane: "Everything got built around it, and some community members became frustrated."[11]

In late 2018, BioMed bought the parcel for $50.5 million—a heady premium over the roughly $10 million KnicKrehm had paid for it. The following year, the realty company advanced a grand plan that included the performing arts complex, but with a commercial and lab building on top of it to finance its construction. Finally, in December 2020, the Cambridge City Council unanimously approved BioMed's proposal for a triple-tiered tower with a large cultural and arts space, including a theater, as its base.

Figure 40 BioMed Realty's design for a triple-tiered office tower with performing arts space on the lower floors, viewed from the winter skating rink in Termeer Square.
Source: ©2022 BioMed Realty, L.P. All rights reserved.

BioMed hoped to begin construction in spring 2022, targeting a 2025 opening. Kendall Square might finally get some real nightlife.

*   *   *

An arts center, coupled with a rebound in the restaurant and retail scene, could be a potential game-changer, improving the vitality of Kendall Square and realizing the *activate the streets* vision. But that still left a fundamental question: Who is it all for?

Mostly the affluent—and that remains an issue that won't be resolved anytime soon, if ever. It helps that Cambridge now requires any residential development with more than nine units to set aside 20 percent of those units for affordable, or "inclusionary," housing, notes Nancy Ryan, board chair of the Cambridge Residents Alliance. But given Kendall Square prices, Ryan says, "it means that 80 percent—the rest of the housing that is being built—is highly unaffordable to ordinary folks."[12]

The question of affordability extends to startups and small businesses, adds Baerkahn. "Most of the new stuff we're building in Kendall, we're building for big companies. My concern is that the bump factor may still be there, but it's primarily the people at the big corporations that are bumping into each other."[13]

As in many areas, Kendall Square has long experienced tension between real estate developers, big and small businesses, the city, and residents. "The greed factor is well underway here," says one current resident, Bob Simha, the previously mentioned long-time head of MIT planning. He says real estate companies and the city have largely worked to maximize their bottom lines—too often at the expense of residents, who have been largely left to fight their own battles. "Every amenity in this area of any consequence [was] produced by the neighborhood twisting the arm of the developers," Simha asserts.

It's a matter of finding the right balance, says Chuck Hinds, president of the East Cambridge Planning Team, a group representing what is still largely a working-class neighborhood of about ten thousand abutting Kendall Square that historically has been a "first stop" for a variety of ethnic groups. "That's where the balance is—making the neighborhood feel comfortable and happy, and they're making money."[14]

Hinds has worked with the group for more than twenty-five years—including two stints as president. A top priority is stopping the creep of high-rise buildings into the residential area. "If you go to the center of the neighborhood and you look around, these tall buildings are starting to make a wall around the center of East Cambridge," he says. "It gives a lot of people a feeling of claustrophobia."

In 2019, the group applied to have East Cambridge designated a Neighborhood Conservation District by the city. Such a designation would not provide blanket protection but would give residents more firepower to curb further development. The proposal was conditionally approved by the Cambridge Historical Commission late that year, thereby granting the status temporarily, pending a year-long study. That study was extended by a year due to the coronavirus pandemic, and was set to be complete in September 2021.

Additional top priorities for the East Cambridge Planning Team include protecting open space and minimizing other impacts of large-scale development. Before the COVID pandemic, the group successfully blocked a new electric power substation planned for East Cambridge. Members galvanized the neighborhood, enlisted the help of politicians and other citizen groups, and threatened to take the fight to the courts. A compromise was reached in early 2020 that involved building the substation underground in Kendall Square near the Biogen and Akamai headquarters and creating a park on top of it. Boston Properties owned the land, and agreed to build the station and park in exchange for the city allowing it to also develop three buildings on the site—two commercial and one residential.[15]

That project was formally approved by the City Council in February 2021. "There is like an invisible wall now from Binney Street to the neighborhood," says lifelong East Cambridge resident Tim Toomey, who has been a councilor since 1992. In short,

Toomey does not expect any new further Kendall Square encroachment into East Cambridge.

But while there has been tension over development, and rising real estate prices have encouraged many longtime families to sell their homes and move farther out, Toomey is philosophic about the changes. The Broad Institute's COVID testing facility on Charles Street, the street on which he lives, was a Budweiser distribution plant when he was growing up: Toomey remembers the brewer's Clydesdale horses coming each summer to put on a show. "So you're seeing a warehouse distribution center for beer go to testing thousands of swabs every day. To me that is a very positive thing," he says. The same holds true for an array of other developments. "I'm looking at what is coming out of those buildings in terms of lifesaving cures. There's so much more benefiting not only the people who live here but around the country and the world."[16]

For his part, Hinds anticipates that the "next battle" will likely be around the Volpe Center, but points to the substation outcome as a win for all concerned. "You know, people think we're cranky and yell a lot and are very bossy, but what comes out of this is projects are better," he says.

*   *   *

One thing that wasn't getting better was the Kendall Square commute. As the Kendall Square Association's oft-cited Medium post lamented: "You can't find the cure for cancer while sitting in traffic."

The group had taken a series of steps to call attention to the square's transportation challenge. Some forty Kendall Square CEOs signed a letter to the governor and top Massachusetts State House and Senate leaders demanding action on the "transportation crisis." In 2019, the KSA signed up twenty Kendall Square companies for an eighteen-month pilot program to test traffic-reduction ideas and collect data on traffic and commuting. "We need to get in there and make sure this doesn't cripple us," KSA president C. A. Webb asserts.[17]

The top priority: the subway. The Kendall/MIT station on the MBTA's Red Line is the linchpin of Kendall Square's commute—handling roughly one-third of everyone working or visiting the square. In 2016, the latest full figures available, average ridership during the 8:00 a.m. weekday commute was put at twenty-four thousand passengers—on a system with a maximum design capacity of twenty thousand. Ridership was expected to double between 2012 and 2040, according to *Transport Kendall*, a 2019 report put out in partnership by the Kendall Square Association, the City of Cambridge, and the Cambridge Redevelopment Authority. On top of all that, Red Line service was notoriously unreliable.

A fleet of new cars—the most modern ones dated to 1994 and some reportedly went all the way back to 1969—was supposed to begin coming into service in 2019. But delays hit even before the pandemic, which only made things worse. It apparently wasn't until December 2020 that the first new car was delivered to the Red Line, and the fleet's full delivery was pushed to winter 2024, a year behind schedule. Whenever the order is finally completed, the modernized fleet is expected to have the capacity to handle up to thirty-one thousand passengers per hour. However, *Transport Kendall* warned, "With the current demand and additional growth expected in Kendall Square, overcrowding could continue to be a challenge in the future even with the design capacity increase."[18]

Another piece of the plan, and to many the most important for the future, hinged on creating an entirely new transit line. This was a proposal to install a modern rail system along the old Grand Junction tracks that had served as a freight connection since 1868, passing through the square and paralleling the MIT campus on their path between North Station in Boston and a planned West Station near Boston University in Allston.

An estimated 42 percent of all jobs in Cambridge and a third of all residents—roughly forty-nine thousand jobs and thirty-three thousand Cantabrigians—lie within a half mile of this route. Trains on the line, which is still in sporadic operation, had previously transported circus elephants to shows in Boston. Turning this into a modern commuter line would provide direct access to the square for people coming from Allston and the western suburbs on the one hand and those streaming into North Station on the other. Located near the TD Garden arena, North Station is a major hub serving four commuter rail lines and Amtrak, with an adjacent subway station for the Green and Orange Lines. "If you connect North Station to Kendall Square, it changes everything," asserts Bryan Koop, an executive vice president and head of the Boston region for Boston Properties, who calls the envisioned line the "brain train."

Grand plans were already in the works around the Grand Junction—including a pathway for cyclists, joggers, skateboarders, pedestrians, and others alongside a section of tracks running from the Boston University Bridge through Kendall Square and into Somerville. A small section of the envisioned fourteen-foot-wide pathway, with newly planted trees, rose gardens, and seating areas, opened in June 2016—opposite Akamai's headquarters and the Whitehead Institute. Full design of the pathway was scheduled for completion by the end of 2020, with construction set to begin on other stretches in 2021.

The effects of the COVID-19 pandemic on all this were unclear in late 2020—and no commitments had ever been made for the rail service.[19] In October 2020, worried the issue had been forsaken in the wake of the pandemic, C. A. Webb from the Kendall Square Association, then MassBio CEO Bob Coughlin, and Tim Murray, CEO of the Worcester Regional Chamber of Commerce, penned an article advocating for the Grand Junction transit line.

"We have the opportunity to lay the track for rail service that would connect, for the first time, our growing innovation economies in Worcester, Metro-West, Harvard's Allston development, Kendall Square, and North Station," the trio argued. "We could transform downtown Worcester into a restaurant and retail hub, open up jobs to the east and west for commuters in Worcester or Boston (and all the towns in between), and usher folks into Kendall's life sciences and technology epicenter."[20]

* * *

The idea that regions could benefit from more tightly connecting innovation districts—through transportation infrastructure, public and private initiatives, and other means—was nothing new. The MIT Regional Entrepreneurship Acceleration Program (MIT REAP) makes this point—and tries to help teach people, companies, and governments how to foster such a climate (see chapter 27).

This "rising tide lifts all innovation boats" approach was also what Mayor Menino had pitched to Susan Hockfield in 2010. In her talk a decade later at MassBio's conference, Hockfield pointed to two areas where she feels Boston holds an edge and could strengthen its innovation cluster even more—one in human capital, the other in its science and technology capabilities. "Neither of these opportunities is unique to Kendall Square," she explains. "But we have elements that make it easier for us."[21]

When it comes to attracting talent, Hockfield and many others point to Massachusetts's legions of top colleges and universities as a unique recruiting ground. Indeed, education technology company Plexuss identified 118 institutions of higher education within ten miles of Boston—and in US News & World Report's 2020 rankings of colleges and universities, Massachusetts was home to seven of the country's top fifty universities and six of the top fifty liberal arts colleges.[22] These provide an almost unrivaled pool of students—and potentially entrepreneurial faculty. "If we're not using those resources to foster our innovation economy, that's just bad on us," Hockfield asserts.

Boston's other big advantage lies in science and technology—more specifically, in the convergence of biology and engineering, Hockfield says. Bringing these disciplines more closely together to fight cancer is the premise behind MIT's Koch Institute, which Hockfield helped spearhead as MIT's president and where she is now based. The future power of this convergence is also central to her 2019 book, *The Age of Living Machines: How Biology Will Build the Next Technology Revolution*. "It's not enough just to have a cool scientific insight. You need really great engineering to turn these biological insights into marketplace products," she told me. "And one of the things that often isn't sufficiently described for the biotech revolution in Kendall Square is that the engineers were a critical part of it. The engagement of both biologists and engineers, really from the beginning and continuing today, is why we have a head start—because biotech is intrinsically multidisciplinary."

So what *is* next? The last few chapters of this book contain insights from business, university, and civic leaders about the challenges and pathways ahead for Kendall Square. Chapter 25 is a compilation of short quotes and predictions from twenty-six of these leaders. Chapter 28 offers more about the convergence of key fields like biology, engineering, and artificial intelligence—as well as a look at other emerging fields that may hold the keys to the square's future.

Hockfield, for one, believes the confluence of biology and engineering is poised to deliver breakthroughs akin to what happened through digital computing. "If we think about the technologies that transformed the twentieth century the most profoundly, it's the digital computer technologies that just make our lives so different," she says. "And all of these fabulous information technologies were made possible because physicists around 1900 decoded the parts list of the physical world. They discovered subatomic particles—electrons, protons, neutrons—as well as X-rays and other forces that connect and organize the components of the parts list of physics. With the electron in hand, the electronics industry was born and then its successors, the computer and information industries. These new technologies and industries were dramatically accelerated by World War II, laying the ground work for the computer revolution."

"But truth be told, biology was far behind; before around 1950, biology had not yet revealed its parts list," Hockfield continues. "We didn't know what the components were. Biologists were describing the *behavior* of biological organisms—they didn't know what the fundamental controlling parts were. But the revolution in molecular biology, beginning in the late 1940s and continuing today, revealed biology's parts list: it led to the discovery of DNA and RNA and the understanding that these molecules carried the information to build proteins, the workhorses of biology. Engineers are now using this parts list to build technologies that to my mind will be the miracle technologies of the twenty-first century."

* * *

If history is any guide, Kendall Square should not count on biotech lasting forever—at least in its current form. "The pendulum shifts every twenty years or so. It will probably shift again," says C. A. Webb. "Will that return Kendall to being kind of the destitute, dusty land of old? I don't think so. But how do we future proof Kendall? What if biotech went bust one day? I mean, we all thought Polaroid was going to be here forever. So how do we send the right signals to make sure that we keep attracting the next generation of whatever technology is? Whatever innovation means in that era, how do we make sure they keep finding their way and being born in Kendall?"

Answering that question is a work in progress. But part of the answer lies in what Webb sees as the essence of the square. "The story of Kendall is not just biotech. That becomes an easy way to understand it; it's what our minds like to do. We like to take a lot of data and cull it down to a soundbite. But the story of Kendall is collaboration. So if you're a business that requires active partnerships and active collaborations in which to thrive, if you are a business which requires steady acquisitions in which to thrive, Kendall Square is one of the key places in the planet that you want to be. Because all you have to do is stumble out your door, and it's all there."

25 VOICES OF THE SQUARE

Where is Kendall Square going—for better and worse? What is its magic power? Its Achilles' heel? What remains to be proved? What advances does the future hold? The square is teeming with amazing people—business people, scientists, entrepreneurs, residents, activists, politicians—whose illuminative (and sometimes amusing) insights, predictions, and cautions proved difficult to fit into a narrative history. Gathered in the following pages are some thoughts that stood out. All but one are derived from interviews for this book (see List of Interviews). The exception is an email from Helen Greiner. Some comments are about what Kendall Square needs; others hold predictions about the future of science, technology, and medicine. Many are from people not quoted elsewhere in the book. And some are from observers outside the square, looking in.

Noubar Afeyan, Founder and CEO, Flagship Pioneering

"The key is rejuvenation. So what is the rejuvenating force? Well, startups are rejuvenating, and universities are rejuvenating—because they're constantly attracting new talent, new people. That regenerative capacity of an ecosystem, which has been missed in my view by people, is a key, key component of what drives it. That is there in spades in Kendall Square. It's self-fulfilling. Startups will do what they do. Universities will do what they do. They'll pump talent out, they'll pump out scientific expertise, scientific results. So I don't see Kendall Square losing out to something else for any reason. It doesn't mean that other places can't exist or can't get there faster. But I don't think it matters. As long as it's vibrant, it's fine."

Bill Aulet, Managing Director, Martin Trust Center for MIT Entrepreneurship; Professor of the Practice, MIT Sloan School of Management

"It's head spinning how much Kendall Square has changed. When I graduated in 1980, you didn't go down to Kendall Square—it was dangerous. Then when I came back to Boston in the late 1980s, it was an interesting, curious place. Today, it's like the Dubai of Massachusetts, with buildings going up everywhere, and it feels like it could be overbuilt. Who knows what tomorrow will bring?"

David Baltimore, Founding Director, Whitehead Institute; Nobel Laureate; Former President, Rockefeller University and Caltech

"A sort of entrepreneurial spirit developed around Kendall Square, and having old buildings was helpful because you can renovate them relatively cheaply and get space to start up things. There's a real issue of whether biotech is moving away from its traditional centers in Cambridge and the Bay Area because it's been so successful that it's limited space and driven up the cost of doing business and driven up living costs."

Sangeeta Bhatia, Serial Entrepreneur; Professor of Health Sciences and Technology and Electrical Engineering and Computer Science, MIT

"To me, there is magic in Kendall Square—the density of ideas and people, the energy, the chance passing on the street/restaurant/elevator, the mix of trainees and luminaries, of biology and engineering, of tech and medicine, it's just special. I started my faculty career in California—which has its own charm—but eventually I realized that when you train in Cambridge, it creates a sort of addiction in your brain that can never be truly satisfied anywhere else in the world. Happily the food and coffee is a lot better than it was in the nineties!"

Joshua Boger, Cofounder and Former CEO, Vertex Pharmaceuticals

"If one wants to stay ahead in life sciences, you have to look at, 'What's the next transformative technology? What's the next tool? What's the next reagent for innovation?' I think if you ask that now, you have to look toward the rapidly advancing field of artificial intelligence, but not like it is being considered by Google and computer companies. Boston will continue to be patient-oriented and medicine-oriented and have a life sciences edge over Silicon Valley. They are looking for answers to problems like self-driving cars—fundamentally a simple problem—that merely take money and engineering. But in human disease—think of Alzheimer's or other major diseases—you need to marry technology with creative medical insights and a systems biology understanding of complex processes. Problems like Alzheimer's are tough, and better suited to the Boston area."

Joe Chung, Cofounder and CEO, Redstar Ventures; Kinto Care; Cofounder and Former Chairman and CTO, Art Technology Group

"It always seems like one of the most important things about doing a startup is you see other people do it, and you're like, 'Fuck, I can do that. Those bozos? Hey, if they can do it, I can do it.' I think that's really important. It's that kind of proof of possibility."

"It's easy to be really skeptical of stuff, and, you know, a lot of the time you're right—because most startups fail. So you've already got the odds on your side if you just poopoo everything. You look pretty smart."

Jim Collins, Serial Entrepreneur, Professor of Medical Engineering and Science, MIT; MacArthur "Genius" Award Winner

"I moved from Boston University to MIT in 2014, in part to become a member of the translation culture at MIT, which is focused on commercializing discoveries and inventions from academia so as to maximize impact. What I found was that moving eight thousand feet from Kenmore Square to Kendall Square was a trip bringing me into a completely different world, a world where biotech and translation was central to the activity. It's an amazing community that is dense with interactions from academia to business folks to clinical folks at a level nobody could expect or anticipate unless you actually experienced it."

Stephanie Couch, Executive Director, Lemeleson-MIT Program

"My vision, my passion, is that we think about a creative corridor. You get off at the Green Line, and you walk to the Red Line. You go down First Street, then up to Third Street, and you're hitting different creative spaces that we can program for STEM and the arts, with different specialties. And if we think of it with one comprehensive vision, we have something we don't have now. If we did this, we would be unlike any other area."

Bob Coughlin, Former CEO, MassBio

"Our research has shown that the breaking point—not the tipping point, the breaking point—for our industry, and what will limit us from being able to continue to grow, is: public transportation, infrastructure, congestion, workforce housing, and workforce development. So what do we need to continue to be the best? We need to focus on those issues so that we're prepared for the future. We have to change our policy platform. Instead of advocating for money for tax incentives, we have to start talking about more money for public transportation, more money for infrastructure, and how we can create more of a housing inventory to bring the costs down. As an industry, we haven't done a good enough job on equality, diversity, and inclusion. Why do we need to do that? One, it's the right thing to do. But two, it's a good business decision. We need to continue to recruit, attract, and retain the best and brightest, no matter where they come from or what color they are or what sexual orientation they are. The more diverse our workforce, the better chance we have of staying on top."

Deborah Dunsire, CEO, Lundbeck; Former CEO, Millennium Pharmaceuticals, Forum Pharmaceuticals, XTuit Pharmaceuticals

"Tech is going to come into biotech in a way that I don't think we've really grasped yet. Things like digital therapeutics will bring a different set of players into the innovation world. The big unknown is how will the US healthcare system and healthcare payment model evolve, and what potentially chilling effects could happen."

Johannes Fruehauf, Cofounder and President, LabCentral; Founder and General Partner, Mission BioCapital

"I think a big risk to our ecosystem in general in Boston right now, for life sciences and pharma, is this whole discussion about pricing of drugs. It's a big societal pain point, and it's a very easy and populist target for politicians on both sides of the political spectrum. I am always urging other leaders in the industry to be very conscious of the signal that we project to the greater population about what the industry does, how it behaves, and why there's a national and societal interest to keep a vibrant pharmaceutical innovation system going. I think that is the biggest structural risk that we're facing right now. If there were to be draconian pricing controls enacted in order to achieve certain populist short-term goals, it could do real damage in the long term."

Helen Greiner, CEO, Tertill; Founder and CEO, CyPhy Works; Cofounder and Former President and Chairman, iRobot

"Not in Kendall:(too expensive." (From an email about where her newest company would be based.)

Anne Heatherington, Senior Vice President and Head of Data Sciences Institute, Takeda

"To benefit from Kendall Square, you have to leave your office and get out into it. The commute to the Square can be awful. And then many people work so hard, it is easy to do your work, eat lunch at your desk, and head home. You could be working anywhere. It wasn't until I joined a small biotech where collaboration outside the company was essential—so-called competitors, data consortia, potential hires, others seeking to learn—that I realized the true power of Kendall Square. It is all there. You have many points of intersection. The proximity to peers, to some of the startups, and then the proximity to the academics as well makes a big difference. But you have to embrace it and get out into it."

Penny Heaton, Global Therapeutic Area Head, Vaccines, Janssen/Johnson & Johnson

"COVID accelerated progression of technologies in a way that would have taken years to do otherwise. We got from a gene sequence to Phase 1 data being published in the *New England Journal of Medicine* in six months. That usually takes two to five years. Now we're going from Phase 1 to Phase 3, potentially in another six months. All that usually takes somewhere between seven to nine years. Some of the things that we've done to speed up development could just become part of the routine. I think we will understand a lot of these technologies in ways that we hadn't before and can apply them more broadly.

"And then on the clinical trial front, we're doing things differently. We're doing 60 percent of the enrollment completely remotely. We have collaborations with testing centers like Walgreens, Walmart, CVS, and other big labs. So if someone comes in to get tested, they can be invited to be in the study. Then if they turn out to be positive, they can be enrolled in any of the fifty states—wherever they are. Consenting is done online. If they're eligible and they consent to be in the study, then we ship what we call a clinic in the box to their home. It has their medications. It has their O2 sat [oxygen saturation] monitor, their thermometer, their swabs for testing, and everything. The study nurse and physicians talk to them on the phone every day to monitor their progress. If needed, a home health nurse can go out, or they have other physicians they can refer them to in the area. So it's a really different way of doing clinical trials. If it works, it could change how we do clinical trials forever."

Danny Hillis, Cofounder, Thinking Machines; Founder, Applied Invention

"I expect that the next big thing is going to be the entanglement of engineering and biology—engineering living things, and using biologically inspired methods to engineer machines. Kendall Square is a natural nexus for that."

Eric Lander, President and Founding Director, Broad Institute of MIT and Harvard; Science Advisor to President Biden and Director, White House Office of Science and Technology Policy

"There are various stages of denial. The first one is institutions say, 'Well, we don't need to do this—we don't really need it.' The second is, 'We're already doing it.' Then the third is, 'All right. all right, but we can do it alone.'"

"You have to have a vision coupled to—there's no other word for it—the arrogance that you can kind of somehow pull it off, because if you get too realistic, you can talk yourself out of anything. Realism is good in appropriate doses, but so is the willingness to think audaciously of what could happen if we all work together. And you can't do it alone."

Mark Levin, Cofounder, Third Rock Ventures, Millennium Pharmaceuticals

"It's an extraordinary time to think about how things fit together. Pattern recognition. This stuff's all coming together—CRISPR, gene therapy, Car-T cells, AI. So that explosion for the future, I'm really excited about. The biggest challenge I think we're all worried about is making sure that we can supply all these great drugs not just to a few people—that everybody will have access. If it costs $200,000 a drug a year, it's just not going to work. So how do we make these affordable, and justifiable, with return for everybody—and make sure it doesn't get over-legislated? We don't seem to have the will politically in Washington to get together and figure this out, as opposed to making it a political hot button. What about sitting down and solving it together? We're just not doing that yet. I'm worried about that."

Travis McCready, Executive Director and National Practice Leader, JLL; Former President and CEO, Massachusetts Life Sciences Center; Former Executive Director, Kendall Square Association

"On one street corner, you're standing in front of one of the most dynamic, exciting, world-changing, scientific, innovative companies that is funded with hundreds of millions of dollars of venture capital. And literally you go across the street, and you're standing in front of public housing. There's something just so deeply American about that, which is, you know, both a shame and a missed opportunity. So I guess that's the next evolution of Kendall—to figure this out."

Lee McGuire, Chair of the Board, Kendall Square Association; Chief Communications Officer, Broad Institute of MIT and Harvard

"Our proximity, both the physical proximity and the intellectual proximity, allows us to really close the gap between early-stage, curiosity-driven science and the therapeutic development that leads to therapies. COVID is showing us again how important that is. The challenge with covid, and any pandemic, is that proximity is part of the problem. Because we are such a physically small community, a lot of what happens in Kendall Square is driven by person-to-person interaction. Those relationships are built up over time but are essential when many are remote—because it's really hard to forge a new relationship in a mostly remote environment. So I think what we are seeing now is that a lot of innovation, rapid iteration, and essential evolution of facing down the pandemic is happening because of relationships built over many, many years among people in Kendall Square, and also people who used to be in Kendall Square and now are somewhere else."

"The question moving forward is once we get past this crisis, how do we make sure that we are able to resume and deepen that relationship-building? We're still not reflective of the full diversity, not just of science but of the world, in Kendall Square. I think that's the next big challenge. If we just go back to the thing that has powered Kendall Square for the last twenty or thirty years, that will only take us so far."

Fiona Murray, Professor of Entrepreneurship and Associate Dean for Innovation and Inclusion, MIT Sloan School of Management

"We used to take being together for granted. One of the things that has changed [due to COVID], and will probably continue to be different for a while, is that we're going to think about time together, in proximity with one another, as a scarce resource, much like money or any other scarce resource—as opposed to being something that is kind of infinite. What that means is that there may be more deliberateness around our time together. But there's this weird tension. It'll have to be more intentional, but if it's intentional, then how do you get all that serendipity that we thought was important?

"I suspect that the people who consider themselves innovators will want to be co-located again. Some stuff actually requires us to be in a physical location, like lab work. But more interesting is the work that isn't like that—the early stage of innovation where we're brainstorming. We're still living on all sorts of social capital and the social fabric of relationships we already have. Another obvious question is how do we bring new people into our network? Especially people who are different than us. And then how do you actually do stuff that requires more of that ideation? I've yet to see that working really well online."

Katie Rae, CEO and Managing Partner, The Engine

"Just think about how much has happened and what's been built in the last ten years, and then you think about the next ten years. From a company-building perspective, and a collisions perspective, and an MIT perspective, Kendall Square is one of the most exciting places to be in the world. Honestly. From the city planning perspective, and an infrastructure perspective, we were tempting fate a bit to think we are going to have all those people coming in along the often-struggling Red Line. We have to change how we get to work; it's going to have to morph. We have to be more bicycle-friendly and all that. Also culture and food. It's less sterile now than it was ten years ago, but we still have a ways to go to make it a truly vibrant place where humans want to be after 7:00 p.m. This means businesses that serve all our needs—personal and business—as well as all the various demographics that work here. I mean, I know that I'd welcome things like a close-by nail salon to pop over to in Kendall Square."

Bill Sahlman, Professor Emeritus, Harvard Business School

"I believe Kendall Square and Allston will thrive. There is more than enough human and financial capital in Boston to support two closely linked areas, particularly if they can address the transportation issue between Harvard Square, Kendall Square, Allston, and Longwood. We have more space over here than they have in Cambridge. If they straighten out the Mass Turnpike, we're going to have almost one hundred acres of commercialize-able space next to an applied engineering and science school, next to Harvard Business School, Boston University, and Northeastern, and across the river from MIT. It takes a community of different kinds of people to make these places work—and instead of paying ninety-two bucks a square foot in Kendall Square, my guess is you'll pay fifty bucks a square foot here. So there's going to be a nexus here. The center of Harvard becomes Harvard Business School effectively. And my guess is they'll figure out clever ways to move people from Kendall Square to Allston."

Phil Sharp, Institute Professor, MIT; Nobel Laureate; Cofounder, Biogen, Alnylam

"Alzheimer's is one of the most debilitating, discouraging, thieves of humanity that I know. We have today not had a single treatment that changes the course of this disease. It occurs in the brain, which makes it extraordinarily difficult to study because you just can't open a brain and look at it and then put it back. But with the tools of biotechnology, it is probable, and likely, that we will have treatments that will slow the progression of this disease.

"I'm optimistic, and I think this will happen in Cambridge. Beyond this is Parkinson's, which is equally debilitating, and then a whole host of diseases like ALS, and others. You go back fifty years, we thought cancer was something we could only treat with radiation and poison. Now we have a large canopy in our medicine cabinet of treatments that add additional quality years to most cancer patients' lives, except for a few rare cancers. I think we're at a stage where we're going to see the same progression in these degenerative chronic diseases. Can we cure them? Maybe so, maybe not. Can we control them and give you quality years? I think we're going to see that. And that's important. Recognizing your wife when you're ninety and then enjoying your grandkids—that's important."

Sophie Vandebroek, Former Chief Technology Officer, Xerox; Former Chief Operating Officer, IBM Research

"It's the convergence. AI is going to be everywhere. It's the new electricity, as people have been calling it. So a closer collaboration between life sciences and the hospitals, the tech firms, as well as universities, to really leverage AI for good, right—make sure that AI can indeed positively make an impact. It's robotics. It's security. It's healthcare. Creating that whole future by having these different areas collaborate even more closely could be something that Kendall Square can really make a difference in."

"From where I live west of Cambridge, if there was no traffic, I should be able to get to Kendall Square in twenty minutes. But unless we left before 7:00 a.m., it would take an hour. If I compare this to Tech City, it's like the Kendall Square of London, it's a whole different feeling because it has fast trains, including a train that comes all the way from the mainland. There's a nice modern station. And then you exit, and everything is walking. There are no cars, but all these companies are there. There are big plazas with restaurants and people getting together. It's really nice to be in that space, because it's optimized for pedestrians. You don't have that feeling in Kendall Square. It's very different. So Kendall Square could benefit from having more areas where no cars can go. Then of course it needs to be complimented with good public transportation."

C. A. Webb, President, Kendall Square Association

"Kendall Square just has such a different sensibility than other dense business districts. You think about Midtown Manhattan or the financial district in London, or a lot of places where there's a business case that drives companies to choose to plant themselves there. There's a feeling of competition, to the extent that the businesses are curious about what's happening to their left and right—but there's really no feeling of collectivism. In Kendall Square, there is this really wonderful sense of collectivism because of the magic that has been created over the last fifteen to twenty years as it's become the dominant center of biotech. It has become the home of most of the big pharma companies. And it has become a definitive tech center in its own right. This place has generated its own identity, sense of community, sense of connectedness, and sense of collectivism that other places have not."

"Everyone came here to be next to MIT and to be within arm's reach of one another. Every company in Kendall has some sort of research partnership, some type of active collaboration, with someone else in Kendall. That's why you're there. But you don't necessarily know what the next partnership needs to be. And so there's a business case to be made for creating a social platform, a social network, so that when you need them, you have them."

Susan Whitehead, Life Member, MIT Corporation; Vice Chairman, Whitehead Institute for Biomedical Research

"One of the things that has really struck me in the biological revolution is that there were a bunch of remarkable leaders—Salvador Luria in particular. So my question in anxious moments is, 'Who's the next Salvador Luria?' I don't know who the successor is. I've been thinking about the difference between scientists and engineers. Engineers really solve problems. Scientists love asking questions—it's just fundamentally different. There are lots of people who are great questioners, but that is different from being a great leader. I really want some leaders to emerge in the life sciences. I don't know who the next one is—and it really matters, particularly if Kendall Square is to remain as Kendall Square."

26 ELEVEN DECISIONS THAT SHAPED KENDALL SQUARE

When considering how Kendall Square became a world-renowned innovation ecosystem, the first thought on virtually everyone's mind is MIT. As investor-cum–innovation commentator Juan Enriquez says, "Without MIT, there would not be a Kendall Square, or there would be a lot different Kendall Square. It's almost impossible to think of how you'd build that somewhere else in the world without the anchor of one of the world's great universities."[1]

MIT's 1912 decision to purchase land in Cambridge along the Charlies River and move from its original Boston home has to go down as the most important decision in Kendall Square history. That decision, in turn, only came about because of a critical court ruling. In June 1905, MIT's board—the Technology Corporation—approved a merger with Harvard that would have it move to Allston, near today's business school, to become the backbone of Harvard's applied science and engineering programs. MIT planned to pay for its share of the merger by selling its Back Bay land. However, the Massachusetts Supreme Judicial Court ruled that because MIT had acquired the property through federal land grant funding, it did not have the right to sell it—and the merger was abandoned.[2] Had that ruling gone the other way, MIT would almost certainly have gone through with the union.

What would have happened if that been the case? Would MIT have been subsumed by the culture of its intellectual predecessor—as many MIT alumni and faculty feared? Or would it have maintained and enhanced its identity and culture for innovation built on applied science and technology? Would Allston have become, in essence, the Boston region's Kendall Square?

Unless some MIT whiz kid of the future invents a time/dimension machine that can explore alternate realities, we will never know. But if I had to make a bet, it would be that MIT's growth would have been stunted—making it less than what it is today. It's extremely difficult to see how both cultures could have thrived in the same institution. The separation made a huge difference—for Kendall Square, yes, but also for science and innovation globally. Two world-leading, competing centers of excellence so close together is a gift that no place else in the world has enjoyed.

But setting aside the two trailblazing events just mentioned, several other key decisions since World War II have shaped and even altered the square's trajectory—for better or worse. Ahead, in chronological order, are nine others that seem the most important and influential. Not all are equally weighted, some are far more speculative than others—and some were decisions *not* to do something. And, of course, many other important decisions were made that are not included here.

Radiation Laboratory: The creation of a major radar research lab at MIT during World War II—the Rad Lab—had a transformational effect on the institute. It also shaped at least a generation of industry around MIT, in the square and along Route 128. (See chapter 8.)

Urban renewal: In 1960, MIT, the City of Cambridge, and real estate developer Cabot, Cabot & Forbes announced plans to revitalize land next to MIT. The subsequent development of Technology Square helped inspired a major urban renewal effort that was central to Kendall Square's evolution. "It is in fact the ignition," asserts Bob Simha, MIT's longtime director of planning.[3] (See chapter 9.)

NASA center: The early 1960s decision to build NASA's Electronics Research Center in Kendall Square dramatically affected the square's development for years—and inspired lasting folklore. The ERC's closure in 1970, less than seven years after it opened, left some eleven acres slated for NASA vacant. What would have happened had NASA stuck to its plans? Would Biogen and the biotech industry that followed it have been forced to find a new center elsewhere in the region? (See chapter 10.)

MIT abandons plans to start a medical school: In the 1960s, MIT was taking steps to form its own medical school, but ultimately decided against the move. Instead, in 1970, it teamed with Harvard University to create the Harvard-MIT Program in Health Sciences and Technology (HST). Students emerge with a medical degree, a PhD, or both. "MIT not having its own medical school and hospital meant it had to work with Harvard," says David Altshuler, a Broad Institute founding member and veteran of the program. "That creates this dynamic that otherwise never have happened. This is what sets the stage."[4]

Recombinant DNA ordinance: Cambridge's February 7, 1977, adoption of the nation's first municipal biosafety ordinance laid out the steps needed to conduct recombinant DNA experiments within city limits. The ordinance became a stabilizing force for the nascent biotech industry. "What it did at a very critical time was it established rules," says former city councilor David Clem, who later became a real estate developer. "That's really why Cambridge became the heart of life science."[5] (See chapter 12.)

MIT revamps its licensing office: In 1985, MIT overhauled its approach to technology licensing. The institute reorganized its Patent, Copyright, and Licensing Office, renaming it the Technology Licensing Office. A main goal was "to become more proactive in moving MIT science and technology into commercialization, and to place greater emphasis on the role of new companies in accomplishing that," notes longtime Sloan School professor Ed Roberts.[6]

Novartis moves its worldwide research headquarters to Cambridge: In May 2002, Novartis stunned the pharmaceutical world with the announcement that it would move its worldwide research nerve center to Cambridge, next to MIT. The move sparked a new era for big pharma, getting it closer to the centers of talent and expertise coming out of universities. Since then, virtually every big drugmaker has opened a significant facility in or near Kendall Square.

No venture capital for Zuckerberg: In 2004, Harvard student Mark Zuckerberg failed to convince Boston venture capitalists to invest in what became Facebook. What would have happened had a few local VCs taken the plunge? Would Kendall Square and the Boston area have become much more consumer- or social media–centric, dampening or muting the biotech influence?[7]

MIT's forms its new AI college: In October 2018, MIT announced it would invest $1 billion toward identifying and addressing the opportunities and pitfalls of artificial intelligence. At the heart of its plan was establishment of the Stephen A. Schwarzman College of Computing, where studies of both the technology and ethics of artificial intelligence would be centered. "The initiative marks the single largest investment in computing and AI by an American academic institution, and will help position the United States to lead the world in preparing for the rapid evolution of computing and AI," the school's announcement proclaimed. The move has the potential to become a transformational decision for Kendall Square.[8] (For more on this, see chapter 28.)

27 LESSONS AND OBSERVATIONS

Bill Aulet, managing director of the Martin Trust Center for MIT Entrepreneurship, has a favorite story of taking part in a Harvard Kennedy School discussion about innovation and the economy. Among his fellow panelists was Massachusetts Governor Charlie Baker. When it came time for audience questions, somebody asked: "How do you build an entrepreneurial ecosystem?"[1]

"Charlie jumped right up before I could answer," Aulet recalls. "He said, 'It's easy.' And I'm thinking, 'What? It's easy?' Without missing a beat, Baker explained: 'You just get Harvard and MIT to set up in your state, and then you wait 150 years.'"

While Baker's answer was tongue-in-cheek, the question is one a multitude of people and governments are clamoring to have answered. Indeed, people come from far and wide to learn about Kendall Square and the Boston innovation ecosystem. Sometimes they reach out informally to various groups, companies, or individuals. But many are also willing to pony up. Each year, up to eight teams pay $300,000 apiece to be part of the MIT Regional Entrepreneurship Acceleration Program. The two-year MIT REAP program provides hardwired rules, workshops, and other assistance to help regional teams replicate some of the magic of Kendall Square and other innovation ecosystems in their home areas. "Every single week, we are hosting a delegation from another country," Fiona Murray, one REAP organizer, told the school's alumni magazine in the precoronavirus days. "They want to know, 'How can we have an ecosystem like this?'"[2]

While writing this book, I encountered many people who shared insights and lessons about Kendall Square's special sauce and what makes innovation hubs successful more generally. And of course, there's a lot in the literature, both historical and contemporary. As might well be expected, there aren't just five steps or even ten steps to building a place like Kendall Square—and getting it right (or mostly right) is an extensive, iterative process that is never finished.

I've tried to distill a multitude of insights from leaders in and around Kendall Square, and a few outside experts, into a manageable view of those that seem most salient. While many have appeared elsewhere in this book, it seems worthwhile to collect them in an easy-to-grasp format. Not everyone agrees with every step, of course. And even when they agree that something is important, they may have different views about *how* important it is. Is a world-class university a must-have, for example? Yes, says *The Rise of the Creative Class* author Richard Florida. Others, like Fiona Murray, say it's important but not essential. I've included both views.

I've arranged things in a sort of hierarchy: three views of the essentials; general principles to keep in mind; additional core ingredients; and policies and programs. Complementing all this is a fun, but insightful, take on Kendall Square from my conversation with Juan Enriquez I've called "A Few Things about MIT That Might Be Transferrable" (box 13). The chapter ends with a short section called networks that posits that a core of collaboration is essential to a system's enduring success. While not all the advice herein will necessarily be replicable, or even relevant to a given ecosystem or situation, it may still yield clues to what might work.

Three Views of the Essentials

"Invention is a flower, innovation is a weed," Ethernet inventor and MIT alum Bob Metcalfe once noted.[3] In short, innovation can be wild and messy. That also means that various innovators and experts might have differing, overlapping, but slightly

Box 13
A Few Things about MIT That Might Be Transferrable

> Although Kendall Square emerged as an economic and innovation hub long before MIT opened its doors in Cambridge in 1916, it's almost impossible to imagine the square today without the institute. MIT's influence is virtually everywhere. But while it isn't possible to 3D-print a top-tier university to anchor an ecosystem, some core elements of what MIT brings to the table might be transferrable. Few people are few more eloquent at summing them up than Juan Enriquez.*
>
> The secret goes way beyond having a great university, he says. "The whole Ivy league is an attempt to British-ize higher education. And MIT was just exactly the opposite. We don't care how the buildings fit together. We just want them to fit together. We don't care how these departments fit together—we don't want any distance between departments. We want to make these hallways narrow instead of spacious. So what that leads to is a series of internal streets that recreate the dynamics of some of the great, interesting, quirky, fun cities in the world. MIT recreated that sense of serendipity, that sense of what's around the next corner. You're constantly pinballing into people from different areas, with different thoughts. And that is just so completely different from an eating club at Princeton."
>
> Kendall Square is an extension or extrapolation of that philosophy in a very small urban area, says Enriquez. "If you tried to get to the businesses near Stanford that are generated by Stanford—good luck walking there. So what's fascinating about this thing [Kendall Square] is how tightly interwoven and interconnected the commercial infrastructure, the startup infrastructure, is to the academic infrastructure."
>
> "The other thing that MIT really leveraged is an absolute excellence and focus on making stuff. That whole maker mentality way before you had maker movements is incredibly important. These folks are not solving abstract problems. These are folks trying to say, 'What's the problem that I can solve with something that I can make?'
>
> "The link to industry and the willingness to link to applied research is also incredibly important. It made it legitimate for professors to cross the line back and forth. That's something that if you wanted to do at Harvard, you tended to do it by going into government and taking a two-year or three-year sabbatical. You didn't tend to do it by walking from your office to startups to a government research lab and back again every day."

*All quotes from Enriquez come from our interview of June 11, 2020, with follow-up via email for clarification.

contradictory views of what innovation means. But that doesn't mean one view is wrong and another is right. Here are three takes on the essential ingredients for an innovation hub—two from Kendall Square insiders, one from an outside observer.

It's All about the Entrepreneurs

"An entrepreneurial ecosystem has one necessary and sufficient condition—and that is entrepreneurs," says Aulet, who besides his role at the entrepreneurship center is professor of the practice at the Sloan School. "Sometimes we make business too complicated and we lose track of the real simplicity and the central element of it. This is what happens when you study things—you get caught up in the minors and you don't know the majors. A good entrepreneurial ecosystem has more entrepreneurs, better entrepreneurs, and entrepreneurs who are connected. You can have all the other fancy things you want, with policies, venture capitalists, educational institutions, and all that. But at the end of the day, it comes down to how many entrepreneurs do you have, how good are they, and how connected are they. And that comes down to talent, training that talent, and connecting that talent."

Entrepreneurs *and* a Top-Tier University

"A world-class university is a necessary prerequisite. Without it, you can have a creative district, you can have an arts district, you can have a place people like to be. But you're not going to have a high-technology cluster or complex," says Richard Florida.[4] But that alone will only get you so far—and that is where entrepreneurs come in, he adds. "A university can be blasting out this technological signal, but you need somehow to develop a cluster of technological and entrepreneurial talent that can harness that signal so it doesn't leak out and go to other places. So really the two prerequisites are incredible technological talent and a set of *technological*-entrepreneurial talent that can capitalize on it."

A Community of Stakeholders

A successful innovation ecosystem consists of a series of overlapping communities, says Fiona Murray, who, along with Aulet and a few other colleagues, has worked with close to sixty regions in the past decade through MIT REAP. The program identifies five key stakeholders: entrepreneurs, universities, risk capital, corporations, and government. "You have these overlapping communities, and you need them to sort of intersect, to overlap, and have some shared interests," Murray sums up.[5]

Things can get a bit tricky because people often have different views of what each of these stakeholder categories means—and of their relative importance, she stresses. For instance, while Murray believes a world-class university like MIT or Stanford is a huge asset that can set the tone for an entire region, she doesn't see it as a prerequisite of a successful innovation ecosystem. "I genuinely think you need universities to play a role," she says. "Do they always have to be in the lead like MIT has been to be like MIT? I don't think so."

There is even wider disagreement over the role of government and innovation. Many people believe government should set the table with some general policies and then get out of the way of entrepreneurs and business. "I don't fully espouse that view," Murray says. For instance, governments can play vastly different roles depending on their level—city, regional, state, national—and where they are. "I think in lots of countries, the government proactively reinforcing and supporting the community is also a form of legitimacy as much as everything else. When government says it's a good idea, it makes a difference."

In the end, a city or region doesn't need all these players in order to succeed, but it sure helps. Says Murray, "The places that do best are the ones that really do have all five stakeholders at the table."

General Principles to Keep in Mind

Location[3]

"There are three things that matter in property: location, location, location." The origins of that adage remain unclear; even the language detective William Safire could not sort it out.[6]

But the saying is so fitting for innovation ecosystems, it should be expressed as an exponential. A Brasilia-like model city just won't cut it—no matter how well-conceived and designed. Kendall Square has been blessed by its crossroads location. In the early days, it offered cheap land just across the Charles River from Boston and linked that major city more directly to Harvard and northern trade centers. Access to rail and sea traffic enhanced its edge. These advantages helped make it a major economic and innovation hub decades before MIT opened its doors in Cambridge in 1916. (See chapters 3–7.)

Today, while real estate prices have risen to exorbitant levels, the square is still an easily accessible crossroads—adjacent to MIT and just a few subway stops from Harvard in one direction and from Boston and its centers of government and finance and incredible research hospital network in the other. In light traffic, it's only a ten-minute taxi or rideshare to Logan International Airport. Over the years, the square has become a scientific and technological crossroads as well. It serves as a base of operations for a potent mix of international corporations, research organizations, and startups. Key to the future, it also stands as a crossroads of scientific and technological disciplines, where biology, chemistry, engineering, data, artificial intelligence, and other fields are blending with increasing sophistication and power, offering the potential for a new wave of important innovations.

Location also has another meaning: it can refer to where a company is in relation to other companies of its ilk. In general, the closer similar companies are to each other and potential customers and partners, the better. This was well-documented by AnnaLee Saxenian in *Regional Advantage*, who showed it helped in building businesses, attracting talent, and sharing best practices that can help companies adapt to changing circumstances. Richard Florida has shown the power of concentration in a different way, noting that innovation seems to arise more in areas with tight clusters of creative and talented people.

Kendall Square has some compelling advantages on this front. "San Francisco is a big hub, but things are a lot more spread out," says former Millennium CEO Deborah Dunsire, who now heads Denmark-based biotech Lundbeck. "You can't just go over and have a lunch with somebody. The commute to Cambridge is getting horrible, but once you're there it's very easy to navigate. You can go visit someone, and be back in your office in a couple of hours."[7]

Initial Plans Almost Always Go Awry

In Kendall Square's case, major visions for the future have almost always been off base, to the point that MIT lecturer Joost Bonsen calls the square "the prediction failure zone." Whether it was Charles Davenport's vision of a luxurious real estate development where MIT now stands, the idea that Polaroid or Lotus would continue their dominance, visions that NASA's Electronics Research Center would create thousands of jobs, initial urban renewal plans, or the square's branding as AI Alley, all too often predictions about the ecosystem's future have come up short. Biotech, a growing presence for over forty years, pretty much took people by surprise.

All this means that plans should be as flexible as possible: be prepared to pivot.

Continuity Is Queen

Building or revitalizing an ecosystem takes decades, demanding continuity of philosophy and leadership. "If you don't have continuity, you just never get to do the right thing because everything takes a long time," says MIT's long-time director of planning, Robert Simha, who helped engineer Kendall Square's post-1960 revitalization.

Investor and Techstars cofounder Brad Feld echoes this view in *Startup Communities*, saying leaders must make a long-term commitment to an area's innovation fabric. "I like to say this has to be at least 20 years from today to reinforce the sense that this has to be meaningful in length," he says. "Optimally, the commitment resets daily; it should be a forward-looking, 20-year commitment."[8]

Additional Core Ingredients

In addition to essential ingredients like the five stakeholders MIT has identified, there is a long list of other factors in building an innovation ecosystem. The most important:

- Discovery science and technology
- Infrastructure and affordable startup space
- Culture of commercialization
- Accessible, value-added funding
- Connections between large corporations and startups
- Licensing expertise
- Startup competitions

A great view across several of these sectors comes from Johannes Fruehauf, cofounder of LabCentral, which provides space and facilities for biotech and medical startups. Launched in Kendall Square in 2013, it has been expanding around the world. Discovery science and technology, he notes, comes from top universities, research hospitals, and institutes like the Broad and Koch. These innovations often inspire company formation—and the startups that arise typically need affordable lab and manufacturing space. The shared equipment at places like LabCentral or Greentown Labs in Somerville offer significant capital efficiencies that can give startups a leg up.

Arguably more important than great science is culture, says Fruehauf: "I'm building these labs in many other cities outside of Cambridge. The contrast actually is very helpful to see what it is that we have here, and what others don't. It's not just buildings. It's also not just great academic research, but it's the confluence of all of this. It also needs this culture that encourages, and a priori does not discourage, thinking of commercializing your invention. We have examples of universities that are excellent in basic research, but that really have a culture that is averse to professors thinking about commercializing their findings. They think it's uncool. They think it's selling out. And I think they're missing the mark."[9]

Funding is another big issue. Access to investment capital is on everyone's list of what makes for a strong cluster. Murray puts it slightly behind entrepreneurs and universities in her hierarchy of stakeholders—emphasizing that if you have the first two, the investment community will soon be there. "You need to get capital involved, but I think it comes along," she says. Importantly, a strong investment community is about much more than just opening wallets. "It's not enough to just provide money to these groups," says Fruehauf. "Often times they also need guidance in terms of board or strategy—experience around the areas that the founders themselves might not be familiar with."

Kendall Square and the Boston region also get high marks when it comes to a shopping list of other key ingredients. One involves the connections between larger companies and startups, which run especially deep in the life sciences. Multinational corporations have long maintained a variety of partnerships and other programs that support startups—to gain an edge on talent and get access to new technology, often as a first step toward acquisition. Corporate culture around partnering varies widely, but a critical attribute when working with young companies is to have a clear orientation toward boosting the startup, says Feld. This means "helping and providing value upfront, instead of entering into the relationship with a startup with a goal of extracting value. When the leaders of a corporation innovation program understand this, good things can happen."[10]

The Boston region also is home to strong technology licensing expertise at universities and nonprofits (see "Spotlight: Lita Nelsen on Technology Licensing and How 'Clusters Feed Themselves'"), as well as in the legal sector. Then there are various startup programs and competitions—from the MIT 100K to Techstars to MassChallenge. "You can have talent, but to create accessible output you really need to build teams—because nobody gets anything done by themselves," says Aulet. "So what these competitions do is they are a forcing function to get people to create teams and to start to move toward developing a business."[11]

Policies and Programs

A multitude of policies and programs seek to encourage innovation and entrepreneurship—some more successfully than others. But in Kendall Square's case, a few stand out as especially significant.

Collaborations and Initiatives Can Make a Difference

In 1960, a unique partnership between the City of Cambridge, MIT, and real estate developer Cabot, Cabot & Forbes catalyzed development of Technology Square. The development became the world headquarters for Polaroid, the home of Project Mac and MIT's artificial intelligence lab, and it helped inspire larger urban renewal plans that shaped the square to this day (chapters 9 and 25).

Beginning around 2004, the state undertook a series of efforts to put out the welcome mat for large pharma companies, says then secretary for economic development Ranch Kimball. These included streamlining permitting and zoning processes, improving collaboration between various state offices involved in life sciences, and facilitating meetings with top university scientists. "All these things start creating a culture that says Kendall Square is not a hostile insider's club; it's a welcoming place for people from around the world," says Kimball.[12]

Then, in mid-2008, under Governor Deval Patrick, the state established the ten-year, $1 billion Massachusetts Life Sciences Initiative. The initiative, which provided $500 million in capital investments, $250 million in tax incentives, and $250 million for grants and other discretionary efforts, changed the game, says former MassBio CEO Bob Coughlin. "The key to our growth has been because of one thing: industry, academia, and government all working in a true partnership," he proclaims.[13] Before the initiative, Coughlin says, "We had the most NIH [National Institutes of Health]-funded hospitals. We had great venture capital firms. We had a lot of the ingredients to be successful, but government wasn't a partner. Look at the growth that we experienced from 2008 to 2012—during a recession. Two or three big pharma companies were here in the mid-2000s. Now, nineteen of the top twenty have a physical presence here. So how the heck does that happen? We did it by showing that government wanted them here. It happened because we changed our mindset."

Susan Windham-Bannister, founding president and CEO of the Massachusetts Life Sciences Center that implemented the intiaitive, feels the same. She says the core concept drew on the work of both Luis Suarez-Villa at the University of California Irvine and Michael Porter at Harvard Business School. Areas that have high innovation capacity, she explains, are able to consistently translate technology from academia into the marketplace through commercialization. "It's a competency that you develop—with all of the underlying and supporting ingredients that you need to do it over and over again," she says. "So the strategy for the life sciences initiative was to build up Massachusetts' innovation capacity by investing in that supporting platform. Those ingredients fall into five buckets—setting aside federal regulation, because we had no control over that. It's your academic culture. It's your capital and culture for entrepreneurship. It's your workforce. It's your infrastructure. And it's your ecosystem. And everything that the Massachusetts Life Sciences Center funded was targeted to one or more of those areas." She point to the MLSC's two $5 million grants to LabCentral mentioned in chapter 22 as a prime example. "The need to access very expensive wet lab space is often what keeps entrepreneurs in the university, so the creation of LabCentral accelerated getting entrepreneurs out of our universities and into the community," she says.[14]

Protect Startups

"Keep a high concentration of startups," Cambridge Innovation Center founder and CEO Tim Rowe advises. "The City of Cambridge passed a law some years ago that incentivized landlords to reserve part of their space for startups. This was needed, because the popularity of the area could easily have seen them squeezed out, as landlords chase high-credit corporate tenants" (see "Spotlight: Innovation Space Zoning—How Kendall Square Hopes to Keep Its Startup Community Vibrant").

Networks: How It All Fits Together

The ecosystems that do best, as Murray relates, bring a multitude of stakeholders to the table. But there is together and there is *together*. The power of deep and trusting relationships that allow for more sharing of insights and strategy, of successful entrepreneurs who mentor their younger counterparts, of the ability to tap into networks to build new teams—these all form an intangible that makes the whole greater than the sum of the parts. Such collaborative networks were lacking in Boston in the Route 128 heyday, and that had a lot to do with the region falling behind its Silicon Valley rivals. Heading into the third decade of the 2000s, Kendall Square and the greater Boston biotech community are thriving in large part because of their strong networks.

"It's something in the water," Phil Sharp once said, marveling at this sense of collaboration in the biotech sphere.[15] This happened in part because a very tight biology community helped create the field—and grew it together. Now the early pioneers will soon head into retirement, if they aren't there already. A few, like Henri Termeer, have passed away. Will they have instilled this same sense of togetherness and openness to the next generation? Will it spread to include more women, people of color, and others who have not typically enjoyed the same access to money and networks? And will it spread to new arenas, such as the convergence of healthcare and artificial intelligence, or some of the other emerging areas showcased in the final chapter?

28 CONVERGENCE AND CONSILIENCE

"One of the things that I worry about is all the bio stuff is amazing, but it needs to be balanced. You don't want any ecosystem that's too imbalanced in any industry, because you have these huge boom and bust cycles."

That's Katie Rae talking. She's CEO and managing partner of The Engine, a venture capital firm conceived and created by MIT that invests in and provides mentorship and infrastructure support to "tough tech" startups. These are young companies pursuing extremely challenging, potentially game-changing innovations that may take many years to come to fruition, and which therefore are often not attractive initially to venture capitalists looking for shorter-term returns. The Engine is all about helping these companies get through the "valley of death" between ideas, proven lab research, and commercialization. Some call this *patient capital*, as The Engine's model allows twelve years or more for some of the longer-term initiatives to provide returns—although it invests in startups on shorter time cycles as well. It doesn't put in a lot of money by today's standards—with "first check" sizes typically running $1 million to $3 million. The hope is that these early investments, plus its mentoring, infrastructure support, and access to both equipment and connections beyond the reach of many startups, will help companies bring their ideas to a point where they can attract capital from more traditional investors.

Rae is a highly influential figure in the Boston startup ecosystem—having cut her teeth at companies from Lycos to Eons to giants like Microsoft, with an early Silicon Valley sojourn thrown in. More recently, she spent more than six years as managing director or chairman of Techstars Boston, and along the way she became one of the area's biggest early-stage investors, with more than one-hundred thirty investments in Boston-area startups.

Armed with her wide experience, Rae has been running The Engine since its 2016 inception. The upstart facility set up on five floors of a refitted space on Massachusetts Avenue in funky Central Square. But it's revving up to move to the outer borders of Kendall Square—an old Polaroid factory at 730–750 Main Street, with LabCentral on one flank across Osborn Street and the candy factory where all the world's Junior Mints are made on the other.[1] This long-term home will have lab and maker facilities, in addition to offices and meeting spaces. And it will offer a safe haven for companies that may need to set up in a suburb but could use a foot in Kendall Square. "So if you're doing

Figure 41 Katie Rae standing in The Engine's workshop, where science and tech startups receive access to specialized equipment.
Source: Tony Luong.

a tough tech company out in Woburn because you need warehouse space, you've got a place to come have your meetings in Cambridge," Rae explains. "We want people to feel like, 'Okay, this is my home too,' even part time. I think everyone has to think about doing that—inviting people who can't afford the rents in Kendall Square into that collision path, because it's so key." If anything, that desire and need for interaction became more apparent during the COVID pandemic, she says. "People love to be together. They love those collisions, you know? There's been a clamor for more, not less, and we want to facilitate that."

Getting back to the need for Kendall Square to be balanced in the type of companies it hosts, Rae says the square is actually doing all right. It just needs to be better, and The Engine plans to help. "I think the cool thing about Kendall is it's still diverse, and it's going to get more diverse in terms of types of companies there," she says. "One of the reasons we are building The Engine's long-term home there is that it brings all kinds of startups—some of it biology, some of it chemistry, some of it quantum, all kinds of different nascent industries essentially—into the heart of Kendall Square, which will allow that kind of community to also grow up there."

* * *

When I met with E. O. Wilson to kick off this book, one big theme we talked about was evolution. A thriving ecosystem, just like a thriving person, isn't static; it keeps evolving and growing, spawning novel species, adapting to changing conditions. One of the key ways that happens in innovation ecosystems is the convergence of different technologies or scientific disciples to inspire ideas and innovations, and sometimes new fields.

A number of people are already working on the next technological iteration of Kendall Square. It seems clear that the dominant current threads, in computing, machine learning, and especially in biotechnology, are not going away in the foreseeable future. In biotech alone, emerging tools like CRISPR gene editing, as well as "older" innovations like genomics, RNA interference, and gene therapy, are just beginning to make their marks. To all appearances, their future is very bright.

But . . . well, you never know. The experts thought that about Polaroid and Lotus, too. Visionaries foresaw an amazing future for AI Alley but largely missed the ascendance—and just about totally missed the transcendence—of biotechnology.

Similarly, potentially powerful forces of change are at work today. When talking to people about new fields of growth that might power not just Kendall Square, but the entire region, two major lines of thought come through. Both involve convergence. Atop many people's list is the convergence of artificial intelligence, healthcare, and biology.

This convergence has been underway for a number of years. Every biotech and pharma company employs a mashup of computing power and data science along with its biology. The Broad Institute, with its powerful genomics platform, utilizes a lot of machine learning and AI. GNS Healthcare, a Kendall Square startup that recently moved a half-mile away to Somerville, uses its "causal AI technology" to figure out which patients respond to a given drug and why—as well as to discover new drug targets for specific patient populations. A host of startups these days champion their use of AI, with wide variance in how they employ it. In short, there are many flavors of AI and myriad ways to bring it to bear on health care. These run from analyzing medical images with unprecedented accuracy to diagnosing disease to finding drug compounds. "The convergence of molecular patient data, computing, and bleeding edge AI mathematics will do more to transform our understanding of complex diseases such as cancer, neurodegeneration, and immune system diseases and our ability to discover and develop drugs and better match them to patients in the real world than any other innovation," says GNS Healthcare cofounder and CEO Colin Hill. "This is the key that unlocks a new age of predictive biology that will change the way we discover, develop, and use new and existing medicines."[2]

One manifestation of this trend can be seen in Takeda's Data Sciences Institute.[3] The institute is based near Central Square, but its 250 statisticians, programmers, real-world data experts, digital tools specialists, and others are spread all over—including Kendall Square and other sites around the world.

The ultimate goal is to analyze and crunch data to design better drug trials and help improve patient outcomes, explains Anne Heatherington, the senior vice president who heads the institute. Her group does this in a number of ways that don't involve AI, such as employing digital tools to gather patient data remotely, as well as traditional mathematical methods. But one longer-term effort seeks to use vast computing power, natural language processing, and artificial intelligence techniques to mine patient records and medical data to better diagnose patients. For some rare diseases, it can take seven years or more to make the right diagnosis, which makes outcomes more problematic, says Heatherington. "If there are algorithms built into the hospital systems, patients could potentially get flagged for particular diseases much, much earlier," she says.

In early 2020, the company announced a collaboration with the MIT Jameel Clinic for Machine Learning in Health—known as the J-Clinic—that would help with this effort and others. The three- to five-year initiative (dollar terms were not announced) is designed to explore issues at the intersection of artificial intelligence and health that Takeda believes could impact its business. It started with ten projects—among them efforts in diagnosing gastrointestinal diseases, drug manufacturing, and biomarkers—each involving a joint team of Takeda and MIT people. "The plan is each of these ten projects will have a two-year lifespan, and then we will kick off another round of projects after that," says Heatherington.[4]

Coming at this convergence from the big tech perspective is IBM, which established its IBM Watson Health headquarters to Kendall Square in 2016. Around the same time, it beefed up IBM Research's local strength in artificial intelligence—putting both groups in the same building along Binney Street. Big Blue then underscored its growing Kendall Square AI activities with several major financial commitments. In 2016, it announced a five-year, $50 million collaboration with the Broad Institute to support an initiative to sequence cancer samples and then analyze related genomic data with Watson. It expanded that in 2019, announcing a separate three-year joint project aimed at helping physicians wield genomics, AI, and clinical data to better predict the likelihood of patients developing serious cardiovascular diseases. Meanwhile, in 2017, it also announced a ten-year, $250 million commitment to fund a new laboratory at MIT—the MIT-IBM Watson AI Lab.[5]

Through this lab, IBM backs about fifty projects per year, addressing a variety of issues that include algorithms, hardware, applications for specific areas like healthcare, and the ethics of AI. The company's team devoted to the effort was housed initially within IBM Watson Health's Kendall Square headquarters. But it was set to move, along with all of IBM Research Cambridge, to Pi Tower at 314 Main Street when that building opened in 2021, putting IBM's researchers directly alongside their MIT partners.

These are just two examples of what large corporations are doing around the convergence of computing, AI, and healthcare. They are hardly alone. MIT, not surprisingly, is at the heart of many collaborations—and has AI-related initiatives with a number of other companies and organizations. As mentioned in chapter 1, one of the latest, announced in March 2021, involves the creation of the Eric and Wendy Schmidt Center at the Broad Institute of MIT and Harvard. The upstart center, backed by a $150 million endowment from the former Google CEO and his wife, will bring artificial intelligence and computer science more broadly to bear against a variety of diseases and medical conditions. It also will follow the Broad tradition of collaborating with a number of other top institutions and companies worldwide.[6]

Well before the Schmidt Center's formation, MIT took things to another level in 2018 when it announced a $1 billion commitment toward addressing the opportunities and potential hurdles posed by the increasing power of AI and computing. The commitment included the establishment of a new college, the Schwarzman College of Computing. The college marks the most significant structural change at MIT in more than seventy years. Fifty new faculty appointments will be created, and, importantly, one of the main goals involves integrating AI into every discipline at MIT. "The College will reorient MIT to bring the power of computing and AI to all fields of study—and, in turn, to allow the future direction of computing and AI to be shaped by insights from all of these other disciplines, including the humanities," the school's announcement proclaimed.[7]

This is a critical point, says Lisa Amini, director of IBM Research Cambridge, who oversees IBM Research's major AI collaborations with universities around the world through what it calls the AI Horizons Network. "It's not just, 'Let's build this great big computer science department, bigger than anyone else's,'" she says. "What I see is a better engagement with and leveraging of the tools of computer science by many different disciplines: engineering, finance, architectural. To me, that's the blockbuster."[8]

She likely won't get any disagreement from the college's inaugural dean, Daniel Huttenlocher, who started in January 2020 after being recruited from Cornell. Huttenlocher earned his master and doctorate degrees at MIT in the 1980s, working out of the AI lab in Tech Square. "I believe we're in a time now that is in many ways similar to the time period in which MIT was founded—when a lot of technology practice had gotten out ahead of our understanding of it. We're there with computing and AI today, much in the ways we were with what we now know as engineering," he told *Slice of MIT*, the school's online alumni publication, not long after taking his new role. An interdisciplinary mindset is critical to addressing many future challenges, he says, and the college will provide more paths for students "that aren't just the silos of the disciplines."[9]

* * *

The integration of artificial intelligence and machine learning with a variety of fields opens the door for a wave of innovations and startups. In the eyes of Jim Collins, a serial entrepreneur and a professor of Medical Engineering and Science at MIT, AI writ large is "going to be one of the two dominant themes of this century." The other, he believes: synthetic biology.

Collins is uniquely positioned to come to this conclusion. He is one of the seminal figures in the field, winner of a MacArthur "genius" award largely for his work on this frontier. One of the companies he cofounded, Synlogic, is based along Binney Street in Kendall Square, seeking to develop a novel class of bioengineered, "living" drugs that target tumors and various diseases or conditions, among them gastrointestinal and immune disorders. But that is just one face of synthetic biology—and as it becomes increasingly melded with the power of artificial intelligence, even more possibilities arise, Collins predicts. "The integration of AI with synthetic biology will allow us to use advanced computational tools to understand and embrace the complexity of biological systems, enabling us to harness the power and diversity of biology for the benefit of our planet. We will be able to endow living cells with novel functions, enabling us to tackle some of the world's great challenges in health, food, energy, climate change, and sustainability."[10]

This kind of convergence is more along the lines of what Katie Rae means when she talks about creating a better balance of industries in Kendall Square, but even it is only part of what she imagines. So what are the tough tech arenas she has in mind?

"So, three big areas," Rae relates. "One is reversal and mitigation of climate change. That's decarbonization of everything. It might be energy generation. It could be fusion. It could be long-term battery storage, or it could be decarbonizing industrial processes. All of those things. Second big bucket is human health and agriculture. How do we discover drugs? How do we truly manufacture tissue so we can cure disease? And the third, we'll call them advanced systems and infrastructure. These are things that connect the world, or improve how we can understand things. It's what's driving improved transit for smart cities. It might be quantum AI on the compute power side, or it could be photonics and semiconductors on the infrastructure side, or it could be space and mobility."

Companies at The Engine speak to all these arenas. Commonwealth Fusion Systems is developing high-field magnets based on new superconducting technology to better insulate the plasma that is key to a fusion system—allowing for a much smaller reactor that might make the process commercially viable. Italian energy company Eni put in $50 million to back Commonwealth. Zapata Computing is developing algorithms and software for quantum computers that hold the potential to solve equations too complicated for today's most powerful supercomputers. Cellino seeks to manufacture human tissue on demand. Quaise is advancing a novel drilling system that could make affordable geothermal energy feasible.

"These are things that are leaps forward in how we as a world figure things out, improve transportation, cure diseases, solve the energy crisis, fight climate change, etcetera," sums up Rae.

*   *   *

Katie Rae's words sparked a series of thoughts. Things like nuclear fusion and quantum computing have eluded successful commercialization for decades. No doubt these dreams will someday become reality, but is that day actually within sight? Could these fledgling companies, or others like them, become the icons of a future iteration of Kendall Square?

I also thought of an E. O. Wilson book, *Consilience: The Unity of Knowledge*, about the idea of fusing science with the arts and humanities. "The greatest enterprise of the mind has always been and always will be the attempted linkage of the sciences and humanities," he wrote.[11] A long-sought-after arts and cultural center had finally been approved for Kendall Square at the end of 2020. What would happen if Rae's breed of entrepreneurs joined the existing Kendall Square crowd—and they in turn rubbed shoulders with artists and musicians? If Kendall also became an arts ecosystem, would that produce collisions that led to innovations people haven't even imagined yet? That is certainly part of what Richard Florida predicted could happen in *The Rise of the Creative Class*.

But most of all I thought of the power of Rae's last sentence. To me, her words evoked a key trait—in many ways, the essence—of Kendall Square. The square has logged over two hundred years as a center of economic growth and innovation, to varying degrees and in varying forms. One word Rae used in particular spoke to the inevitability of that continuing, more or less along the lines it had done for two centuries, but not exactly. More of the same, only different—and probably not quite what you imagined. To me, this was a beautiful word to sum up the likely future of Kendall Square.

Etcetera.

ACKNOWLEDGMENTS

Any effort like this book cannot be done without the help of many people. A collective thank you to all, including the more than one hundred who gave their time in interviews—many of them sitting for my questions more than once. They, and others, helped me check facts, work through complicated situations, dig up photos and documents, and much more. Any errors are theirs—not! It's more like, what I have gotten right is thanks to them. I am deeply grateful to all for their help.

Nicholas Negroponte brought this idea to me. He had to ping me twice. Thank you, Nicholas, for persisting and for helping me get going on this wonderful project! Great thanks as well to Doron Weber and the Sloan Foundation for supporting this endeavor.

In the course of my research and writing, several people truly went above and beyond. Charlie Sullivan, executive director of the Cambridge Historical Commission, was incredibly generous with his time and expertise, reviewing large sections of the manuscript, providing source material, and generally sharing his unrivaled knowledge of Cambridge. In the true spirit of collaboration and assistance, Karen Weintraub shared materials she and her husband, Michael Kuchta, had gathered for their own great book, *Born in Cambridge*, despite the fact that our efforts overlapped a small bit in several places. Vic McElheny, as he has done many times in the past, shared his encyclopedic knowledge of the history of science and technology, and helped place events in their larger context. Phil Sharp opened many doors, including inviting me to a special dinner celebrating Biogen's fortieth anniversary early in my efforts. I also enjoyed special access to three special leaders at MIT: Susan Hockfield, Sangeeta Bhatia, and Nancy Hopkins, whose vitally important work to encourage and support women founders in biotech is described in chapter 21, "Forty Missing Companies."

Three people provided invaluable help with research and editing. First was my former *Technology Review* colleague Tracy Staedter. Then came Madeleine Turner, who dug up documents, maintained the bibliography, and more. The third was John Carey, with his great eye for detail. Great thanks to all.

Throughout this process, Gita Manaktala and Amy Brand at the MIT Press provided steadfast support, helping me sort through the myriad challenges that come with such a book. The team behind them was incredible, including Erika Barrios, Suraiya Jetha, Sean Reilly, and Kathleen Caruso.

Lastly, I thank my family—my children, Kacey and Robbie, and my wife, Nancy Walser. You readers know how it is: just by being in my life they give me encouragement, hope, strength, joy, and a lot more.

LIST OF INTERVIEWS

Afeyan, Noubar—February 20, 2020; July 14, 2020; March 19, 2021; May 4, 2021

Altshuler, David—June 18, 2020

Amini, Lisa—September 14, 2020

Andrews, Tom—May 21, 2020

Aulet, Bill—October 27, 2020; December 11, 2020

Baerkahn, Jesse—June 24, 2020

Baltimore, David—January 30, 2020

Baty, Gordon—March 13, 2020

Bhatia, Sangeeta—June 16, 2020

Boger, Joshua—May 15, 2020

Booth, Bruce—September 1, 2020

Bonsen, Joost—December 20, 2018; January 18, 2019

Bosley, Katrine—April 14, 2021

Bourque, Janice—June 9, 2020

Broderick, Shawn—August 18, 2020

Chung, Joe—February 21, 2020

Clem, Chet—July 26, 2019

Clem, David—February 14, 2019; July 26, 2019; February 27, 2020; July 22, 2021

Collins, James—January 29, 2021; February 4, 2021

Conrades, George—March 11, 2020

Cooney, Charles—March 6, 2020

Cooper, Phil—January 12, 2020

Couch, Stephanie—November 17, 2020

Coughlin, Robert—August 6, 2020

Dietz, Ryan—June 2, 2021

Dunsire, Deborah—June 9, 2020

Durant, John—November 29, 2018

Emmens, Matthew—August 4, 2020

Enriquez, Juan—June 11, 2020

Evans, Tom—April 9, 2020

Fagnan, Jeff—September 16, 2020

Florida, Richard—November 20, 2020

Fox, Marvin—July 26, 2017 (Interview with Karen Weintraub, shared with author)

Fruehauf, Johannes—March 8, 2019; October 15, 2020

Gilbert, Wally—January 27, 2020

Haglund, Karl—July 26, 2019

Harthorne, John—September 14, 2020

Heatherington, Anne—August 26, 2020

Heaton, Penny—August 13, 2020

Hillis, Daniel—November 22, 2019

Hinds, Chuck—August 28, 2020

Hockfield, Susan—September 28, 2018; September 15, 2020

Hoffman, Ellen—November 26, 2018

Hopkins, Nancy—June 24, 2020

Hughes, Tom—April 27, 2020

Hunt, Aisling—April 9, 2019

Jewart, Eric—September 1, 2020

Kane, Bill—March 25, 2020; May 19, 2020; July 22, 2021

Kapor, Mitch—January 25, 2019 (Interview with Karen Weintraub and Michael Kuchta, shared with author); November 29, 2019 (Interview with author)

Kay, Alan—January 14, 2020

Kimball, Ranch—January 13, 2021

Kleespies, Gavin—September 13, 2019

Koop, Bryan—April 2, 2020; April 30, 2021

Krim, Robert—July 23, 2019

Lander, Eric—November 30, 2020

Langer, Robert—June 8, 2020; June 11, 2021

Lang, Ilene—December 20, 2019

Larrea, Hasier—May 24, 2019

Larson, Kent—March 8, 2019; April 27, 2021

Larson, Michela—December 5, 2019; April 9, 2020

Leighton, Tom—March 6, 2020

Levering, Alexandra—April 10, 2019; April 3, 2020

Levin, Mark—May 18, 2020

Lockwood, Jeffrey—October 11, 2019; April 27, 2020

Lodish, Harvey—February 5, 2020

Maraganore, John—April 28, 2020; April 29, 2020; January 5, 2021

Marsh, Steve—November 16, 2018; October 2, 2020

McElheny, Victor—February 5, 2019

McCready, Travis—July 30, 2020

McGuire, Lee—July 1, 2020

Morgan, R. Gregory—May 17, 2018

Murray, Fiona—March 26, 2019; November 6, 2020

Nassi, Ike—February 27, 2020

Negroponte, Nicholas—April 25, 2018

Nelsen, Lita L.—March 9, 2020

Noftsker, Russell—November 22, 2019; December 4, 2019

Novack, Kenneth—February 4, 2020

Olmsted, Andy—April 2, 2020

Plump, Andy—August 31, 2020

Rae, Katie—September 18, 2020

Roberts, Edward—December 16, 2019

Rossi, Derrick—May 26, 2021

Rowe, Tim—April 9, 2019; April 1, 2020

Rowland, Robert—November 25, 2019

Ruiz, Israel—November 28, 2018

Ryan, Nancy—March 26, 2019

Safdie, Moshe—October 12, 2018

Sahlman, Bill—December 10, 2019

Sayare, Mitch—February 4, 2020

Schwarz, Erica—April 10, 2019

Sharp, Phillip—April 25, 2018; October 30, 2020

Simha, O. Robert—November 30, 2018; October 30, 2019

Solomon, Cynthia—January 29, 2020

Springer, Timothy—July 1, 2021

Stalder, Carrie—June 12, 2020

Steinert, Heidi—February 20, 2020

Stowe, Barbara—May 7, 2018; April 9, 2020

Sturtevant, Reed—December 11, 2019

Sullivan, Charles—September 3, 2019; November 9, 2020

Termeer, Belinda—May 14, 2020

Toomey, Tim—February 4, 2021

Webb, C. A.—February 7, 2019; June 12, 2020

Wilson, E. O.—February 22, 2019

Vandebroek, Sophie—August 7, 2020

Vasella, Daniel—June 30, 2020

Vinter, Steve—July 28, 2020

Walts, Alan—May 14, 2020

Weiss, Rainer—January 30, 2020

Whitehead, Susan—February 17, 2020

Windham-Bannister, Susan—May 24, 2021

Woods, Nancy Bellows—August 19, 2019 (Interview with Madeleine Turner)

NOTES

Introduction 1.0, or Preface

1. Wilson quotes from interview of February 22, 2019.

Introduction 2.0

1. Quoted in Logan, "MIT to Start."
2. K. Larson interview, March 8, 2019, with follow-up email.
3. Bonsen interview, December 20, 2018.
4. Simha interview, November 30, 2018.
5. Chakrabarti and Bracken, "Cambridge Elections."
6. Webb interview, February 7, 2019.
7. Marsh, Inside the Dome meeting, October 9, 2018.
8. Hockfield quotes from interview of September 28, 2018, with some clarifications by email.

Chapter 1

1. Boston Consulting Group, "Protecting and Strengthening Kendall Square."
2. Levin interview, May 18, 2020. *The Soul of a New Machine* followed engineers at Data General racing to build a next-generation minicomputer.
3. Bluebird was set to take over the entire building in mid-2021, when Sanofi Genzyme planned to move to a new East Cambridge development called Cambridge Crossing. Bristol Myers Squibb, mentioned ahead, will move to the same development around 2023. For Bluebird's approval and pricing, see the company website: https://www.bluebirdbio.com/; Lovelace, "Bluebird Bio."
4. Enriquez, "Vassar and Main."
5. Broad Institute communications, "Broad Institute Launches."
6. This science competition started in 1942 and was taken over by Intel in 1998 and by Regeneron Pharmaceuticals in 2016. It is now called the Regeneron Science Talent Search.
7. While the official address is 700 Main, different building entrances have other addresses, among them 708 Main Street, 28 Osborn St., and 2 Osborn.
8. Murray interview, March 26, 2019. All Murray quotes are from this interview.
9. I learned about MassBio's COVID company list and Volpe's tweet from Leung, "Massachusetts Miracle." Madeleine Turner then analyzed the list to identify Kendall Square companies. See https://

www.massbio.org/covid-19-resource-center/massachusetts-life-sciences-companies-working-to-address-covid-19/. Volpe's tweet was from November 16, 2020.

10. Buderi, "Danny Hillis." Hillis asked to amend the quote slightly for this book, adding the phrase "in Palo Alto," and substituting "Kendall Square" for "Cambridge."

11. McCready interview, July 30, 2020. All McCready quotes and details are from this interview.

Chapter 2

1. Description of the MIT Media Lab's City Science zone from multiple visits, K. Larson and Bonsen interviews, and email with K. Larson.

2. Zoning had been approved in 2013.

3. There actually was a pharmacy at 238 Main Street when I was a fellow at MIT in 1986–1987. I have not been able to determine when it opened, but it closed in 2005. MIT leased the space to Fidelity, which could afford a higher rent, sources say.

4. 292 Main Street, MIT building E38, is known as the MIT Press Building: its ground floor most recently housed the MIT Press Bookstore. It's also called the Suffolk Building, after the Suffolk Engraving & Electrotyping Company that occupied it starting around 1920. Next door is E37, which includes the graduate student housing tower. Its facade at 264 Main Street incorporates another historic edifice, the J. L. Hammett building. Hammett was the country's first manufacturer of school supplies. Sullivan interview, November 9, 2020. See also report to the Cambridge Historical Commission by executive director Charles Sullivan: https://www.cambridgema.gov/~/media/Files/historicalcommission/pdf/Landmark_reports/kendall_2015_memo.pdf?la=en.

5. Murray interview, November 6, 2020. See also Carson, "Opening for Innovation"; MIT InnovationHQ website: https://innovation.mit.edu/innovation-infrastructure/ihq/. Programs and organizations to be housed in the building include the MIT Innovation Initiative, the Deshpande Center, the Legatum Center for Development and Entrepreneurship, Venture Mentoring Services, I-Corps, and the MIT Sandbox, which provides seed funding for student-spawned entrepreneurial ideas.

6. Schwarzman, the CEO of the private equity firm the Blackstone Group, provided a $350 million foundational gift. As of late 2020, completion was expected sometime in 2023. See Lohr, "M.I.T. Plans College." We'll return to this in chapter 26.

7. Kendall Square Initiative details and other plans are from interviews with Steve Marsh and Israel Ruiz, plus email correspondence with Marsh and MITIMCo. I also attended the Inside the Dome meeting on October 9, 2018. See Chandler and MIT News Office, "A New Era"; Logan, "MIT to Start" MIT-IBM Watson AI Lab details from Vandebroek and Amini interviews; MIT-IBM Watson AI Lab website: https://mitibmwatsonailab.mit.edu.

8. Bonsen interview, December 20, 2018. All Bonsen quotes are from this interview.

9. Larson interview, March 8, 2019. All Larson quotes are from this interview and follow-up emails.

10. Hockfield interview, September 28, 2018. All Hockfield quotes are from this interview.

11. Marsh quotes from Inside the Dome meeting, October 9, 2018. Other details from Marsh interview, November 16, 2018.

12. Ruiz interview, November 28, 2018

13. Correa, quoted in Mitchell and Vest, *Imagining MIT*, 90.

14. The three quotes in this section are from Simha, "A Critical Look."

15. Simha email, July 20, 2021.

16. Enriquez, "Vassar and Main."

17. Webb interview, February 7, 2019. All Webb quotes are from this interview.

18. Kendall Square Association, "You Can't Find."

Chapter 3

1. Sullivan interview, September 3, 2019. Unless noted, all Sullivan quotes are from this interview.

2. For more on early Cambridge, see Maycock and Sullivan, *Building Old Cambridge*; Maycock, *East Cambridge*.

3. Maycock and Sullivan, *Building Old Cambridge*; Maycock, *East Cambridge*; Wikipedia, "Francis Dana."

4. Quoted in Haglund, *Inventing the Charles River*, 13.

5. "West Boston Bridge Editorial."

6. For more on the West Boston Bridge, see Maycock, *East Cambridge*, 17, 60; Haglund, *Inventing the Charles River*, 13.

7. The Middlesex Canal fully opened in 1803, kicking off land speculation that in part led to development of canals. Sullivan interview, September 3, 2019, and follow-up emails; Wikipedia, "Middlesex Canal."

8. Maycock, *East Cambridge*, 25 (map), 60; Sullivan interview, September 3, 2019. See also Haglund, *Inventing the Charles River*, 13, 388.

9. Maycock, *East Cambridge*, 60.

10. Maycock, *East Cambridge*, 60. Maycock summarizes the state of Kendall Square after the creation of the West Boston Bridge. The Sullivan interviews were also very helpful. Some details came from Wikipedia, "Embargo Act of 1807" and "War of 1812."

Chapter 4

1. The hotel was at the corner of Massachusetts Avenue and Pearl Street.

2. Mid-Continent Railway Museum, "Davenport & Bridges." Many details about Davenport's business are from this article.

3. Mid-Continent Railway Museum, "Davenport & Bridges."

4. Mid-Continent Railway Museum, "Davenport & Bridges." See also Gilman, *Cambridge of Eighteen Hundred and Ninety-Six*, 321; MacDonald, "Center Aisle Train Car."

5. Harvard Trust Company, *Edward Kendall, 1821–1915*, 6.

6. MacDonald, "Center Aisle Train Car."

7. Davenport's brother Alvin teamed with Albert Bridges's twin brother, Alfred, to run another operation the business set up in Fitchburg, Massachusetts, a mill town forty-five miles northwest of Cambridge. The operation appears to have burned down in December 1849. (Many years later, in the Harry Potter series, Fitchburg became the fictional home of the Fitchburg Finches professional Quidditch team.)

8. Mid-Continent Railway Museum, "Davenport & Bridges."

9. Cambridge Historical Commission, "Davenport Car Works."

10. Quoted in Mid-Continent Railway Museum, "Davenport & Bridges."

11. Mid-Continent Railway Museum, "Davenport & Bridges."

12. Harvard Trust Company, *Charles Davenport, 1812–1903*, 10.

13. Harvard Trust Company, *Charles Davenport, 1812–1903*, 10. Davenport's former partner Albert Bridges was apparently not part of the new business and went on to establish a successful railway supply business with his twin, Alfred. See MacDonald, "Center Aisle Train Car," for a factory description.

14. Mid-Continent Railway Museum, "Davenport & Bridges." See also Charles Davenport biographical sketch, city of Watertown, Massachusetts, website, at https://www.watertown-ma.gov/DocumentCenter/View/639/Charles-Davenport?bidId=.

15. For a picture of Kendall Square's development in the 1800s, I rely mainly on Maycock, *East Cambridge*, 60–62; Sullivan interview, September 3, 2019.

16. Gilkerson, "Kendall Square."

17. Davenport joined two other men in creating the Union Railroad in 1848. The endeavor was chartered to erect a line between Somerville and Brookline, across the marshes and mud flats of East Cambridge and Cambridgeport. The tracks were to cross the Charles River at the Cottage Farm Bridge, where the Boston University Bridge is today, and connect with Boston and Worcester Railroad—providing a crucial link from the western environs to the deep-water docks of Charlestown and east Boston via the Grand Junction Railroad. Boston and Worcester Railroad agreed to help build the line. Service began in 1855, but a storm wiped out a bridge shortly thereafter, and Boston and Worcester Railroad refused to help rebuild it. A lawsuit ensued, and the line would not be reconstructed until 1869, when Boston and Albany Railroad took over operation (Davenport had apparently sold or given up his stake in the Union Railroad to the Grand Junction). It remains the only rail line between northern and southern railroads in Boston.

18. Sullivan interview, September 3, 2019, with clarifications via email.

19. The Charles River Dam was moved downstream in 1978. The two locks near the Museum of Science are kept permanently open.

20. Gilman, *Cambridge of Eighteen Hundred and Ninety-Six*, 313.

21. For companies and major industries, I rely chiefly on Gilman, 321ff; Maycock, *East Cambridge*; Sullivan interview, September 3, 2019. For populations at different times, see Paige, *History of Cambridge*, 452; Wikipedia, "Cambridge, Massachusetts."

22. Kleespies interview, September 13, 2019; Sullivan comments on draft manuscript. See also Wikipedia, "Houghton Mifflin Harcourt"; Dornbusch, "Riverside Press."

23. Sullivan interview, September 3, 2019.

24. Kleespies interview, September 13, 2019.

25. Dornbusch, "Boston Woven Hose & Rubber."

26. Boston Woven Hose was one of three companies whose whistles signaled the curfew. Toomey interview, February 4, 2021; Charlie Sullivan communication; and "For First."

27. Boston Woven Hose's story was gleaned from the Kleespies and Sullivan interviews of September 3, 2019, and November 9, 2020; Foss, "Boston Woven Hose and Rubber," parts 1 and 2; and Dornbusch, "Boston Woven Hose & Rubber."

28. Logan, "One Kendall Square Complex." One Kendall Square is a misnomer that has no doubt confused many people, as it's more than a half mile from the true Kendall Square, where Broadway, Main Street, and Third Street intersect.

29. One example is the former Blake and Knowles Foundry at 101 Rogers Street. The original foundry was constructed in 1883. Blake and Knowles and its successor, Worthington Pump Works, ultimately occupied three city blocks between Bent and what's now called Linskey Way, spanning Binney Street. The companies made steam-powered pumps for various industrial purposes. Reportedly, every naval ship in World War I utilized more than a thousand Worthington pumps, Charlie Sullivan relates. For

several years, the fourth floor of 101 Rogers served as Xconomy's headquarters. The building has been rebranded the Foundry, and in 2020 it was being renovated to include space for a variety of programs, including art, dance, and workforce education, as well as entrepreneurship. Couch interview, November 17, 2020.

30. Harvard Trust Company, *Edward Kendall, 1821–1915*.

Chapter 5

1. "Cambridge Board of Trade Celebrates." As far as I can determine, Kendall did not make a financial contribution to the station—so it seems the use of *sponsor* here meant *namesake*. The VIP train ride account is mostly drawn from this article, with a few details from Harvard Trust Company, *Edward Kendall, 1821–1915*. The Charles Sullivan interview of September 3, 2019, provided background.

2. In 1932, the Charles stop—now called Charles/MGH—would be added in Boston between Kendall and Park.

3. Quotes and description from "Large Crowd Participates."

4. "Large Crowd Participates."

5. The sale was referenced in chapter 4.

6. Harvard Trust Company, *Edward Kendall, 1821–1915*.

7. Harvard Trust Company.

8. Harvard Trust Company.

9. "Subway Stations."

10. For details of Edward Kendall's life, I rely chiefly on Harvard Trust Company, *Edward Kendall, 1821–1915*. See also "Deacon Kendall Dies at Holden."

Chapter 6

1. Haglund, *Inventing the Charles River*, 88.

2. Quoted in Mid-Continent Railway Museum, "Davenport & Bridges."

3. Sullivan interview, September 3, 2019.

4. Details on Davenport and the Union Railroad, the Charles River Embankment Company, and Davenport's efforts to develop the area can be found in Haglund, *Inventing the Charles River*, 88–92, 400.

5. MIT bought the hotel in 1937, undertook renovations, and opened it in 1938 as a residence initially named Graduate House, but soon renamed Ashdown House. Ashdown House moved to Albany Street in 2008. The former hotel was renovated again and renamed Maseeh Hall in 2011 after MIT alum Fariborz Maseeh, who supported its restoration. Thanks to Charlie Sullivan for helping sort this out. The story of Ashdown House is on its website.

6. Prescott, *When M.I.T. Was "Boston Tech"*, 197.

7. Prescott, 194.

8. Prescott, 197. For the story of the MIT and Harvard merger, I rely chiefly on this account. See also Haglund, *Inventing the Charles River*, 207ff. Another source with interesting details is Wildes, "Electrical Engineering." Henry Higginson's background is from Wikipedia, "Henry Lee Higginson." Some details on MIT's founding are from Puleo, *City So Grand*, 158.

9. See the MIT Libraries website: https://libraries.mit.edu/mithistory/institute/offices/office-of-the-mit-president/henry-smith-pritchett-1857-1939/.

10. Wildes, "Electrical Engineering," 4–20.

11. Prescott, *When M.I.T. Was "Boston Tech"*, 249.

12. See Haglund, *Inventing the Charles River*, 207–208.

13. Quoted in Haglund, 208.

14. Haglund, 209.

15. Prescott, *When M.I.T. Was "Boston Tech"*, 250.

16. Scottish word for shy or bashful.

17. Carnegie founded the Carnegie Technical Schools in 1900. In 1912, it became the Carnegie Institute of Technology. In 1967, the school merged with the Mellon Institute of Industrial Research to become Carnegie Mellon University. See Wikipedia, "Carnegie Mellon University."

18. Carnegie's letter to Maclaurin, quoted in Prescott, *When M.I.T. Was "Boston Tech"*, 250; emphasis in original.

19. Prescott, 251.

20. Prescott, 251.

21. Prescott, 251.

22. Prescott, 263.

23. Details of Eastman's gift from Prescott, 262–264.

24. Prescott, 264. Other details of the transaction are from this account.

25. "Deacon Kendall Dies at Holden."

26. Prescott, *When M.I.T. Was "Boston Tech"*, 299.

27. Prescott, 308.

28. Prescott, 312. Italics in original.

29. The Harvard Business School was started in 1908 inside the humanities department but became a separate entity two years later. The forerunner of MIT's Sloan School of Management was established in 1914, but a master's degree was not offered until 1925.

30. Haglund, *Inventing the Charles River*, 210.

31. Details of MIT's design drawn from Prescott, *When M.I.T. Was "Boston Tech"*; and Haglund, *Inventing the Charles River*, 207–215.

Chapter 7

1. Maycock, *East Cambridge*, 62

2. "Factories, Not Colleges."

3. Stone, *History of Massachusetts Industries*, 773. Stone's book, published in 1930, sounds eerily similar in its description of Cambridge's growth to the *Boston Daily Globe* article from January 1927.

4. Stone, *History of Massachusetts Industries*, 773. Thanks to Charles Sullivan for the likely definition of *manufactures*.

5. Stone, *History of Massachusetts Industries*, 773–774.

6. Stone, 774.

7. Population info from Seaburg, Dahill and Rose, *Cambridge on the Charles*, 100; City of Cambridge website: https://www.cambridgema.gov/cdd/factsandmaps/demographicfaq. The city hit a peak population of 120,740 in 1950; the Census Bureau estimated its 2018 population at just under 119,000.

8. Broad Canal was used for fuel shipments until 1982 per an email from Charles Sullivan, October 15, 2019.

9. Stone, *History of Massachusetts Industries*, 774.

10. Sullivan interview, September 3, 2019.

11. Stats from Prescott, *When M.I.T. Was "Boston Tech"*, chap. 4.

12. The account of Lever Brothers in Cambridge and Kendall Square is drawn chiefly from Stone, *History of Massachusetts Industries*, 779, 844ff. See also Dornbusch, "Lever Brothers." Charles Sullivan provided lots of help. Lever Brothers transferred its US headquarters to New York's Park Avenue in 1949 but kept the factory open another ten years.

13. An MIT alum, Donald Des Granges, designed this building; Shreve, Lamb, and Harmon, designers of the Empire State Building, were consulting architects. In 2016, 50 Memorial Drive was rebranded the Morris and Sophie Chang Building in honor of the Changs. Morris Chang is an MIT alum who headed the Texas Instruments worldwide semiconductor business before becoming founding chairman of the Taiwan Semiconductor Manufacturing Company.

14. The Kennedy Biscuit Company named its fig paste–filled biscuit for the city of Newton. It named other biscuits after other towns in Massachusetts. Sullivan interview, November 9, 2020. Later, the company merged with the New York Biscuit Company to create Nabisco. See also Moravek, "Candy Land."

15. For candy makers, see Moravek, "Candy Land;" MacDonald, "Automated Candy Production."

16. Lockwood interview, October 11, 2019. Much of the $175 million Novartis spent renovating the Necco facility went to removing sticky residue from the floors and sugar spores from inside the walls.

17. See MacDonald, "Automated Candy Production"; Logan, "In Biotech Hub Cambridge"; Kirsner, "A Factory in Cambridge."

18. The Riverside Press closed in 1971 after 120 years of continuous service in the city. Its brand name and equipment were sold to Rand McNally and operations shifted to Michigan. For Houghton Mifflin's story, see Dornbusch, "Riverside Press."

19. The company was purchased by Charles Little and James Brown in 1837 and formally known as Little, Brown and Company starting in 1837. For more, see Stone, *History of Massachusetts Industries*, 776–779.

20. See Stone, *History of Massachusetts Industries*, 843–844.

21. Gilman, *Cambridge of Eighteen Hundred and Ninety-Six*, 337–338.

22. Stone, *History of Massachusetts Industries*, 811; Gilman, 337–339; Wikipedia, "Athenaeum Press."

23. In the interest of full disclosure, I was an investor in SensAble Technologies, which eventually went out of business with no return on investment. My company, Xconomy, was headquartered in the basement for about three years until late 2016.

24. Sullivan interview, September 3, 2019.

25. Sullivan interview, September 3, 2019.

26. For the GenRad story, I rely primarily on two histories by company insiders: Van Veen, *General Radio Story*; and Thiessen, *History of General Radio Company*.

27. Stone, *History of Massachusetts Industries*, 826.

28. Stone, 824ff; Wikipedia, "Arthur D. Little."

29. Stone, 849–853. See also Dornbusch, "Warren Brothers"; Petroski, *Road Taken*.

30. Both quotes from Stone, 850.

31. Raytheon reported more than $25 billion in revenues in 2018. In 2019, it and United Technologies announced an all-stock "merger of equals" that would give Raytheon shareholders 43 percent of the combined company. For Raytheon's story, see MacDonald, "Radar and Microwave Ovens." See also Wikipedia, "Raytheon Company."

32. See Morgan, O'Connor, and Hoag, *Draper at 25*; MacDonald, "Guidance Systems."; Wikipedia, "Draper Laboratory."

33. Sullivan interview, September 3, 2019.

Chapter 8

1. Wiesner interview, May 22, 1992. Quoted in Buderi, *Invention That Changed the World*, 129.

2. Quoted in Buderi, *Invention That Changed the World*, 34.

3. Buderi, 34.

4. Quoted in Buderi, 45.

5. For the story of the NDRC, the British mission, and the Rad Lab formation, I rely chiefly on my book, *Invention That Changed the World*, 27–51. See also Newton, Patterson, and Perkins, *Five Years at the Radiation Laboratory*, esp. 6–15.

6. Quoted in Buderi, *Invention That Changed the World*, 49.

7. Quoted in Buderi, 212.

8. Buderi, 169. For other details of the success of the lab's ASV systems, see 167–169.

9. Buderi, 251.

10. For the Rad Lab's impact on MIT and beyond immediately after World War II, including RLE's formation, see Buderi, 246–257.

11. Buderi, 255.

12. https://www.rle.mit.edu/about/.

13. Quoted in Rand, *Cambridge, U.S.A.*, 101. This book was created around a collection of *New Yorker* articles.

14. Quoted in Buderi, *Invention That Changed the World*, 255.

15. See Buderi, 258ff.

16. Buderi, 256.

17. Buderi, 256; Ante, *Creative Capital*, 113. ARD was the first publicly funded venture firm.

18. See Morgan, O'Connor, and Hoag, *Draper at 25*.

19. For the section on Whirlwind, Lincoln Laboratory, and SAGE, I rely chiefly on Buderi, *Invention That Changed the World*, 354–406. See also Ante, *Creative Capital*, 147ff; MacDonald, "Venture Capital."

20. Ante, 197. For DEC's valuation and stock price, see 196–197. For more on the success of High Voltage Engineering, see 143.

21. Maycock, *East Cambridge*, 63.

Spotlight: The F&T

1. Stalder email, June 8, 2020. Details of Venture Cafe from this and several follow-ups, plus our June 12, 2020, interview. Tim Rowe also provided details in various emails.

2. Fox interview with Karen Weintraub, July 26, 2017, transcript provided to author. All quotes and many details from Marvin come from this interview.

3. Cooney interview, March 6, 2020. Details about meal tickets and booths from David Clem interviews, February 14, 2019, July 26, 2019, and July 22, 2021; and Chet Clem interview July 26, 2019.

4. Weiss interview, January 30, 2020. All Weiss's quotes, and descriptions of tables and scientific experiments, are from this interview.

5. Lettvin was lead author of the influential 1959 paper, "What the Frog's Eye Tells the Frog's Brain." His poem is published in Waugh, "F&T fans."

6. Quote is from Mather email to author, July 22, 2021. Mather shared the prize with George Smoot.

7. Weiss and two Caltech scientists, Kip Thorne and Barry Barish, were awarded the 2017 Nobel Prize in Physics "for decisive contributions to the LIGO detector and the observation of gravitational waves."

8. Baltimore interview, January 30, 2020.

9. Baltimore interview.

Chapter 9

1. Quoted in Miara, "Reinvention of Kendall Square." See also Winling, *Building the Ivory Tower*.

2. Simha, *MIT Campus Planning*, 78.

3. Simha, "Brief History."

4. Simha interview, October 30, 2019.

5. Technology Square story drawn chiefly from Simha, *MIT Campus Planning*, 78–79; Simha interview, October 30, 2019. See also the Cabot, Cabot & Forbes website: http://ccfne.com/?page_id=16. MIT real estate acquisition strategy from Simha interview and Simha, "A Brief History."

6. Simha interview, October 30, 2019.

7. Simha interview.

8. Simha, *MIT Campus Planning*, 79.

9. Simha interview, October 30, 2019.

10. Chiou et al., "Marriage of Convenience," 6–7.

11. Project MAC section drawn chiefly from Chiou et al.; Wikipedia, "MIT Computer Science and Artificial Intelligence Laboratory." The original funding was funneled through the Office of Naval Research on behalf of ARPA.

12. Between 1999 and 2002, the Blockhouse was demolished. Four new buildings were later added and the seven structures that exist today renumbered as 100–700. See Wikipedia, "Technology Square (Cambridge, Massachusetts)."

13. He dropped out twice—first after his freshman year, and then again in 1932, after returning to Harvard in 1929.

14. The early offices of Polaroid were described in McElheny, *Insisting on the Impossible*, 47, 58; and in emails and notes from McElheny.

15. General Polaroid details from multiple sources, including Wikipedia, "Polaroid Corporation." The company filed for bankruptcy in 2002—but it had long since ceded its status as a Kendall Square icon.

16. Quoted in MacDonald, "Polaroid."

17. See "Athenaeum Press Sold." Cuneo Press operated in the building at least into the early 1970s.

18. Rude, "Surprising Link."

19. Rude; Saxon, "Richard S. Morse"; and emails from Ken Morse.

20. Arthur D. Little moved to West Cambridge in 1953. Lever Brothers shifted its US corporate headquarters to Park Avenue in 1949. Warren Brothers moved in the early 1950s, selling its building to Godrey L. Cabot Inc. (now Cabot Corp.), a maker of specialty chemicals. Ultimately, all those buildings were acquired by MIT. Cabot is now headquartered in Boston. One board member is Juan Enriquez, who authored a *WIRED* piece about "the world's most innovative intersection" (Enriquez, "Vassar and Main"). He is a Cabot family member.

Chapter 10

1. Butrica, "The Electronics Research Center: NASA'S Little Known." Kennedy made his famous speech about this in September 1962, and the effort became the Apollo program. Neil Armstrong and Buzz Aldrin walked on the moon on July 20, 1969.

2. Butrica, "Electronics Research Center: NASA's Little Known."

3. Butrica, "Electronics Research Center."

4. Details related to NASA and the Kendall Square Urban Renewal Project come from a variety of sources. These include the Rowland interview, November 25, 2019; Simha interview, October 30, 2019; and the two documents by Butrica. Also essential are Cambridge Redevelopment Authority, "Kendall Square Urban Renewal Plan" and "Background of the Kendall Square Urban Renewal Project." See also Simha, *MIT Campus Planning*, 84–86, "A Brief History," and "Cambridge Redevelopment Authority." Tercyak has also written about this: "Initial Years," "Six Pivotal Episodes," and "MBTA Role." See also Maycock, *East Cambridge*; Fitzpatrick, "Duck Pin"; Winling, *Building the Ivory Tower*. Various accounts have slightly different acreage numbers. I use CRA figures.

5. Simha, *MIT Campus Planning*, 84.

6. Simha interview, October 30, 2019.

7. Simha, *MIT Campus Planning*, 84. These were Section 112 credits, based on Section 112 of the Housing Act of 1949. There are slight discrepancies over the amount of the credits Cambridge received. I went with Tercyak, "Six Pivotal Episodes."

8. Maycock, *East Cambridge*, 63.

9. NASA launched the center's L-band mid-Atlantic satellite for navigation, communications, and air traffic control in August 1969 as the experimental ATS-E (now called the ATS-5) satellite.

10. NASA's historical paper did not reach a conclusion. "Was the Electronics Research Center simply the victim of Nixon's revenge against the Kennedy's? [sic] Or was closing the center a necessary part of NASA's third consecutive budget cut, a cut that also eliminated the Voyager mission to Mars, the NERVA II nuclear rocket, and much of Apollo Applications (the sequel to Apollo)?" it asked. Butrica, "Electronics Research Center: NASA's Little Known."

11. Cambridge Redevelopment Authority, "Background of the Kendall Square Urban Renewal Project."

12. See Tercyak, "Six Pivotal Episodes," "MBTA Role," and "Initial Years."

13. Maycock, *East Cambridge*, 63.

14. The Gateway Center; see Maycock, 64.

15. Safdie interview, October 12, 2018. All Safdie quotes come from this interview. See also Maycock, *East Cambridge*, 64–67. Plans called for some housing in Parcel 3 that was never built; it instead became the Marriott Residence Inn.

16. Simha interview, October 30, 2019.

Chapter 11

1. Kapor interview, November 29, 2019. Unless otherwise noted, all Kapor quotes are from this interview. Other Lotus details are from this interview and several other sources, including Sahlman, *Lotus Development Corporation*; "Lotus Development History"; Wikipedia, "Lotus Software."

2. See Huang, "Entrepreneur Walk of Fame"; Buderi, "Kendall Square."

3. Kapor interview with Karen Weintraub and Michael Kuchta, January 25, 2019. Great thanks to them for sharing interviews done for their own book.

4. Sahlman, *Lotus Development Corporation*. Lotus's first product was a graphics presentation program called the Executive Briefing System that Kapor had begun earlier for Apple. With Sachs in the fold, he focused anew on the spreadsheet.

5. Kapor interview with Weintraub and Kuchta, January 25, 2019.

6. Kapor interview with Weintraub and Kuchta.

7. Sahlman, *Lotus Development Corporation*, 14.

8. Edited portions of Kapor interview with Weintraub and Kuchta. Other company culture details from Wikipedia, "Lotus Software," and conversations with Kapor and Reed Sturtevant.

9. After stepping down as CEO, Kapor remained chairman for about a year. He left completely in 1986.

10. A variety of offices and condominiums were also erected in this period around Canal Park in East Cambridge.

11. Sturtevant interview, December 11, 2019.

12. Lemann, "CIA Still at Tech Square."

13. Hillis interview, November 22, 2019. All Hillis quotes and details are from this interview. Some lab details pieced together from other sources, including Noftsker interviews, and Levy, *Hackers*.

14. Levy, *Hackers*, 98.

15. Additional Thinking Machines details from an array of articles, including Davis, "Road to Reasoning;" Taubes, "Rise and Fall;" Markoff, "American Express to Buy 2 Top Supercomputers," "Computer Said to Be the Fastest;" Bulkeley, "Reorganization of Thinking Machines Filed."

16. Rosenberg, "AI Alley's Longest Winter." Some details about Gold Hill and Palladian also come from this article.

17. Golden, "'AI Alley.'" Market size facts and details about the Gold Hill offices and IBM's AI lab are also from this article.

18. Golden, "'AI Alley.'"

19. Noftsker had grown up in Carlsbad, New Mexico. He got his solo pilot's license at age sixteen, bought his own single-engine, sixty-five-horsepower Luscombe with $1,400 saved from odd jobs, and flew the roughly two hundred miles between his home and Las Cruces almost every week when attending college at New Mexico State. Thanks to an introduction to Minsky by a childhood friend who worked at Project Mac, he had been hired in 1965 as AI lab administrator. He then hired many of the hackers and programmers who became legendary names in computing and AI, including Richard

Stallman, Bill Gosper, and Richard Greenblatt. He left MIT in 1973, settling near Los Angeles, where he cofounded a company that made computer controls for precision welding machines.

20. See Wikipedia, "Richard Greenblatt (programmer)."

21. More details of Lisp Machines and Symbolics and the controversy around them are in Levy, *Hackers*. See also Wikipedia, "Lisp Machines"; Stallman, "My Lisp Experiences."

22. Funding came from three leading VCs: Memorial Drive Trust, affiliated with Arthur D. Little; General Doriot's American Research and Development; and Patricof Associates. When Symbolics was founded, Bob Adams, an MIT grad and part of the founding team behind Scientific Data Systems, was named president. Noftsker served as secretary. Noftsker later took over as president, then the company's top position. Still later, he became the first CEO and board chairman.

23. Noftsker email, December 19, 2019.

24. Before going public, the company raised $16.5 million in 1983 through a private placement.

25. Details of Palladian's formation from Sahlman, *Palladian Software*. Computer Pictures, which Cooper had previously founded, made software that simplified the analysis of business data through graphics and also automated the creation of some of the graphics. Cooper's Sloan thesis was titled "Artificial Intelligence: A Heuristic Search for Commercial and Management Science Applications."

26. Cooper interview, January 12, 2020. All Cooper's quotes are from this interview, along with the AI Alley billboard story and some Palladian details. Cooper says AI Alley represented the fusion of two lines of thought. He had learned of Boston's Pi Alley, where leftover type from newspaper composition—called *pied type*—was reputedly collected. The other inspiration was Allen's Alley, a routine in the long-running Fred Allen radio show popular in the 1930s and 1940s. As a comedy gag, Allen and his wife would stroll down the alley, talking to the colorful residents. Relates Cooper: "So I knew about Pi Alley where all the type was thrown. And then I thought about if you walked through Kendall Square and knocked on all the different doors, you'd meet all these characters."

27. Lohr, "Charles Bachman." See also Smith, "Charles Bachman." Bachman Information Systems, founded in 1983, went public in 1991; shares surged for a time.

28. Movie references from Wikipedia, "Thinking Machines Corporation."

29. The bankruptcy filing was largely driven by the long-term lease Thinking Machines held, Hillis notes, which was seen as an albatross around its neck and not the asset such leases would become as the square filled up.

Chapter 12

1. Kifner, "'Creation of Life' Experiment."

2. Cooke, "Mayor Vellucci."

3. Property tax information from City of Cambridge.

4. Gilbert interview, January 27, 2020.

5. Information about the early days of molecular biology at Harvard and MIT, and of biotech, come from multiple sources. These include interviews with Sharp and Gilbert and communications from Victor McElheny; Luria, *Slot Machine*; Durant, "Refrain from Using."

6. Baltimore interview, January 30, 2020.

7. Cold Spring Harbor had been run by James Watson since 1968, even though he remained on Harvard's faculty.

8. Durant, "Refrain from Using," 148–149.

9. McElheny email, February 4, 2019.

10. Baltimore interview.

11. Durant, "Refrain from Using," 146.

12. Durant, 146.

13. The letter appeared in *Nature*, *Science*, and *Proceedings of the National Academy of Sciences*.

14. Many things have been written about the Asilomar conference. I rely chiefly on Durant, "Refrain From"; Anerbach, "Man Created Risks." See also Feldman and Lowe, "Consensus from Controversy." This was the second Asilomar meeting. The first, entitled Biohazards in Biological Research, was held January 22–24, 1973.

15. Durant, "Refrain from Using," 150.

16. Gottlieb and Jerome, "Biohazards at Harvard."

17. Vellucci interview with Rae Goodell, May 9, 1977. He also told Goodell: "I shook up the whole news media around the world. I got Harvard and MIT scientists to go down to marketplace and peddle their goods like a street hawker and a peddler and I cut them down to size. From now on there's a new era in the world of science."

18. Hall, *Invisible Frontiers*, 44.

19. Quoted in Durant, "Refrain from Using," 151. Some details from Feldman and Lowe, "Consensus from Controversy."

20. The former councilors Clem refers to are Cornelia "Connie" Wheeler and Daniel "Dan" Hayes Jr. Wheeler represented the progressive group known as the CCA, for Cambridge Civic Association. Hayes was an independent. Also a former city mayor, he chaired the review board. The inclusion of those two, says Clem, "was the glue that kept this committee going. It took the standard conflict between liberals and independents out of this commission because it was viewed as having credibility by both sides." D. Clem interview, July 22, 2021.

21. Feldman and Lowe, "Consensus from Controversy." See also Durant, "Refrain from Using."

22. Additional steps called for by the ordinance included lead scientists being available for public hearings; and proof, if required, that lab staff members had been trained in biosafety. See Feldman and Lowe, "Consensus from Controversy"; Durant, "Refrain from Using."

23. D. Clem interview, February 14, 2019.

24. Durant, "Refrain from Using," 156–157.

25. Sharp interview, April 25, 2018. Unless noted, all Sharp quotes are from this interview.

26. Sharp in talk at MIT, November 19, 2019.

27. Sharp, onstage chat on November 19, 2019, after a screening of the documentary film, "From Controversy to Cure," about the Cambridge biotech boom. Some details of Inco's investment in Genentech from Hall, *Invisible Frontiers*.

28. Good details about the early maneuverings leading to formation of Biogen can be found in Hall, *Invisible Frontiers*, 192ff. I also relied on interviews with Sharp and Gilbert.

29. Lubrizol, a specialty chemical maker now owned by Berkshire Hathaway, invested in several biotech companies.

30. The Paris meeting story, quotes, and account of how Biogen got its name are from Hall, *Invisible Frontiers*, 209–210. Other details of Biogen's early days from Hall, as well as Dick and Jones, "Commercialization of Molecular Biology." See also Roberts, *Celebrating Entrepreneurship*.

31. Quoted in Roberts, *Celebrating Entrepreneurship*, 139.

32. Gilbert interview, January 27, 2020. Details about the rules giving the SAB more power and about initial investments and valuation also come from this interview.

33. Biogen was originally set up as a patent-holding company, most likely on the assumption that it would monetize its science by licensing its technologies to other companies. However, later that year, legal incorporation was shifted to the Netherlands Antilles "in order to take advantage of a favorable treaty network," according to Ken Novack, an attorney who long represented Biogen as outside counsel. Registration was later shifted to the United States. Novack email, February 2, 2020. See also Hall, *Invisible Frontiers*, 212ff.

34. Early funding details from Gilbert interview, January 27, 2020.

35. Dick and Jones, "Commercialization of Molecular Biology."

36. Details of Geneva lab from Gilbert interview, January 27, 2020; Dick and Jones, "Commercialization of Molecular Biology."

37. Cloning alpha interferon was a significant achievement. On hearing the news, Weissmann rushed back from his vacation and called Biogen's attorney, Jim Haley, in California. Haley arrived in Europe on Christmas Day to begin patent filing work. More results were shared at a previously scheduled meeting of Biogen's scientific advisory board on January 12, 1980. The company decided to announce the news publicly at a hastily arranged seminar at MIT on January 16—followed by a press conference in Boston. Sharp interview, April 25, 2018; Rasmussen, *Gene Jockeys*, 111; Schmeck, "Natural Virus-Fighting."

38. Grand Metropolitan once owned Pillsbury and Burger King. In 1997, it merged with Guinness to form Diageo.

39. Quoted in Roberts, *Celebrating Entrepreneurship*, 140.

40. Feldman and Lowe, "Consensus from Controversy."

41. I rely on Feldman and Lowe for their account of the amended permit process and experience.

42. Quoted in Roberts, *Celebrating Entrepreneurship*, 140.

Chapter 13

1. Statistics on early biotech companies from Dick and Jones, "Commercialization of Molecular Biology."

2. Founding dates and influence of early biotech companies determined from multiple sources, including interviews with biotech leaders and company websites.

3. Quoted in Feldman and Lowe, "Consensus from Controversy."

4. Wyeth-AHP paid $666 million for its stake. The Genetics Institute story is pieced together chiefly from FundingUniverse, "Genetics Institute"; and Feldman and Lowe, "Consensus from Controversy." See also Timmerman, "Genetics Institute Alumni."

5. Sayare interview, February 4, 2020. Early Immunogen story comes largely from this interview, supplemented by details from the company's website.

6. Lodish interview, February 5, 2020. All Lodish quotes are from this interview. Details of Genzyme's early days and BIA's role in it come from this interview; Roberts, *Celebrating Entrepreneurship*, 142–143; and Fearer, "Improbable Circle of Life." A few details are from the Steinert interview, February 20, 2020.

7. Lynch, "Genzyme 'Adventure' Ends."

8. Baltimore interview with Charles Weiner and Rae Goodell, May 3, 1977.

9. Details on Jack Whitehead, Technicon, and Whitehead Institute founding from a variety of sources, primarily: Baltimore interview, January 30, 2020; Susan Whitehead interview, February 17, 2020; Lodish interview, February 5, 2020; Knox, "Financing Science's Taj Mahal"; Knox, "MIT Debates Gift"; Klein, "Fracas over Funds"; "MIT Approves Ties"; Wikipedia, "Whitehead Institute"; Whitehead Institute website: https://wi.mit.edu/history.

10. Whitehead interview, February 17, 2020. Lewis Thomas's story comes from this conversation and follow-up email.

11. Lodish interview, February 5, 2020.

12. Baltimore interview, January 30, 2020.

13. Baltimore interview, January 30, 2020.

14. In addition, the agreement specified that while the Whitehead Institute would own patents stemming from research at the facility, it would split royalties and other revenue with MIT after various expenses were deducted.

15. Baltimore interview, January 30, 2020.

16. Knox, "Financing Science's Taj Mahal."

17. Wikipedia, "Whitehead Institute."

18. Startup figures from Lisa Girard, Whitehead's director of strategic communications.

19. Afeyan interview, February 20, 2020.

Chapter 14

1. English (Larson's first chef), Adams, Goin, and Lynch all won James Beard awards while opening a slew of well-known restaurants. For the last few years, the manager of Michela's was Christopher Myers, who became another renowned Boston restaurateur. Larson closed Michela's in 1994 for personal reasons. The Blue Room closed in 2017.

2. Brand, *Media Lab*, 82.

3. Negroponte interview, April 25, 2018.

4. Atari's lab was headed by Cynthia Solomon, who had started as Minsky's secretary in 1962 but soon demonstrated her computer chops, codesigning Logo with Papert and Wally Feurzeig. The lab grew to twenty-two full-time staffers and ten consultants. It closed in April 1984, after the bottom fell out of the home video game market. Three staffers, including Minsky's daughter, Margaret, were among the first Media Lab graduate students when it opened in 1985.

5. Sculley email, February 24, 2020. All Sculley quotes are from this email.

6. Details of Apple's Advanced Technology Lab come chiefly from Nassi interview, Febrary 27, 2020. Supporting details from Sculley email.

7. D. Clem interviews, February 14, 2019, July 26, 2019, and February 27, 2020, and Wikipedia, "Digital Equipment Corporation."

8. Greif email, May 9, 2020. Other details about the Lotus lab and IBM Research are from this email.

9. Many historys and articles cover the basics of ARPANET, BBN, WC3, and the early internet. For some history and one take on what's ahead, see Roush, "What Is the Future of the Internet?"

10. Leighton interview, March 6, 2020. Unless noted, all Leighton quotes come from this interview. The Akamai story is drawn chiefly from interviews with Leighton and Conrades; Raskin, *No Better Time*; Edelman, Eisenmann, and Van den Steen, *Akamai Technologies*; and Van den Steen, *Akamai's Edge (A)* and *Akamai's Edge (B)*. See also Spinrad, "New Cool." Leighton had also worked as a consultant for Thinking Machines, teaching parallel computing to its staff during some summers.

11. The protocol's authors were Leighton, Lewin, David Karger, Eric Lehman, Matthew Levine, and Rina Panigrahy.

12. The team's initial company name was Cachet Technologies.

13. A short precompetition, essentially an elevator pitch, leads up to the main event. In Akamai's year, the precompetition prize was $1K, with each of the ten category winners receiving one hundred dollars. Prize money for the main event was supposed to be distributed as $30K for first place and $10K each to the next two places. But when two teams tied for first, an anonymous donor added $20K to the pot—enabling the winners to each get $30K, with $10K going to third place. Then one winner, Direct Hit, donated its prize to the finalists out of the money awards levels—as it had just secured a big venture round. See Raskin, *No Better Time*. Leighton thinks Akamai got $5K. The competition was rebranded *the 100K* in 2006, thanks to an endowment from Leighton and Anne Lewin, Danny Lewin's widow.

14. Quoted in Raskin, *No Better Time*.

15. Quoted in Raskin.

16. BBN sale to GTE from Conrades interview, March 11, 2020. See also Schiesel, "GTE Discloses 3 Big Deals."

17. As a venture partner, Conrades was not a full partner in the firm—a position usually called a *managing director* or *general partner*. *Venture partners* are typically experienced businesspeople who help manage and advise portfolio companies but don't work full time. Many are semiretired or taking a breather between executive positions.

18. Conrades interview, March 11, 2020. All Conrades quotes are from this interview.

19. A trailer had been shown in movie theaters the previous November. But the March 1999 trailer release was a much bigger deal. For more on *Phantom Menace* and Akamai, see Trenholm, "20 Years Ago."

20. Associated Press, "Akamai Technologies Stock Surges." Information on peak stock price and valuation from Edelman, Isenmann, and Van den Steen, *Akamai Technologies*. In April 2001, *Forbes* estimated Lewin's wealth at $285.9 million, with Leighton just behind at $284.3 million, as cited in Raskin, "Sept. 11 Secrets."

21. 9/11 Commission, *9/11 Commission Report*, 5. See also Raskin, "Sept. 11 Secrets," for more on Lewin's actions that day.

Spotlight: Lita Nelsen

1. Garde, "Biotech's Behind-the-Scenes."

2. Garde, "Biotech's Behind-the-Scenes."

3. Nelsen interview, March 9, 2020. Unless noted, all Nelsen's quotes come from this interview, as does bulk of her personal story, with some follow-up emails.

4. Bauter Engel, "Exit Interview."

5. MIT participated in IPAs, but did not do much with it, says Nelsen.

6. Bauter Engel, "Exit Interview."

Chapter 15

1. Rowe email, January 8, 2021.

2. CIC's story was drawn chiefly from these sources: Rowe interviews, April 9, 2019, April 1, 2020 (and follow-up emails); Olmsted interview, April 2, 2020; Kerr, Kerr, and Brownell, *CIC (A)* and

CIC (B); Roush, "Cambridge Innovation Center Turns 10"; and Pan, "CIC Health." See also the CIC website and annual impact reports.

3. Olmstead interview, April 2, 2020. All Olmsted quotes are from this interview.

4. Roush, "Cambridge Innovation Center Turns 10."

5. Rowe interview, April 9, 2019. Unless otherwise noted, all Rowe quotes are from this interview.

6. Olmstead helped lead Research Dataware through two acquisitions.

7. Ember was acquired in 2012 by Austin-based Silicon Laboratories for $72 million and milestones. Venture investors had put in an estimated $89 million. See also Huang, "Ember CEO."

8. Over the years, various competitors popped up—including Silicon Valley–based Plug and Play, which opened in 2006, and New York–based WeWork, which launched in 2010.

9. Kerr, Kerr, and Brownell, *CIC (A)*, 5.

10. Roush, "New Business Association." See also Kerr, Kerr, and Brownell, *CIC (A)*.

11. Quoted in Roush, "Cambridge Innovation Center Turns 10." For more on fire, see Kerr, Kerr, and Brownell, *CIC (A)*.

12. Rowe interview, April 1, 2020.

13. Mamlet email, April 10, 2020.

14. Hunt interview, April 9, 2019.

Spotlight: Innovation Space Zoning

1. Rowe interview, April 9, 2019. All Rowe quotes are from this interview.

2. Kane interview, March 25, 2020. All Kane quotes are from this interview.

3. I rely chiefly on interviews (and follow-up emails) with Alexandra Levering, April 10, 2019 and April 3, 2020; interview with Tom Evans, April 9, 2020; and City of Cambridge, "Ordinance # 1378." (Note that the Kendall Square Urban Renewal Area was renamed the Kendall Square Urban Redevelopment Area in 2021.) Also important is Cambridge Zoning Ordinance 14.000—Mixed Use Development District: Kendall Square, 14.32.5 (City of Cambridge, "Article 14.000"). This area covers the core of Kendall Square around the Marriott Plaza. Similar, but not identical, rules govern other parts of Cambridge, including several that overlap with Kendall Square. A variety of other information about past and current plans for Kendall Square, including the K2 Plan, can be found on the Cambridge Redvelopment Authority website: https://www.cambridgeredevelopment.org/kendall-square-3.

4. Evans interview, April 9, 2020. All Evans quotes are from this interview. The Link's space is managed by TSNE, a nonprofit formerly known as Third Sector New England.

5. Boston Properties, "Innovation Space Compliance Report 2019"; Levering interviews; Koop interview, April 2, 2020.

6. Emails with CRA, Alexandria Real Estate Equities, and LabCentral.

Chapter 16

1. Stowe interview, May 7, 2018, with follow-up email, July 28, 2021.

2. Bosley interview, April 14, 2021. All Bosley quotes are from this interview.

3. Bourque interview, June 9, 2020.

4. Sale price was $278.8 million. At that time, Tech Square comprised the four early buildings, plus three others under construction. MIT sold 90 percent of the property to Alexandria Real Estate

Equities in 2006 for $540 million. ARE later bought the remaining 10 percent. See Brehm, "MIT to Sell Share in Technology Square"; Seiffert, "Alexandria Aims to Repeat Success."

5. Novartis was formed in 1996 around the agrochemical and pharmaceutical arms of Ciba-Geigy and Sandoz, following their merger.

6. Vasella interview, June 30, 2020. Unless otherwise noted, all Vasella quotes come from this interview.

7. Novartis's drug approval numbers and NIBR plans are compiled from announcements, early coverage, and subsequent interviews. These include: "Novartis Steps Up US Research Investment"; Fuhrmans and Zimmerman, "Novartis to Move Global Lab"; Halber, "Novartis Opening in Tech Square"; Lockwood interviews; Hughes interview, April 27, 2020. See also Henry, "Sweet Research Life at Novartis."

8. Quoted in Halber, "Novartis Opening in Tech Square." This article gives figures on Cambridge biotechs at the time.

9. Fishman email, February 25, 2020.

10. Hughes interview, April 27, 2020. All Hughes quotes are from this interview.

11. Story from Lockwood interview, October 11, 2019.

12. Story from Hughes email, February 16, 2021.

13. Contest details from Lockwood interview, April 27, 2020.

14. Quoted in Aoki, "Candy Coated Power-Cleaning." The article also has renovation details.

15. Lockwood interviews. See also Moravek, "Candy Land."

16. Lockwood interview, October 11, 2019.

Spotlight: Bob Langer

1. Patent, research paper numbers, and citation figures from Langer's bio and email correspondence. See also Prokesch, "Edison of Medicine."

2. Kimball interview, January 13, 2021.

3. Langer interview, June 8, 2020. Unless otherwise noted, all Langer quotes are from this interview.

4. Prokesch, "Edison of Medicine."

Chapter 17

1. Morse's remarks and event details from Morse email, June 14, 2020.

2. Levin interview, May 18, 2020. Unless otherwise noted, all Levin quotes are from this interview.

3. Fisher, "Rocky Road from Startup to Big-Time Player"; interviews with Biogen veterans; "Biogen Inc."

4. Fisher.

5. Fisher.

6. One important deal involved hepatitis B vaccine technology licensed to SmithKline Beecham, which proved a watershed after its 1989 approval. For landmark alpha interferon approval for hairy cell leukemia, see Henderson, "Cancer Drug Interferon."

7. Quoted in Fisher, "Rocky Road from Startup to Big-Time Player."

8. Avonex beat out a competing beta interferon drug called Betaseron, originally developed at Cetus but licensed to Schering (different from Schering Plough). Lawyers were involved. See Fisher.

9. Maraganore interview, April 28, 2020. All Maraganore quotes are from this interview.

10. D. Clem interview, February 14, 2019. Clem quotes and his story are from this interview, with follow-ups from additional conversations.

11. The two residential sites were bought by Twining Properties.

12. Details of Genzyme's success are drawn primarily from Toffel and Sesia, *Genzyme Center (A, B, C)*. The "leaking money" quote is from Wilke, "Genzyme Says It Has Potential Hot Seller."

13. Genzyme signed the lease in August 2000, but the term didn't start until April 30, 2003. Construction was completed that November. Meanwhile, Genzyme stayed at One Kendall. The details here are primarily from Clem and Kane interviews and emails.

14. Genzyme's lease expired July 31, 2018, with options on two successive ten-year terms. Details from D. Clem interviews, February 14, 2019, and July 26, 2019; Kane interview; Toffel and Sesia, *Genzyme Center (A, B, C)*. See also Campbell, "'Green' Building."

15. Quoted in Toffel and Sesia, *Genzyme Center (A)*.

16. Boger interview, May 15, 2020. All Boger quotes are from this interview.

17. Details of Vertex's lease and subleases from Kane interview and BioMed Realty emails. Vertex signed a lease in January 2001. It became effective January 1, 2003, and expired April 30, 2018.

18. Details of Vertex in Kendall Square from interviews with Boger, Kane, and D. Clem, February 14, 2019, and July 26, 2019; and email with Vertex communications. Employment stats from City of Cambridge: https://www.cambridgema.gov/cdd/factsandmaps/economicdata/top25employers.

19. In 2013, Ibsen leased a floor and a half, moving in the following year. Also in 2014, Baxter took the rest of the building. Baxter transferred its lease to spin-off company Baxalta. Then Baxalta was bought by Shire, which in turn was acquired by Takeda in early 2019. The lease expires in early 2027.

20. In 2004, University Residential Communities was formed by a group of Harvard and MIT faculty seeking to develop intergenerational housing in Cambridge for university-related people. The effort was led by former MIT president Paul Gray, with Bob Simha as executive director. The initial project was established at the 160-unit complex at 303 Third Street, where over forty condominium units were under agreement. The plan ultimately fell through during the 2008 recession. However, Simha relates, eleven purchasers who held out took ownership of their condos. The rest of the units in that block, now called Third Square, are rented as apartments.

21. Millennium story from Levin interview, May 19, 2020, and follow-up emails; Maraganore interviews, April 28–29, 2020; Thomke and Nimgade, *Millennium Pharmaceuticals, Inc. (A)*; and Shah, "Mark Levin." See also Johannes and Moore, "Bayer, Millennium Enter Alliance." Some additional LeukoSite details are from Springer interview, July 1, 2021.

22. Millennium's board included John Doerr of Kleiner Perkins, Tony Evnin from Venrock, and Bill Hellman of Greylock. Gillis later became managing director of Arch Venture Partners, another successful biotech venture firm.

23. Dunsire cut Millennium's inflammation research to focus on oncology. She bet big on another compound acquired through LeukoSite: a monoclonal antibody for treating Crohn's disease and ulcerative colitis. This became Entyvio, a multibillion drug for Takeda. Dunsire interview, June 9, 2020.

24. Carroll, "Ex-Biogen Exec."

25. Alnylam's founders included Sharp and Westphal, John Clarke of Cardinal Partners, and four other scientists: Thomas Tuschl and David Bartel, both then at the Whitehead Institute; Paul Schimmel, then at MIT; and Phillip Zamore of the University of Massachusetts Amherst.

26. Sharp had reached out about RNA interference the previous year, and Maraganore had arranged for Sharp, Tuschl, and Bartel to meet with Millennium, which took a nonexclusive license for MIT's inventions around the field.

27. Alnylam's drug for transthyretin amyloidosis is patisiran; the AHP treatment givosiran; the primary hyperoxaluria type 1 drug is lumasiran. The latest approvals, for lumasiran and inclisiran, came against the backdrop of a major partnership with private equity firm Blackstone Group, signed in April 2020, that could bring Alnylam up to $2 billion. The deal, designed to carry Alnylam to independence without further financing, included Blackstone receiving half the biotech's royalties from inclisiran. The drug's approval was delayed due to a December 2020 FDA ruling over a manufacturing issue, but was widely expected to go ahead in 2021. See company website: https://www.alnylam.com; Vinluan, "Alnylam Gets Quick FDA OK"; Gardizy, "Alnylam Wins FDA Approval"; Al Idrus, "Alnylam Scores $2B"; Liu, "FDA Rebuffs Inclisiran"; O'Riordan, "Inclisiran Approved in Europe."

28. All details from Maraganore interview, January 5, 2021.

29. Andrews interview, May 21, 2020.

30. Quotes and information on early MassBio days from Bourque interview, June 9, 2020, with follow-up emails. The organization's first home under her was inside Feinstein Kean Healthcare, a public relations firm founded by Marcia Kean and Peter Feinstein, who had headed corporate communications for Biogen. MassBio moved twice under Bourque—first to the Athenaeum Building, then to One Broadway. It is now based in Technology Square.

Chapter 18

1. Details on building history from Broad Institute; McGuire interview, July 1, 2020, and follow-up emails; and Eric Lander interview, November 30, 2020, plus follow-up emailed notes, November 29, 2020, and January 25, 2021.

2. Sharp interview, April 25, 2018.

3. Lander communication to author via emailed notes. For information on Joseph O'Donnell and Boston Concessions, later named Boston Culinary Group, see "Joseph J. O'Donnell, MBA 1971."

4. See Wikipedia, "Human Genome Project." The race was officially called a tie between government-funded HGP and Celera Genomics.

5. Altshuler interview, June 18, 2020. All Altshuler quotes and background details are from this interview.

6. Lander communication with author via emailed notes.

7. The Celera story is complicated. For a start, see Wikipedia, "Human Genome Project." I filled in some information through Afeyan's interview, July 14, 2020, and Lander's comments and interview of November 30, 2020. More details are in "Spotlight: Flagship Pioneering—Kendall Square Company Creator."

8. See Wade, "Scientists Complete Rough Draft," and Wikipedia, "Human Genome Project." The entire human geneome was finally sequenced in mid-2021.

9. Kolata, "Power in Numbers."

10. Lander interview, November 30, 2020. Unless noted, all Lander quotes are from this interview.

11. Broad, *The Art*. Broad's memoir has a different version of this and related events, but he and Lander agree generally.

12. Athenahealth, among others, set up at the Arsenal site, which is roughly 6.5 miles from the Broad.

13. Broad, *The Art*, 163. For information on funding, see 95–96. As of 2021, the Broads had given or pledged more than $1 billion to the institute, including $150 million announced in March 2021, part of which will help support the new Eric and Wendy Schmidt Center mentioned in chapters 1 and 28.

14. Nhuch, "Broad Institute Expands."

15. Ted Stanley made his fortune in collectibles. Details of the Stanley Center gift and the 75 Ames Street building, which opened in 2014, from Broad Institute press releases: "$650 Million Commitment"; and Nhuch, "Broad Institute Expands." See also Roberts, "Ted Stanley."

16. Kolata, "Power in Numbers."

17. Begley, "Why Eric Lander Morphed."

18. Broad Foundation, "Broad Institute Director Eric Lander Wins."

19. Broad Institute, "Eric Lander to Take Academic Leave."

20. McGuire interview, July 1, 2020.

21. The Broad had a research exemption under state COVID rules, allowing it to stay open during the pandemic. However, it decided to suspend a lot of work until new safety protocols were in place.

22. McGuire interview, July 1, 2020. See also Broad Institute, "Broad Institute Provides COVID-19."

23. Begley, "Why Eric Lander Morphed." Most episode details are from this and Hall, "Embarrassing, Destructive Fight." See also Regalado, "Scientist's Contested History"; Lander, "Heroes of CRISPR."

24. Hall, "Embarrassing, Destructive Fight."

25. Subsequent rulings favored the Broad. It won patents on Zhang's discovery in 2014, while the University of California did not receive patents for Doudna's work. The UC system lost before a patent office appeal board in 2017 and in a federal appeals court decision the following year. In September 2020, the US Patent Office ruled for the Broad in a related dispute as well. See Begley and Cooney, "Two Female CRISPR Scientists."

26. Begley, "Why Eric Lander Morphed."

27. Hall, "Embarrassing, Destructive Fight."

28. Quoted in Hall.

29. @eric_lander, October 7, 2020.

30. Quoted in Begley and Cooney, "Two Female CRISPR Scientists."

Chapter 19

1. Timmerman, "Biogen Idec CEO." Details about Biogen's return to Cambridge are pieced together from this article; McBride, "New CEO George Scangos"; Timmerman, "George Scangos"; and Ross and Weisman, "Biogen Idec Expanding Cambridge Footprint."

2. Ross and Weisman, "Biogen Idec Expanding Cambridge Footprint."

3. Altshuler interview, June 18, 2020.

4. Vertex-cum-Alnylam space details come from interviews with Maraganore, Boger, Emmens, Clem, and Kane, and email with Vertex communications. A few points are from Weisman, "New Complex"; Donnelly, "Vertex Moving." Starting in early 2015, Vertex sublet its three floors to startup space provider Mass Innovation Labs (later called SmartLabs). Alnylam's lease plan was announced in April 2015—though Alnylam couldn't move in until mid-2018. See also Vermes, "Booming Alnylam."

5. Maraganore interview, January 5, 2021.

6. Details from multiple press accounts, plus company materials. Shire sold its oncology franchise to Servier ahead of selling its remaining assets to Takeda. For overview of Takeda's activities, see Saltzman, "Takeda, already." As noted in chapter 17, Aveo Pharmaceuticals experienced an FDA setback after it took occupancy of 650 East Kendall. Baxter took over the lease in 2014, then gave way to Baxalta when it spun out of Baxter in 2015. Thanks to Bill Kane for helping sort this out. State employment rankings from MassBio, *2020 Industry Snapshot*.

7. Alnylam had an option on the building—but waived its claim, opting for the nearby former Vertex building. Other details about the Lyme/BioMed development are gleaned from Kane's interviews. As this played out, another glitzy building went up near the Genzyme Center. The glass-walled structure, at 450 Kendall Street, opened in 2015, marking the fourth office or lab complex to occupy the ten-acre parcel BioMed Realty had acquired from Lyme Properties. Life sciences venture capital firm MPM Capital and Eli Lilly were the main tenants of the five-story building in 2020, along with a ground-floor restaurant fittingly called Glass House.

8. Emmens interview, August 4, 2020. All Emmens quotes are from this interview and email clarifications.

9. Emmens interview, August 4, 2020. See also "Shire to Buy Transkaryotic"; "Shire to Acquire TKT."

10. Plump interview, August 31, 2020. All Plump quotes are from this interview.

11. In 2020, Takeda had more than 6,750 employees in Massachusetts, at some twenty sites. Details about Takeda and its strategy are chiefly from Plump interview, August 31, 2020, and supporting Takeda documents.

12. By then, Greif had become familiar with people from IBM Research's headquarters in Yorktown Heights, New York. Leading its software group was someone she knew from MIT: Jeff Jaffe. In 2010, he would return to Kendall Square, as CEO of the World Wide Web consortium.

13. The story of Microsoft Research New England's early days comes chiefly from Roush, "New Microsoft Lab"; Buderi, "Microsoft Research Lab."

14. Buderi, "Microsoft Research Lab."

15. 2019, Chayes was named an associate provost at UC Berkeley and dean of the School of Information. Borgs became a professor. For details of Microsoft in Kendall Square in 2020, I rely chiefly on Jewart interview, September 1, 2020. The building's other main tenant in 2020 was InterSystems Corp., a privately held database management software and technology company founded by MIT grad Phillip "Terry" Ragon. In 2009, Ragon and his wife, Susan, pledged $100 million to create a medical research institute. Four years later, the Ragon Institute of MGH, MIT, and Harvard moved from Boston to 400 Technology Square. In 2020, the Ragon Institute refocused much of its work on the coronavirus.

16. Kerr, Kerr, and Brownell, *CIC (A)*, 5. Other details about Android's move to CIC are also from this source.

17. Cambridge Center was rebranded Kendall Center in 2015, with individual buildings given street addresses, replacing the original number system (e.g., One Cambridge Center).

18. Vinter interview, July 28, 2020. Unless otherwise noted, all Vinter quotes are from this interview.

19. Most details on Google's early growth in Cambridge come from Vinter interview, July 28, 2020. See also Kirsner, "Inside Google's Cambridge, MA Offices"; Roush, "Google's Open House." Roush includes a bit on early projects as well.

20. ITA cofounder and CEO Jeremey Wertheimer would serve for years as Google's co-site lead with Vinter and Brian Cusack. Information on ITA comes chiefly from Vinter interview, July 28, 2020; Roush, "ITA Software." See also Huang, "ITA Software Bought by Google."

21. 325 Main is being erected on the site of the four-story building previously known as Three Cambridge Center. Details on Google's expansion from Koop interview, April 2, 2020; Logan, "It's Official."

22. Later, in 2008, venture firm Flybridge Capital Partners started Stay in MA. This program paid student attendance fees up to one hundred dollars for local high-tech events to help students connect to the Boston scene and possible jobs. See Buderi, "Flybridge Unveils Scholarship Program."

23. Roush, "Google's Open House."

Chapter 20

1. For ARD and Ionics, see MacDonald, "Venture Capital;" Ante, *Creative Capital*, 118–119.

2. See Buderi, "Startup Profile."

3. Roush, "TechStars." Other Techstars details come from this article. See also the Techstars website: https://www.techstars.com; Feld, "I Got My Bathroom."

4. Buderi, "Polaris to Open." Other Dog Patch details are from this article. Dog Patch later shifted to the same space Microsoft loaned to Techstars when it moved to Kendall Square.

5. See Rowe, "Tragedy of the Commons"; Stangler, "Non-compete Agreements." A new noncompete law went into effect October 1, 2018. Although noncompetes were not eliminated, they became much more expensive to enforce. For a summary, see Budoff, "New Massachusetts Non-compete Law."

6. For some perspective on Facebook and Boston venture firms at the time, see Buderi, "Boston VCs Grok Social Media."

7. Florida interview, November 20, 2020. All Florida quotes are from this interview.

8. VC moves taken from news accounts, company websites, press releases, various interviews, and personal experience. See Buderi, "Boston Venture Firms on the Move"; Kirsner, "Latest Venture" and "Venture Firm Matrix Partners."

9. Armony email, August 13, 2020.

10. Quoted in Alspach, "Boston Suddenly State's Tech Startup Capital."

11. MassChallenge details principally from Harthorne interview, September 14, 2020, supplemented by information on the organization's website: https://masschallenge.org.

12. For more details and sources related to CIC, see chapter 15. For LabCentral and BioLabs, see chapter 22. For Alexandria Launch Labs, see Alexandria Real Estate Equities, Inc., "Alexandria LaunchLabs."

13. Alspach, "Boston Suddenly State's Tech Startup Capital."

14. See press release, "Polaris Partners to Relocate." Despite moving its headquarters to Boston, Polaris kept the Waltham office until 2017. It closed Dog Patch in Kendall Square in 2014.

15. CB Insights quotes and details from Buderi, "Cambridge Officially the New Center."

16. CB Insights deal numbers from Alspach, "Boston Suddenly State's Tech Startup Capital." A graphic in the article credits the National Venture Capital Association for some numbers but doesn't identify which numbers.

17. Booth interview, September 1, 2020. All Booth quotes are from this interview, and most Atlas details.

18. Fagnan interview, September 16, 2020. All Fagnan quotes are from this interview.

19. Under the terms, the winner allocated half the $100,000 to an approved charity; the rest was credited to them personally. If the fund reported a 20 percent gain over its lifetime—that is, a profit of $20,000—the winner and the selected charity would each receive $10,000. The winner was Zaqary

Whitnack, a Sacramento-area media and design specialist. Great thanks to Sarah Downey for making this clear.

Spotlight: Flagship Pioneering

1. Booth interview, September 1, 2020. Some Stromedix details also from this interview. See also Weintraub, "Biogen Buys Stromedix."

2. The Flagship story is from interviews with Afeyan, February 20 and July 14, 2020, plus news accounts. All Afeyan quotes are from the second interview, unless otherwise noted, with some details from emailed follow-ups.

3. For more on PerSeptive's acquisition by Perkin-Elmer and the Celera spinoff, see Wikipedia, "Applera." Applera was the successor to PE's life science division.

4. For Rubius, see Vinluan, "Rubius Upsizes IPO." For VC exit rankings, see Murphey, "Top Biotech Venture Capital Funds"; Sheridan, "Among Biotech VCs."

5. Afeyan emailed comments, August 31, 2020.

Chapter 21

1. Bhatia interview, June 16, 2020. Unless noted, all Bhatia quotes are from this interview, as are details of her background.

2. The author has been a member of what was originally called the Boston Biotech Working Group since its inception. All members are volunteers and receive no compensation for their efforts.

3. Hockfield interview, September 15, 2020. Unless noted, all Hockfield quotes are from this interview.

4. The Koch Institute is named for industrialist David Koch, an MIT graduate and board member who survived prostate cancer. He contributed $100 million to the center during Hockfield's presidential term.

5. The group of institutions known as the *MIT-9* included MIT, Harvard, Yale, Princeton, Pennsylvania, Stanford, Michigan, UC Berkeley, and Caltech.

6. I no longer am connected to Xconomy. The company was sold to Informa in August 2016. I stepped down as CEO at the end of 2018 and served as chairman until October 2019.

7. The School of Science had six departments at the time: math, physics, chemistry, biology, brain and cognitive science, and neuroscience. The women later found two other female faculty who had joint appointments in the School of Science, making seventeen in total.

8. Charles M. Vest, "Introductory Comments," in "A Study on the Status of Women Faculty in Science at MIT," special edition, *MIT Faculty Newsletter,* March 1999, https://fnl.mit.edu/wp-content/uploads/2020/06/fnl114xy.pdf.

9. The study originally planned to look at physics as well, but it was dropped due to the low commercialization activity of faculty members.

10. Begley, "Three Star." See also Johnson, "Bias in Biotech Funding."

11. In January 2021, Joe Biden named Zuber as co-chair of the President's Council of Advisors on Science and Technology (PCAST).

12. Begley, "Three Star."

13. Details on the Dolphin Tank plans, including Lodish and Bhatia quotes about it, are from the working group's Zoom meeting, February 1, 2021.

14. MIT Faculty Newsletter, "Women in Biotech."

Chapter 22

1. While the official address is 700 Main Street, several businesses used different doors and addresses, among them 708 Main Street, 28 Osborn Street, and 2 Osborn. In May 2019, Chicago-based Harrison Street Real Estate and Boston-based Bulfinch agreed to pay MIT's investment arm $1.1 billion for a long-term lease on the three buildings—One Portland Street, 610 Main Street, and 700 Main. The cluster, called the Osborn Triangle, is occupied by Pfizer, Novartis, and LabCentral, plus a few smaller tenants. See Logan, "MIT-Owned Lab Buildings."

2. In 1870, $80,000 was worth about $1.5 million in 2019 dollars. Stillson's wrench story is from Wikipedia, "Daniel Chapman Stillson"; Wikipedia, "Pipe Wrench"; and the Walworth website: http://walworth.com/en/historia/.

3. Watson, *Birth and Babyhood of the Telephone*, 27.

4. Watson, 25–27. Article therein shows the image of the newspaper article.

5. Renovation and building history, including some details of Edwin Land's work, are from Cambridge Historical Commission, "Davenport Car Works." Some details of the Kaplan business are from "Joseph Downs Collection."

6. A variety of sources were consulted about Edwin Land. Specifics are found in further notes, but general sources include McElheny, *Insisting on the Impossible*; multiple emails from McElheny.

7. There is apparently no written account of Land's private lab, and his papers were shredded by his personal assistant upon his death in 1991.

8. McElheny email, July 2, 2019.

9. Mole's Hole description and quotes from McElheny emails.

10. For more on Adams and Polaroid, and his friendship with Land, see McElheny, *Insisting on the Impossible*, 203ff.

11. Details of Morse from multiple emails from McElheny; Woods interview (with Madeleine Turner); Bonanos, "Instant Artifact."

12. For more on Polaroid buying property and its later sale (discussed ahead), see Cambridge Historical Commission, "Davenport Car Works."

13. For more on Rogers, see McElheny, *Insisting on the Impossible*, 220ff.

14. McElheny, 154; supplemented by multiple emails with author.

15. LabCentral occupies the entire north side of the complex, called 700 North.

16. LabCentral, *2018 Impact Report*, *2019 Impact Report*.

17. Unless otherwise noted, all Fruehauf quotes are from his interview of March 8, 2019.

18. Cequent raised just under $18 million and also had a collaboration agreement with Novartis. See McBride, "MDRNA and Cequent Pharma."

19. As of 2020, BioLabs operated eight US sites and was eyeing international expansion. The original Cambridge operation moved to the Tufts University space in Boston's Chinatown; it's now called Tufts Launchpad BioLabs. A separate BioLabs facility opened in another part of Kendall Square in partnership with Ipsen.

20. Some details of LabCentral's 2017 expansion are from Weisman, "LabCentral Will More than Double."

Chapter 23

1. Termeer Square dedication posted at https://termeerfoundation.org/inspiredby/termeersquare.

2. Bosley interview, April 14, 2021.

3. Sharp interview, October 30, 2020.

4. Saxenian, *Regional Advantage*.

5. Saxenian, 2.

6. Saxenian, 59–60.

7. Saxenian, 59.

8. Fagnan interview, September 16, 2020.

9. Maraganore interview, April 29, 2020.

10. Sharp interview, October 30, 2020.

11. Hughes interview, April 27, 2020, supplemented by email. All Hughes quotes are from this interview.

12. Levin interview, May 18, 2020.

Spotlight: Mapping the Moderna Network

1. This high-level picture of the network behind Moderna is drawn chiefly from interviews and/or emails with Rossi, Afeyan, Dietz, Springer, and Langer; Garde and Saltzman, "The Story of mRNA"; Elton, "Does Moderna"; Huang, "Moderna"; Lash, "Venture Creation Story"; and the company's S1 filing with the SEC.

Chapter 24

1. Hockfield interview, September 15, 2020. Unless otherwise noted, all Hockfield quotes and the story of her visit with Menino are from this interview.

2. Stockman and Barker, "How Premier."

3. Lemieux et al., "Phylogenetic Analysis Highlights Impact." See also Krueger, "Biogen Conference"; Lemieux et al., "Phylogenetic Analysis Highlights Role." The latter study never mentioned Biogen by name, but it referred to "an international business conference" held February 26–27. See Saltzman, "Biogen Conference;" Zimmer, "One Meeting."

4. Enriquez interview, June 11, 2020.

5. Andrews interview, May 21, 2020. All Andrews quotes are from this interview.

6. Herman, "Pandemic Hasn't Hampered Health Care."

7. See Lopatto, "Pandemic Economy"; Statt, "Microsoft Had Stellar Three Months." IBM took a hit to sales and profits but improved its operating margins. See Novet, "IBM Improves." Quarterly earnings reports released by Alphabet, Apple, Facebook, and Microsoft as of mid-2021 showed their ongoing strength.

8. Rowe email, January 8, 2021.

9. O'Toole email, October 5, 2020.

10. Marsh interview, October 2, 2020. All Marsh quotes are from this interview. Details about the Kendall Square Initiative and Volpe Center are from this interview and another with Marsh from November 16, 2018.

11. Kane interview, March 25, 2020. Details about the Constellation Center come from this interview and from Kirsner, "Star Waiting to Be Born." The property's legal owner was Constellation Charitable Foundation, a nonprofit set up by KnicKrehm. Details of the approval are from Levy, "Kendall Square Lab Tower."

12. Ryan interview, March 26, 2019. All Ryan quotes are from this interview.

13. Baerkahn interview, June 24, 2020.

14. Hinds interview, August 28, 2020. All Hinds quotes, and most details of the East Cambridge Planning Team actions, come from this interview.

15. Hinds interview, August 28, 2020; Koop interview, April 30, 2021. For more on the substation, see Kendall Square: MXD Substation Plan at https://www.mxdsub.site.

16. Toomey interview, February 4, 2021. All Toomey quotes and background are from this interview. As part of another Kendall Square development, a park set to open in fall 2021 on Rogers Street, between Second and Third, will be named after Toomey.

17. Webb interview, June 12, 2020. See also Kendall Square, "You Can't"; Chesto, "Kendall Businesses to Try Experiment."

18. The *Transport Kendall* report can be downloaded at https://www.transportkendall.org. For Red Line car delays, see Vaccaro, "Long Before Pandemic" and "New Red, Orange Line."

19. Core details about Grand Junction Park and the multiuse pathway are from *Transport Kendall*.

20. Webb, Coughlin, and Murray, "In Allston."

21. Hockfield interview, September 15, 2020. All Hockfield quotes are from this interview.

22. See Goldy-Brown, "Cities with Most Colleges"; Bowker, "These Mass. Colleges."

Chapter 26

1. Enriquez interview, June 11, 2020.

2. For the MIT-Harvard merger, I rely chiefly on Prescott, *When M.I.T. Was "Boston Tech"*, 69, 192–201, 249–272; also Haglund, *Inventing the Charles River*, 207ff. See also Wildes, "Electrical Engineering."

3. Simha interview, October 30, 2019

4. Altshuler interview, June 18, 2020.

5. D. Clem interview, February 14, 2019.

6. Roberts, *Celebrating Entrepreneurship*, 23. Other details from 23–24.

7. See Keeley, "Facebook 'Shock.'"

8. "MIT Reshapes Itself." See also "FAQ on College of Computing"; Lohr, "M.I.T. Plans College"; Matheson, "Building Site Identified."

Chapter 27

1. Aulet interview, October 27, 2020. Unless otherwise noted, all Aulet quotes are from this interview.

2. Quoted in Maxwell, "Capturing Kendall Square."

3. Metcalfe, "Invention Is a Flower."

4. Florida interview, November 20, 2020. All Florida quotes are from this interview.

5. Murray interview, November 6, 2020. Unless otherwise noted, all Murray quotes are from this interview.

6. See Safire, "Location, Location, Location."

7. Dunsire interview, June 9, 2020.

8. Feld, *Startup Communities*, 43.

9. Fruehauf interview, October 15, 2020. All Fruehauf quotes are from this interview.

10. Feld, *Startup Communities*, 156.

11. Aulet interview, December 11, 2020.

12. Kimball interview, January 13, 2021.

13. Coughlin interview, August 6, 2020. All Coughlin quotes are from this interview. In early 2021, after more than thirteen years, Coughlin stepped down as MassBio CEO. For the life science initiative, see Timmerman, "Gov. Patrick Travels West"; Executive Office for Administration and Finance, "Life Sciences Initiative."

14. Windham-Bannister interview, May 24, 2021, with follow-up email, July 25, 2021.

15. Panel discussion at MIT, November 19, 2019.

Chapter 28

1. Junior Mints are made at 810 Main St. by Cambridge Brands, a Tootsie Roll Industries subsidiary.

2. Hill email, July 26, 2021.

3. Heatherington interview, August 26, 2020. All Heatherington quotes are from this interview.

4. Takeda details from Heatherington interview, August 26, 2020. See also LoTurco, "MIT School of Engineering and Takeda."

5. Some IBM Watson Health and MIT-IBM Watson AI Lab details from interviews with Vandebroek, August 7, 2020, and Amini, September 14, 2020. See also McCluskey, "With Watson Health"; Ohnesorge, "IBM Confirms Layoffs"; MIT-IBM AI Lab website: https://mitibmwatsonailab.mit.edu.

6. See Broad Institute communications, "Broad Institute Launches"; and Rosen, "Broad Institute Launches $300m Initiative."

7. "MIT Reshapes Itself."

8. Amini interview, September 14, 2020.

9. Taylor, "Conversation with the New Computing Dean."

10. Collins interview, February 4, 2021, with follow-up email, July 28, 2021.

11. Wilson, *Consilience*.

BIBLIOGRAPHY

9/11 Commission. *The 9/11 Commission Report*. Edited by Thomas H. Kean. Washington, DC: National Commission on Terrorist Attacks upon the United States, 2004. https://www.9-11commission.gov/report/911Report.pdf.

Alexandria Real Estate Equities, Inc. "Alexandria LaunchLabs, the Premier Platform Accelerating Early-Stage Life Science Company Growth, Opens First Cambridge Location at the Alexandria Center at One Kendall Square." Press release, December 14, 2018.

Al Idrus, Amirah. "Alnylam Scores $2B from Blackstone to Propel Pipeline Prospects." Fierce Biotech, April 13, 2020. https://www.fiercebiotech.com/biotech/alnylam-scores-2b-from-blackstone-to-propel-pipeline-prospects.

Alspach, Kyle. "Boston Suddenly Finds Itself the State's Tech Startup Capital." BetaBoston, *Boston Globe*, April 6, 2014. https://www.betaboston.com/news/2014/04/06/boston-suddenly-finds-itself-the-states-tech-startup-capital/.

Anerbach, Stuart. "And Man Created Risks: The Risks Of Creation." *Washington Post*, March 9, 1975.

Ante, Spencer E. *Creative Capital: Georges Doriot and the Birth of Venture Capital*. Cambridge, MA: Harvard Business Press, 2008.

Aoki, Naomi. "Candy Coated Power-Cleaning Clears Way for Labs at Old Necco Factory." *Boston Globe*, July 9, 2003.

Aoki, Naomi. "Many Joining Biotech Move to Cambridge." *Boston Globe*, February 19, 2003.

Arsenault, Mark. "How the Biogen Leadership Conference in Boston Spread the Coronavirus." *Boston Globe*, March 10, 2020.

Aspan, Maria. "How The Engine's Katie Rae Is Addressing Racism in VC." *Fortune*, June 17, 2020.

Associated Press. "Akamai Technologies Stock Surges on First Trading Day." *New York Times*, October 30, 1999.

"Athenaeum Press Sold; Plant Under Cuneo Will Continue to Print Books for Ginn." *New York Times*, November 18, 1950.

Babür, Oset. "Jody Adams." The Thirty-One Percent. January 2, 2019. https://thethirtyonepercent.com/home/2019/1/2/jody-adams.

Baltimore, David. Interview by Charles Weiner and Rae Goodell, MIT Archives, Recombinant DNA Controversy Oral History Collection, May 3, 1977.

Bauter Engel, Jeff. "Exit Interview: Lita Nelsen on MIT Tech Transfer, Startups & Culture." Xconomy, May 31, 2016. https://xconomy.com/boston/2016/05/31/exit-interview-lita-nelsen-on-mit-tech-transfer-startups-culture/.

Beals, Gerald. "The Biography of Thomas Edison." Accessed August 1, 2019. http://www.thomasedison.com/biography.html.

Begley, Sharon. "As Twitter Explodes, Eric Lander Apologizes for Toasting James Watson." STAT News, May 14, 2018. https://www.statnews.com/2018/05/14/apology-eric-lander-james-watson/.

Begley, Sharon. "Controversial CRISPR History Sets off an Online Firestorm." STAT News, January 19, 2016. https://www.statnews.com/2016/01/19/crispr-history-firestorm/.

Begley, Sharon. "Three Star Scientists Announce Plan to Solve Biotech's 'Missing Women' Problem." STAT News, January 29, 2020. https://www.statnews.com/2020/01/29/biotechs-missing-women/.

Begley, Sharon. "Why Eric Lander Morphed from Science God to Punching Bag." STAT News, January 25, 2016. https://www.statnews.com/2016/01/25/why-eric-lander-morphed/.

Begley, Sharon, and Elizabeth Cooney. "Two Female CRISPR Scientists Make History, Winning Nobel in Chemistry." STAT News, October 7, 2020. https://www.statnews.com/2020/10/07/two-crispr-scientists-win-nobel-prize-in-chemistry/.

Bell, Alexander Graham. *The Deposition of Alexander Graham Bell*. Boston: American Bell Telephone Company, 1908.

"Biogen Inc." Encyclopedia.com. Accessed April 20, 2020. https://www.encyclopedia.com/social-sciences-and-law/economics-business-and-labor/businesses-and-occupations/biogen-inc.

Blanding, Michael. "The Past and Future of Kendall Square." *MIT Technology Review*, August 18, 2015.

Boeing. "Boeing Completes Acquisition of Aurora Flight Sciences." Press release, November 8, 2017.

Boeing. "Boeing to Establish New Aerospace & Autonomy Center." Press release, August 1, 2018.

Bonanos, Christopher. "Instant Artifact: Remembering Meroë Morse." Press release, August 1, 2018. PolaroidLand. http://www.polaroidland.net/2012/10/21/instant-artifact-remembering-meroe-morse/.

Booth, Rosemary, Jack Dennis, Martha Goodway, David Litster, Gerald O'Leary, Bjorn Poonen, Roger Roach, O. Robert Simha, Jane Sanford Stabile, and Lawrence Stabile. "MIT Volpe Construction Plan Will Damage Faculty Housing Initiative." *MIT Faculty Newsletter*, November 2020.

Boston Consulting Group. "Protecting and Strengthening Kendall Square." Presentation, March 3, 2010.

"Boston Daily Advertiser (Full Paper)." *Boston Daily*, October 19, 1876.

Boston Properties. "Innovation Space Compliance Report 2019." Cambridge, MA: Cambridge Redevelopment Authority, 2019.

Bowker, Brittany. "These Mass. Colleges Were Ranked among the Best in the Nation by US News & World Report." *Boston Globe*, September 14, 2020.

Bradt, Steve. "Boeing Will Be Kendall Square Initiative's First Major Tenant." *MIT News*, August 1, 2018.

Brand, Stewart. *The Media Lab: Inventing the Future at MIT*. New York: Viking, 1987.

Bray, Hiawatha. "How Massachusetts Made the Apollo 11 Moon Landing Possible." *Boston Globe*, July 13, 2019.

Brehm, Denise. "MIT to Sell Share in Technology Square." *MIT News*, January 20, 2006.

Broad, Eli. *The Art of Being Unreasonable: Lessons in Unconventional Thinking*. New York: Wiley, 2012.

Broad Foundation. "Broad Institute Director Eric Lander Wins $3 Million Breakthrough Prize." Press release, February 22, 2013.

Broad Institute. "$650 Million Commitment to Stanley Center at Broad Institute Aims to Galvanize Mental Illness Research." Press release, July 22, 2014.

Broad Institute. "Broad Institute Launches the Eric and Wendy Schmidt Center to Connect Biology, Machine Learning for Understanding Programs of Life." Press release, March 25, 2021.

Broad Institute. "Broad Institute Provides COVID-19 Screening for Students, Faculty, and Staff at More than 100 Colleges and Universities." Press release, September 2, 2020.

Broad Institute. "Eric Lander to Take an Academic Leave to Serve as White House Science Advisor." Press release, January 15, 2021.

Buderi, Robert. "Boston VCs Grok Social Media—So Can We Please Not Tell That Facebook Story Anymore?" Xconomy, May 12, 2009. https://xconomy.com/boston/2009/05/12/boston-vcs-grok-social-media-so-can-we-please-not-tell-that-facebook-story-anymore/.

Buderi, Robert. "Boston Venture Firms on the Move: A Roundup of Who's Heading West (to CA), and Who's Heading East (to Cambridge)." Xconomy, May 19, 2011. https://xconomy.com/boston/2011/05/19/boston-venture-firms-on-the-move-a-roundup-of-whos-heading-west-to-ca-and-whos-heading-east-to-cambridge/.

Buderi, Robert. "Cambridge Officially the New Center of MA Venture Deals—the Data." Xconomy, November 19, 2012. https://xconomy.com/boston/2012/11/19/cambridge-officially-the-new-center-of-massachusetts-venture-deals/.

Buderi, Robert. "Danny Hillis on Boston vs. CA, Last Days with Minsky, and His New Co." Xconomy, May 11, 2016. https://xconomy.com/boston/2016/05/11/danny-hillis-on-boston-vs-ca-last-days-with-minsky-and-his-new-co/.

Buderi, Robert. "EMC Opens Research Arm in Cambridge, Joins MIT Media Lab as Sponsor." Xconomy, June 24, 2009. https://xconomy.com/boston/2009/06/24/emc-opens-research-arm-in-cambridge-joins-mit-media-lab-as-sponsor/.

Buderi, Robert. "Flybridge Unveils Scholarship Program to Help Students Network and Stay in MA." Xconomy, December 16, 2008. https://xconomy.com/boston/2008/12/16/flybridge-unveils-scholarship-program-to-help-students-network-and-stay-in-ma/.

Buderi, Robert. *The Invention That Changed the World*. New York: Simon and Schuster, 1996.

Buderi, Robert. "Just Turned 25, Pegasystems See Big Ride Ahead." Xconomy, May 1, 2008. https://xconomy.com/boston/2008/05/01/just-turned-25-pegasystems-see-big-ride-ahead/.

Buderi, Robert. "Kendall Square Wants an Entrepreneurial Walk of Fame—and So Should Every Innovation Hub." Xconomy, August 11, 2010. https://xconomy.com/national/2010/08/11/kendall-square-wants-an-entrepreneurial-walk-of-fame-and-so-should-every-innovation-hub/.

Buderi, Robert. "Microsoft Research Lab Opens Quietly Next to MIT, Director Says Area's Intellectual Climate Like 'Dry Timber' Ready to Ignite." Xconomy, July 29, 2008. https://xconomy.com/boston/2008/07/29/microsoft-research-lab-opens-quietly-next-to-mit-director-says-intellectual-climate-like-dry-timber-waiting-to-ignite/.

Buderi, Robert. "Polaris to Open Dog Patch Labs Incubator in Cambridge." Xconomy, September 10, 2009. https://xconomy.com/boston/2009/09/10/polaris-to-open-dog-patch-labs-incubator-in-cambridge/.

Buderi, Robert. "Startup Profile: Xconomy Part 2." Xconomy, June 28, 2007. https://xconomy.com/boston/2007/06/28/startup-profile-xconomy-part-2/.

Buderi, Robert. "Where Innovation Happens—A Still-Forming Map of Boston's Growing Tech Lab Cluster." Xconomy, June 9, 2009. https://xconomy.com/boston/2009/06/09/where-innovation-happens-a-still-forming-map-of-bostons-growing-tech-lab-cluster/.

Budoff, Jennifer R. "New Massachusetts Non-compete Law Goes into Effect October 1, 2018." Mintz, August 14, 2018. https://www.mintz.com/insights-center/viewpoints/2018-08-14-new-massachusetts-non-compete-law-goes-effect-october-1-2018.

Bulkeley, William M. "Reorganization of Thinking Machines Filed." *Wall Street Journal*, November 9, 1995.

Butrica, Andrew. "The Electronics Research Center: NASA'S Little Known Venture into Aerospace Electronics." National Aeronautics and Space Administration. Accessed October 2, 2019. https://history.nasa.gov/ercaiaa.html.

Butrica, Andrew. "Electronics Research Center." National Aeronautics and Space Administration. Accessed July 15, 2021. https://history.nasa.gov/erc.html.

Butterfield, Fox. "High Technology Boom Building Up Cambridge." *New York Times*, October 20, 1985.

"Cambridge Board of Trade Celebrates Opening of the New Subway." *Cambridge Chronicle*, March 23, 1912.

Cambridge Community Development Department. "Eastern Cambridge Planning Study." Accessed December 29, 2020. https://www.cambridgema.gov/-/media/Files/CDD/Planning/Studies/ECAPS/ecaps_report_2001.pdf.

"Cambridge Council Allows Harvard DNA Research." *New York Times*, February 8, 1977.

Cambridge Historical Commission. "Davenport Car Works, 700 Main Street." LabCentral, April 1, 2013. https://labcentral.org/news-events/articles/davenport-car-works-700-main-street-4113/.

Cambridge Redevelopment Authority. "Background of the Kendall Square Urban Renewal Project (appendix, Letters and Closeout Agreement)." Accessed November 5, 2019. https://static1.squarespace.com/static/51f173a6e4b04fc573b07c0c/t/5231142de4b0359b56292129/1378948141594/Background+of+the+Kendall+Square+Urban+Renewal+District.pdf.

Cambridge Redevelopment Authority. "The Kendall Square Urban Renewal Plan and Cambridge Center Master Plan 1979–Present." Accessed November 5, 2019. https://www.cambridgeredevelopment.org/redevelopment-history-of-kendall.

Campbell, Robert. "'Green' Building Is Bright and Beautiful." *Boston Globe*, January 18, 2009.

Carraggi, Mike. "MA Stay-at-Home Advisory Issued, Nonessential Businesses to Close." *Patch*. March 23, 2020. https://patch.com/massachusetts/boston/ma-stay-home-order-issued-nonessential-businesses-close.

Carroll, John. "Ex-Biogen Exec Steve Holtzman Hears the Call of Third Rock's Ambitious Decibel." *EndPoints News*, July 13, 2016. https://endpts.com/ex-biogen-exec-steve-holtzman-hears-the-call-of-third-rocks-ambitious-decibel/.

Carson, Joelle. "Opening for Innovation." *MIT Spectrum*, Spring 2020.

Carter, Marc. "MIT's Folding Hiriko City Car Set to Hit the Market Next Year for $16,000." Inhabitat, August 9, 2012. https://inhabitat.com/mits-folding-hiriko-city-car-set-to-hit-the-market-next-year-for-16000/.

Casale, John. "Charles Williams Jr., Part One: Experimental Apparatus Made to Order." Telegraph-History.org. Accessed July 31, 2019. http://www.telegraph-history.org/charles-williams-jr/index.html.

Casale, John. "Charles Williams Jr., Part Two: Human Voice Sent via Telegraph." Telegraph-History.org. Accessed July 31, 2019. http://www.telegraph-history.org/charles-williams-jr/part2.html.

Chakrabarti, Meghna, and Amy Bracken. "Cambridge Elections Bring Out Conflicting Views On Housing." WBUR, November 13, 2017.

Chandler, David L., and MIT News Office. "A New Era Set to Begin in Kendall Square." *MIT News*, May 17, 2016.

Chesto, Jon. "A 1-Acre Lot in Kendall Square Sells for More than $50 Million." *Boston Globe*, August 29, 2018.

Chesto, Jon. "Kendall Businesses to Try a Different Kind of Experiment—Fixing Traffic." *Boston Globe*, October 9, 2019.

Chiou, Stefanie, Craig Music, Kara Sprague, and Rebekah Wahba. "A Marriage of Convenience: The Founding of the MIT Artificial Intelligence Laboratory." MIT course project, December 5, 2011. http://web.mit.edu/6.933/www/Fall2001/AILab.pdf.

"Citizens' Trade Association Again Discusses This Question—Voice Vote Favors Four Stations." *Cambridge Tribune*, April 27, 1907.

City of Cambridge. "Article 14.000—Mixed Use Development District: Kendall Center." Accessed April 1, 2020. https://library.municode.com/ma/cambridge/codes/zoning_ordinance?nodeId=ZONING_ORDINANCE_ART14.000MIUSDEDIKECE.

City of Cambridge. "Ordinance #1378, in Amendment to the Ordinance Entitled 'Zoning Ordinances of the City of Cambridge.'" Kendall Square Urban Renewal Area, Cambridge Redevelopment Authority, Urban Renewal Plan, December 21, 2015, Amended and Restated. https://static1.squarespace.com/static/51f173a6e4b04fc573b07c0c/t/5890c11e893fc01d6e8afab3/1485881644884/KSURP+Amendment+10+Final+Ordinance+12.21.15.pdf.

Cooke, Robert. "Mayor Vellucci 'Welcomes' Biogen to Cambridge." *Boston Globe*, February 24, 1982.

Crescenzo, Sarah de. "Foghorn Therapeutics Inks Chromatin-Targeting Cancer Deal with Merck." Xconomy, July 8, 2020. https://xconomy.com/boston/2020/07/08/foghorn-therapeutics-inks-chromatin-targeting-cancer-deal-with-merck/.

Dabek, Frank. "City Council Approves New Development Ban." *The Tech*, January 26, 2000.

Davis, Bob. "Road to Reasoning: Superfast Computers Mimic the Structure of the Human Brain." *Wall Street Journal*, February 19, 1986.

"Deacon Kendall Dies at Holden." *Cambridge Chronicle*, January 9, 1915.

Dick, Brian, and Mark Jones. "The Commercialization of Molecular Biology: Walter Gilbert and the Biogen Startup." *History and Technology* 33, no. 1 (January 2, 2017): 126–151.

Donnelly, Julie M. "Pfizer Inc. Plans to Sell Cambridgepark Facility." BioSpace, April 2, 2012.

Donnelly, Julie M. "Vertex Moving to Boston's Fan Pier." *Boston Business Journal*, January 24, 2011.

Dornbusch, Erin. "Boston Woven Hose & Rubber." Industry in Cambridge. Accessed September 30, 2019. https://cambridgehistory.org/industry/bostonwoven.html.

Dornbusch, Erin. "Lever Brothers." Industry in Cambridge. https://historycambridge.org/industry/leverbrothers.html.

Dornbusch, Erin. "Riverside Press." Industry in Cambridge. Accessed January 11, 2021. https://cambridgehistory.org/industry/riversidepress.html.

Dornbusch, Erin. "Warren Brothers." Industry in Cambridge. Accessed October 22, 2019. https://cambridgehistory.org/industry/warrenbrothers.html.

Durant, John. "Refrain from Using the Alphabet: How Community Outreach Catalyzed the Life Sciences at MIT." In *Becoming MIT*, edited by David Kaiser, 145–163. Cambridge, MA: MIT Press.

Edelman, Benjamin, Thomas E. Isenmann, and Eric J. Van den Steen. *Akamai Technologies*. Cambridge, MA: Harvard Business Publishing, revised June 2010.

"Edwin Land and Polaroid Photography." American Chemical Society. Accessed August 11, 2019. https://www.acs.org/content/acs/en/education/whatischemistry/landmarks/land-instant-photography.html.

Ehrlich, Paul R., David S. Dobkin, and Darryl Wheye. "Island Biogeography." Stanford University, 1988. https://web.stanford.edu/group/stanfordbirds/text/essays/Island_Biogeography.html.

Elmer-Dewitt, Philip. "Machines from the Lunatic Fringe." *TIME*, June 24, 2001.

Elton, Catherine. "Does Moderna Therapeutics Have the NEXT Next Big Thing?" *Boston Magazine*, February 26, 2013.

Enriquez, Juan. "Vassar and Main: The World's Most Innovative Intersection." *WIRED*, November 2, 2015.

Enriquez, Juan, and Kathryn Taylor. "Battles of Tomorrow Being Waged in Kendall Square." *Boston Globe*, July 26, 2018.

Executive Office for Administration and Finance. "Life Sciences Initiative." Accessed January 15, 2021. https://budget.digital.mass.gov/bb/h1/fy10h1/prnt10/exec10/pbudbrief23.htm.

"Factories, Not Colleges, Make Cambridge Famous." *Boston Daily Globe*, January 16, 1927.

"FAQ on the Newly Established MIT Stephen A. Schwarzman College of Computing." *MIT News*, October 15, 2018.

Farrell, Michael B. "Menino Savors Preview of Innovation District Hub." *Boston Globe*, May 23, 2013.

Fearer, Matt. "An Improbable Circle of Life: For One Whitehead Scientist, a Commitment to Basic Research Hits Home." *Paradigm*, 2014.

Feld, Brad. "I Got My Bathroom." *Feld Thoughts* (blog), January 25, 2008. https://feld.com/archives/2008/01/i-got-my-bathroom.html.

Feld, Brad. *Startup Communities*. Grand Haven: Brilliance Audio, 2016.

Feldman, Maryann and Nichola Lowe. "Consensus from Controversy: Cambridge's Biosafety Ordinance and the Anchoring of the Biotech Industry." *European Planning Studies*, April 1, 2008. https://doi.org/10.1080/09654310801920532.

Feuerstein, Adam. "If Biotech Had a Mount Rushmore, Whose Heads Would You Chisel There?" STAT News, February 4, 2020. https://www.statnews.com/2020/02/04/biotech-mount-rushmore-whose-heads/.

Fidler, Ben. "After Failed Schizophrenia Trial, Forum Pharma to Shutter This Week." Xconomy, June 27, 2016. https://xconomy.com/boston/2016/06/27/after-failed-schizophrenia-trial-forum-pharma-to-shutter-this-week/.

"First Step on Volpe Parcel Planned for 2019." *MIT News*, February 5, 2019.

Fisher, Adam. *Valley of Genius: The Uncensored History of Silicon Valley (As Told by the Hackers, Founders, and Freaks Who Made It Boom)*. New York: Twelve, 2018.

Fisher, Lawrence M. "The Rocky Road from Startup to Big-Time Player: Biogen's Triumph Against the Odds." *Strategy+business*, July 1, 1997.

Fitzpatrick, Garret. "Duck Pin, We Have a Problem." *MIT Technology Review*, August 21, 2012.

"For First Time Tonight at 9:30, Law Requires all Children under 16 to Be Off Streets as Factory Whistles Blast." *Cambridge Chronicle*, July 16, 1953.

Foss, Alden S. "Boston Woven Hose and Rubber Company: Eighty-Four Years in Cambridge." Accessed August 26, 2019. https://cambridgehistory.org/research/boston-woven-hose-part-1/.

Foss, Alden S. "Boston Woven Hose and Rubber Company: Eighty-Four Years in Cambridge (Part 2)." Accessed September 30, 2019. https://cambridgehistory.org/research/boston-woven-hose-part-2/.

"Four Station Men Concluded Their Case, Yesterday—C. H. Morse Asserts That Kendall Square Station Is Assured." *Cambridge Chronicle*, May 9, 1908.

Fuhrmans, Vanessa, and Rachel Zimmerman. "Leading the News: Novartis to Move Global Lab to U.S.—Swiss Drug Maker Follows Other European Companies Shifting Strategy Abroad." *Wall Street Journal*, May 7, 2002.

Funding Universe. "Genetics Institute, Inc. History." http://www.fundinguniverse.com/company-histories/genetics-institute-inc-history/.

Garde, Damian. "Biotech's Behind-the-Scenes Baroness Redefined How Universities Launch Start-ups." STAT News, May 3, 2016. https://www.statnews.com/2016/05/03/mit-biotech-lita-nelsen/.

Garde, Damian, and Jonathan Saltzman. "The Story of mRNA: How a Once-Dismissed Idea Became a Leading Technology in the Covid Vaccine Race." STAT News and *Boston Globe*, November 10, 2020. https://www.statnews.com/2020/11/10/the-story-of-mrna-how-a-once-dismissed-idea-became-a-leading-technology-in-the-covid-vaccine-race/.

Gardizy, Anissa. "Alnylam Wins FDA Approval for First Drug to Treat Rare Kidney Disease." *Boston Globe*, November 24, 2020.

Gardizy, Anissa. "ElevateBio raises $525 Million to Fuel Growth in Cell and Gene Therapy." *Boston Globe*, March 15, 2021.

Gardizy, Anissa. "Foghorn Therapeutics Raises $120 Million in IPO." *Boston Globe*, October 23, 2020.

Gardizy, Anissa. "Seven Months Ago CIC Health Didn't Exist. Today It's Running the State's Mass Vaccination Effort." *Boston Globe*, January 24, 2021.

Gavin, Christopher. "Boston Has the Worst Rush-Hour Traffic in the U.S., Report Says." *Boston Globe*, February 12, 2019.

Gilkerson, Ann. "Kendall Square: A Brief History and Building Survey." Unpublished paper prepared for the Cambridge Historical Commission, October 31, 1983.

Gilman, Arthur. *The Cambridge of Eighteen Hundred and Ninety-Six: A Picture of the City and Its Industries Fifty Years after Its Incorporation.* Cambridge, MA: Riverside Press, 1896.

Golden, Daniel. "'AI Alley': A Haven for Hackers." *Boston Globe*, March 8, 1987.

Goldy-Brown, Sarah. "U.S. Cities with the Most Colleges." Plexuss. Accessed October 21, 2020. https://plexuss.com/news/article/us-cities-with-the-most-colleges.

Gottlieb, Charles and Ross Jerome. "Biohazards at Harvard: Scientists Will Create New Life Forms—but How Safe Will They Be?" *Boston Phoenix*, June 8, 1976.

Haglund, Karl. *Inventing the Charles River*. Cambridge, MA: MIT Press, 2003.

Halber, Deborah. "Novartis Is Opening Research Center in Tech Square." *MIT News*, May 8, 2002.

Hall, Stephen. "The Embarrassing, Destructive Fight over Biotech's Big Breakthrough." *Scientific American*, February 4, 2016.

Hall, Stephen. *Invisible Frontiers: The Race to Synthesize a Human Gene.* Oxford: Oxford University Press, 2002.

Halton, Clay. "Internet Bubble Definition." Investopedia. July 15, 2019. https://www.investopedia.com/terms/i/internet-bubble.asp.

Harvard Business School Club of the GCC. "In Conversation with Stephane Bancel on Jan. 20, 2021." January 21, 2021. https://www.youtube.com/watch?v=tCwYlVUxOUI.

Harvard Trust Company. *Charles Davenport, 1812–1903*. Leaders of Cambridge Industry. Cambridge, MA: Harvard Trust Company, n.d.

Harvard Trust Company. *Edward Kendall, 1821–1915*. Leaders of Cambridge Industry. Cambridge, MA: Harvard Trust Company, n.d.

Hawkinson, John A. "Novartis to Build $600M Complex." *The Tech*, October 29, 2010.

Henderson, Nell. "Cancer Drug Interferon Wins Approval for Commercial Use." *Washington Post*, June 5, 1986.

Henry, Celia M. "The Sweet Research Life at Novartis." *Chemical & Engineering News*, July 11, 2005.

Herman, Bob. "The Pandemic Hasn't Hampered the Health Care Industry." Axios, August 3, 2020. https://www.axios.com/health-care-industry-coronavirus-pandemic-second-quarter-78f19a89-e7d8-4c91-927b-e6b49f7ee60a.html.

Hillman, Michelle. "BioMed Buys Seven Lyme Buildings for $531M." *Boston Business Journal*, April 18, 2005.

Hilts, Philip J. "'Gene Splicers' Fear Recedes, Awe Remains." *Washington Post*, November 4, 1981.

Hockfield, Susan. *The Age of Living Machines: How Biology Will Build the Next Technology Revolution*. New York. W. W. Norton & Company, 2019.

Huang, Gregory. "The Bob Langer and Polaris Family Tree: From Acusphere to Momenta to Visterra." Xconomy, April 19, 2011. https://xconomy.com/boston/2011/04/19/the-bob-langer-and-polaris-company-tree-from-acusphere-to-momenta-to-visterra/.

Huang, Gregory. "Ember CEO: Silicon Labs Acquisition for $72M Is 'Right Thing for the Company.'" Xconomy, May 21, 2012. https://xconomy.com/boston/2012/05/21/ember-ceo-silicon-labs-acquisition-for-72m-is-right-thing-for-the-company/.

Huang, Gregory. "Entrepreneur Walk of Fame Opens in Kendall Square: Gates, Jobs, Kapor, Hewlett, Packard, Swanson, and Edison Are Inaugural Inductees." Xconomy, September 16, 2011. https://xconomy.com/boston/2011/09/16/entrepreneur-walk-of-fame-opens-in-kendall-square-gates-jobs-kapor-hewlett-packard-swanson-and-edison-are-inaugural-inductees/.

Huang, Gregory. "ITA Software Bought by Google for $700M, Shifting Balance of Power in Travel Search." Xconomy, July 1, 2010. https://xconomy.com/boston/2010/07/01/ita-software-bought-by-google-for-700m-shifting-balance-of-power-in-travel-search/.

Huang, Gregory. "Moderna, $40M in Tow, Hopes to Reinvent Biotech With 'Make Your Own Drug.'" Xconomy, December 6, 2012. https://xconomy.com/boston/2012/12/06/moderna-40m-in-tow-hopes-to-reinvent-biotech-with-new-protein-drugs/.

INRIX. "Congestion Costs Each American Nearly 100 Hours, $1,400 a Year." Press release, March 9, 2020.

Johannes, Laura, and Stephen Moore. "Bayer, Millennium Enter Drug-Discovery Alliance." *Barron's*, September 24, 1998.

Johnson, Carolyn. "Bias in Biotech Funding Has Blocked Companies Led by Women." *Washington Post*, January 29, 2020.

Joseph, Andrew. "Amid Takeover Rumors, Baxalta Opens New Kendall Square Research Center." STAT News, December 2, 2015. https://www.statnews.com/2015/12/02/baxalta-kendall-research-center/.

"The Joseph Downs Collection of Manuscripts and Printed Ephemera." Winterthur Library. Accessed July 31, 2019. http://findingaid.winterthur.org/html/html_finding_aids/col0805.htm.

"Joseph J. O'Donnell, MBA 1971." Harvard Business School, January 1, 2005. https://www.alumni.hbs.edu/stories/Pages/story-bulletin.aspx?num=2010.

Kaiser, David, ed. *Becoming MIT*. Cambridge, MA: MIT Press.

Katz, Bruce, and Julie Wagner. "The Rise of Innovation Districts." Brookings, May 2014. https://www.brookings.edu/essay/rise-of-innovation-districts/.

Keeley, Laura. "Facebook 'Shock' Has Boston Firms Searching for Next Zuckerberg." *Bloomberg News*, July 7, 2011.

Kendall Square Association. "You Can't Find the Cure for Cancer While Sitting in Traffic." *Kendall Square* (blog), October 11, 2018. https://medium.com/@kendallnow/you-cant-find-the-cure-for-cancer-while-sitting-in-traffic-9f863343ab72.

Kendall Square Mobility Task Force. *Actions to Transform Mobility*. Boston, MA: Kendall Square Mobility Task Force, 2015. https://static1.squarespace.com/static/5ba10aba1aef1dd45cbb3440/t/5cd5a1d8f9619af48ffd9d59/1557504474958/TransportKendall.pdf.

Keown, Alex. "It's Official: Shire Takes Over Ariad's Old Space in Kendall Square." BioSpace, October 10, 2017.

Kerr, William R., Sari Kerr, and Alexis Brownell. *CIC: Catalyzing Entrepreneurial Ecosystems (A)*. Cambridge, MA: Harvard Business Publishing, 2017.

Kerr, William R., Sari Kerr, and Alexis Brownell. *CIC: Catalyzing Entrepreneurial Ecosystems (B)*. Cambridge, MA: Harvard Business Publishing, 2017.

Kifner, John. "'Creation of Life' Experiment at Harvard Stirs Heated Dispute." *New York Times*, June 17, 1976.

Kirsner, Scott. "Apple Is Growing the Cambridge Research Team Focused on Improving Siri Speech Recognition." BetaBoston, *Boston Globe*, September 8, 2014. http://www.betaboston.com/news/2014/09/08/apple-is-growing-the-cambridge-research-team-focused-on-improving-siri-speech-recognition/.

Kirsner, Scott. "Could COVID-19 Split Apart the Cluster of Innovators That Gives Boston an Edge?" *Boston Globe*, July 21, 2020.

Kirsner, Scott. "A Factory in Cambridge Makes 14 Million Junior Mints a Day. Why Is No One Allowed Inside?" *Boston Globe*, May 4, 2018.

Kirsner, Scott. "Genzyme Nears the End of an Era." *Boston Globe*, February 7, 2011.

Kirsner, Scott. "Inside Google's Cambridge, MA Offices." *Innovation Economy* (blog), May 13, 2008. http://www.innoeco.com/2008/05/inside-googles-cambridge-ma-offices.html.

Kirsner, Scott. "Latest Venture Capital Firm Scouting Space in Cambridge: Charles River Ventures." *Boston Globe*, May 13, 2011.

Kirsner, Scott. "A Star Waiting to Be Born in Kendall Square." *Boston Globe*, February 6, 2015.

Kirsner, Scott. "Venture Firm Matrix Partners Leases Additional Space in Kendall, Will Move HQ There." *Boston Globe*, December 12, 2012.

Klein, Stanley. "A Fracas over Funds Roils at MIT." *New York Times*, November 15, 1981.

Klowden, Kevin, Aaron Melaas, Charlotte Kesteven, and Sam Hanigan. *2020 State Technology and Science Index*. Santa Monica, CA: Milken Institute, 2020. http://statetechandscience.org/State-Technology-and-Science-Index-2020.pdf.

Knox, Richard A. "He's Financing Science's Taj Mahal: Instant Prestige, Controversy Go with His Dream." *Boston Globe*, July 26, 1981.

Knox, Richard A. "MIT Debates Gift of $127m Biocenter." *Boston Globe*, September 17, 1981.

Kolata, Gina. "Power in Numbers." *New York Times*, January 2, 2012.

Krueger, Hanna. "Biogen Conference in Boston Likely Linked to as Many as 300,000 COVID-19 Cases Worldwide, Researchers Say." *Boston Globe*, December 10, 2020.

LabCentral. *2017 Impact Report*. Cambridge, MA: LabCentral, 2017. https://labcentral.org/uploads/assets/2017_impact_report_sp.pdf.

LabCentral. *2018 Impact Report*. Cambridge, MA: LabCentral, 2018. https://labcentral.org/uploads/assets/impact_report_RBG.pdf.

LabCentral. *2019 Impact Report*. Cambridge, MA: LabCentral, 2019. https://labcentral.org/uploads/assets/2019_ImpactReport.pdf.

LabCentral. "Gallery 1832." LabCentral. Accessed February 6, 2019. https://labcentral.org/news-events/gallery-1832/.

Lander, Eric S. "The Broad Institute." Report to MIT President, March 3, 2004. http://web.mit.edu/annualreports/pres04/03.03.pdf.

Lander, Eric S. "The Heroes of CRISPR." *Cell* 164, no. 1–2 (January 14, 2016): 18–28.

"Large Crowd Participates in Initial Subway Trip." *Cambridge Chronicle*, March 30, 1912.

Lash, Alex. "Venture Creation Story: The Early Days Of Moderna." *Start Up*, June 18, 2013.

Lash, Alex. "'Wrong to Toast': Broad's Lander Sorry for Tribute to DNA Pioneer Watson." Xconomy, May 14, 2018. https://xconomy.com/boston/2018/05/14/wrong-to-toast-broads-lander-sorry-for-tribute-to-dna-pioneer-watson/.

Lécuyer, Christophe. *Making Silicon Valley*. Cambridge, MA: MIT Press, 2005.

Lemann, Nicholas. "The CIA Is Still at Tech Square." *Harvard Crimson*, October 6, 1973.

Lemieux, Jacob E., Katherine J. Siddle, Bennett M. Shaw, Christine Loreth, Stephen F. Schaffner, Adrianne Gladden-Young, Gordon Adams, et al. "Phylogenetic Analysis of SARS-CoV-2 in the Boston Area Highlights the Role of Recurrent Importation and Superspreading Events." medRxiv, August 25, 2020. http://dx.doi.org/10.1101/2020.08.23.20178236.

Lemieux, Jacob E., Katherine J. Siddle, Bennett M. Shaw, Christine Loreth, Stephen F. Schaffner, Adrianne Gladden-Young, Gordon Adams, et al. "Phylogenetic Analysis of SARS-CoV-2 in Boston Highlights the Impact of Superspreading Events." *Science* 371, no. 6529 (February 5, 2021). https://doi.org/10.1126/science.abe3261.

Leonhardt, David. "Boston and Pittsburgh, America's Most Successful Sports Cities." *New York Times*, June 4, 2015.

Leung, Shirley. "The Massachusetts Miracle Is Alive and Well. There Are a Lot of Potential Modernas,'" *Boston Globe*, December 26, 2020.

Levy, Marc. "Kendall Square Lab Tower with Public Arts Space Approved, Gas Transfer Station Removal Included." Cambridge Day, December 22, 2020. https://www.cambridgeday.com/2020/12/22/kendall-square-lab-tower-with-public-arts-space-approved-gas-transfer-station-removal-included/.

Levy, Marc. "Massive CambridgeSide Project Is Approved, Transforming Mall into Mini-Neighborhood (Updated)." Cambridge Day, December 17, 2019. https://www.cambridgeday.com/2019/12/17/massive-cambridgeside-project-is-approved-transforming-mall-into-mini-neighborhood/.

Levy, Steven. *Hackers: Heroes of the Computer Revolution*. Newton: O'Reilly Media, 2010.

Liu, Angus. "FDA Rebuffs Novartis' Cholesterol Drug Inclisiran in an 'Inspection-Related' Approval Delay." Fierce Pharma, December 21, 2020. https://www.fiercepharma.com/pharma/fda-rejects-novartis-cholesterol-drug-inclisiran-as-covid-19-delays-manufacturing-inspection.

Liu, Yi-Ling. "Pharma Valley, China's Equivalent of Kendall Square, Is Expanding Rapidly." STAT News, December 3, 2018. https://www.statnews.com/2018/12/03/pharma-valley-chinas-kendall-square/.

Logan, Tim. "Apple Plans to Lease Space in Kendall Square." *Boston Globe*, December 19, 2018.

Logan, Tim. "Developers Are Scrambling for the Next Kendall Square. Where Will It Be?" *Boston Globe*, February 20, 2019.

Logan, Tim. "In Biotech Hub Cambridge, There's Still Room for Junior Mints." *Boston Globe*, August 21, 2019.

Logan, Tim. "In This New Office Building, Kendall Square May Get Some Culture." *Boston Globe*, February 26, 2020.

Logan, Tim. "It's Official: Google Will Occupy a New Tower in Kendall Square." *Boston Globe*, February 13, 2019.

Logan, Tim. "MIT-Owned Lab Buildings Change Hands for Reported $1.1 Billion." *Boston Globe*, May 17, 2019.

Logan, Tim. "MIT to Start Work on a 'New Front Door' to the University." *Boston Globe*, October 9, 2018.

Logan, Tim. "One Kendall Square Complex Sold for $725 Million." *Boston Globe*, July 26, 2018.

Logan, Tim. "With Real Estate Deal, Drug Maker Boosts Its Bottom Line." *Boston Globe*, April 22, 2019.

Lohr, Steve. "Charles W. Bachman, Business Software Innovator, Dies at 92." *New York Times*, July 16, 2017.

Lohr, Steve. "M.I.T. Plans College for Artificial Intelligence, Backed by $1 Billion." *New York Times*, October 15, 2018.

Lok, Corie. "Nancy Hopkins Named Xconomy's 2018 Lifetime Achievement Award Winner." Xconomy, July 24, 2018. https://xconomy.com/boston/2018/07/24/nancy-hopkins-named-xconomys-2018-lifetime-achievement-award-winner/.

Lopatto, Elizabeth. "In the Pandemic Economy, Tech Companies Are Raking It In." The Verge, July 30, 2020. https://www.theverge.com/2020/7/30/21348652/pandemic-earnings-antitrust-google-facebook-apple-amazon.

LoTurco, Lori. "MIT School of Engineering and Takeda Join to Advance Research in Artificial Intelligence and Health." *MIT News*, January 6, 2020.

"Lotus Development Corporation History." FundingUniverse. Accessed December 9, 2019. http://www.fundinguniverse.com/company-histories/lotus-development-corporation-history/.

Lovelace, Berkeley, Jr. "Bluebird Bio CEO Defends $1.8 Million Price for Gene Therapy—'It's Really Thinking about It Differently.'" CNBC, June 14, 2019.

Luria, Salvador Edward. *A Slot Machine, a Broken Test Tube*. New York: HarperCollins, 1984.

Lynch, Brendan. "Longtime CEO Henri Termeer's Genzyme 'Adventure' Ends." *Boston Herald*, July 1, 2011.

MacArthur, Robert H., and Edward O. Wilson. *The Theory of Island Biogeography*. Princeton, New Jersey: Princeton University Press, 1967.

MacDonald, Katie. "Automated Candy Production, 250 Massachusetts Avenue." Innovation in Cambridge. Accessed January 11, 2021. https://cambridgehistory.org/innovation/Automatic%20Candy.html.

MacDonald, Katie. "Center Isle Train Car (Davenport Building)." Innovation in Cambridge. Accessed February 4, 2019. https://cambridgehistory.org/innovation/Center%20Isle%20Train%20Car.html.

MacDonald, Katie. "Frozen Orange Juice—National Research Corp. 70 Memorial Drive." Innovation in Cambridge. Accessed October 22, 2019. https://cambridgehistory.org/innovation/National%20Research%20Corp.html.

MacDonald, Katie. "Guidance Systems, 555 Technology Square." Innovation in Cambridge. Accessed January 11, 2021. https://cambridgehistory.org/innovation/Draper%20Labs.html.

MacDonald, Katie. "Polaroid—748 Memorial Drive." Innovation in Cambridge. Accessed July 31, 2019. https://cambridgehistory.org/innovation/Polaroid.html.

MacDonald, Katie. "Radar and Microwave Ovens, Raytheon, 292 Main Street." Innovation in Cambridge. Accessed January 11, 2021. https://cambridgehistory.org/innovation/Microwaves.html.

MacDonald, Katie. "Venture Capital—146 Sixth Street (Location of Ionics)." Innovation in Cambridge. Accessed August 17, 2020. https://cambridgehistory.org/innovation/Venture%20Capital.html.

Markoff, John. "American Express to Buy 2 Top Supercomputers." *New York Times*, October 30, 1991.

Markoff, John. "Computer Said to Be the Fastest." *New York Times*, June 5, 1991.

MassBio. *2020 Industry Snapshot*. Cambridge, MA: MassBio, 2020. https://www.massbio.org/wp-content/uploads/2020/07/MASSBIO-2020-INDUSTRY-SNAPSHOT-FINAL_7-28-20.pdf.

Matheson, Rob. "Building Site Identified for MIT Stephen A. Schwarzman College of Computing." *MIT News*, December 19, 2018.

Matheson, Rob. "MIT Corporation Life Member and Biotech Pioneer Henri A. Termeer Dies at 71." *MIT News*, May 17, 2017.

Matheson, Rob. "MIT Signs Agreement to Redevelop Volpe Center." *MIT News*, January 18, 2017.

Maxwell, Jill Hecht. "Capturing Kendall Square, from Scotland to Singapore." *MIT Sloan Alumni Magazine*, Winter 2015.

Maycock, Susan E. *East Cambridge (Survey of Architectural History in Cambridge)*. Revised edition. Cambridge, MA: MIT Press, 1989.

Maycock, Susan E., and Charles Sullivan. *Building Old Cambridge: Architecture and Development*. Cambridge, MA: MIT Press, 2016.

"Mayor Wardwell Rejects the Elevated's Plans—He Favors Five Stations—His Letter in Fall." *Cambridge Tribune*, July 27, 1907.

McBride, Ryan. "Broad Institute Scientists' Forma Therapeutics Raises $25M, Aims to Knock Out Underpinnings of Cancer." Xconomy, January 6, 2009. https://xconomy.com/boston/2009/01/06/broad-institute-scientists'-forma-therapeutics-raises-25m-aims-to-knock-out-underpinnings-of-cancer/.

McBride, Ryan. "MDRNA and Cequent Pharma in $46M Merger Deal." Xconomy, April 1, 2010. https://xconomy.com/seattle/2010/04/01/mdrna-and-cequent-pharma-in-46m-merger-deal/.

McBride, Ryan. "New CEO George Scangos Says Biogen Idec's R&D Has to Improve." Xconomy, July 1, 2010. https://xconomy.com/boston/2010/07/01/new-ceo-george-scangos-says-biogen-idecs-rd-has-to-improve/.

McCluskey, Priyanka Dayal. "With Watson Health, IBM Bets Big on Health Care." *Boston Globe*, August 29, 2016.

McElheny, Victor. *Insisting on the Impossible*. New York: Perseus Books, 1998.

McElheny, Victor. "Precipitators of MIT Commitment to Molecular Biology in 1960s." March 14, 2019. Unpublished memo shared with author via email.

Metcalfe, Bob. "Invention Is a Flower, Innovation Is a Weed." *Technology Review*, November 1, 1999.

Miara, Jim. "The Reinvention of Kendall Square." *Urban Land Magazine*, February 17, 2012.

Mid-Continent Railway Museum. "Davenport & Bridges." April 29, 2006. https://www.midcontinent.org/rollingstock/builders/davenport-bridges1.htm.

MIT2016: Celebrating a Century in Cambridge. "Innovation Ecosystem." May 17, 2016. YouTube video, 4:09. https://www.youtube.com/watch?time_continue=1&v=LHCoRam4HQ8.

"MIT Approves Ties with an Institute." *New York Times*, December 5, 1981.

Mitchell, William John, and Charles M. Vest. *Imagining MIT: Designing a Campus for the Twenty-First Century*. Cambridge, MA: MIT Press, 2007.

MIT Faculty Newsletter. "A Study on the Status of Women Faculty in Science at MIT." *MIT Faculty Newsletter*, special edition, March 1999.

MIT Faculty Newsletter. "Women in Biotech." *MIT Faculty Newsletter*, special edition, March/April 2021.

MIT-IBM Watson AI Lab. "Inside the Lab." MIT-IBM Watson AI Lab. Accessed August 6, 2020. https://mitibmwatsonailab.mit.edu/about/.

"MIT Reshapes Itself to Shape the Future." *MIT News*, October 15, 2018.

"MIT to Name Signature Building on the Charles River in Honor of Morris and Sophie Chang." *MIT News*, December 2, 2015.

Moravek, Natalie. "Candy Land: The History of Candy Making in Cambridge." Cambridge Historical Society. Accessed July 15, 2021. https://historycambridge.org/candy/overview.html.

Morgan, Bruce. "A Flair for the Business of Medicine." *Tufts Now*, April 11, 2012.

Morgan, Christopher, Joseph O'Connor, and David Hoag. *Draper at 25: Innovation for the 21st Century*. Cambridge, MA: Charles Stark Draper Laboratory, Inc, 1998.

Murphey, Richard. "Top Biotech Venture Capital Funds of 2018, 2019 and 2020." Bay Bridge Bio. Accessed April 21, 2020. https://www.baybridgebio.com/blog/top_vcs_2018.html.

Murphy, Jack. *Genzyme Center*. Cambridge, MA: Cambridge Community Development Department, July 2004. https://www.cambridgema.gov/~/media/Files/CDD/ZoningDevel/GreenBuildings/CaseStudies/greenbldg_genzyme.pdf.

Nanos, Janelle. "Facebook to Open New Office in Kendall Square, Adding Hundreds of Jobs." *Boston Globe*, August 30, 2017.

National Human Genome Research Institute. "International Human Genome Sequencing Consortium Announces 'Working Draft' of Human Genome." Press release, June 2000.

Neisloss, Liz. "Why Go To The Office? Few Want To—Some Need To." GBH, July 20, 2020.

"New Biotech Facility in the Heart of the City." *Cambridge Chronicle*, November 7, 2001.

Newton, Charles, Therma E. Patterson, and Nancy Joy Perkins. *Five Years at the Radiation Laboratory*. Cambridge, MA: MIT, 1946.

Nhuch, Michelle. "Broad Institute Expands into New Building at 7 Cambridge Center." Broad Institute. Press release, March 9, 2006.

"Novartis Links with Vertex in $800M Product Development Deal." Pharma Letter, September 5, 2000.

"Novartis Steps Up US Research Investment, Opening Cambridge, Massachusetts Biomedical." PR Newswire. Press release, May 6, 2002.

Novet, Jordan. "IBM Improves Gross Margins in Q2 under New CEO." *CNBC*, July 20, 2020.

Nunes, Jesse. "Key Moments in Genzyme's History." *Boston Globe*. Accessed February 24, 2020. http://archive.boston.com/business/healthcare/gallery/genzymehistory/.

Ohnesorge, Lauren. "IBM Confirms Layoffs in Watson Health." *Boston Business Journal*, June 4, 2018.

O'Riordan, Michael. "Inclisiran Approved in Europe for Lowering LDL Cholesterol." TCTMD, December 11, 2020.

Paige, Lucius Robinson. *History of Cambridge, Massachusetts. 1630–1877: With a Genealogical Register*. Cambridge, MA: H. O. Houghton, 1877.

Pan, Deanna. "CIC Health Expands COVID-19 Vaccination Operations to Reggie Lewis Center in Roxbury." *Boston Globe*, February 26, 2021.

Petroski, Henry. *The Road Taken: The History and Future of America's Infrastructure*. New York: Bloomsbury Publishing, 2016.

Pitta, Julie. "Where Lisp Slipped." *Forbes*, October 16, 1989.

"Polaris Partners to Relocate Corporate Headquarters to Boston's Fan Pier." Business Wire, October 7, 2013.

Prabhune, Meenakshi. "A Review of Gene Editing Nobel Prizes: Is CRISPR Headed for One?" Synthego. Accessed July 20, 2020. https://www.synthego.com/blog/gene-editing-nobel-prize. [Title updated to "Nobel Prize Awarded to Jennifer Doudna and Emmanuelle Charpentier for CRISPR Discovery."]

Prescott, Samuel C. *When M.I.T. Was "Boston Tech," 1861–1916*. Cambridge, MA: MIT Press, 1954.

Prokesch, Steven. "The Edison of Medicine." *Harvard Business Review*, March–April 2017. https://hbr.org/2017/03/the-edison-of-medicine.

Puleo, Stephen. *A City So Grand: The Rise of an American Metropolis, Boston 1850–1900*. Boston: Beacon Press, 2011.

Rand, Christopher. *Cambridge, U.S.A*. Oxford: Oxford University Press, 1964.

Raskin, Molly Knight. *No Better Time: The Brief, Remarkable Life of Danny Lewin, the Genius Who Transformed the Internet*. Cambridge, MA: Da Capo Press, 2013.

Raskin, Molly Knight. "Sept. 11 Secrets: The Amazing Life—and Death—of an Internet Pioneer on Flight 11." *Salon*, September 11, 2013. https://www.salon.com/2013/09/11/sept_11_secrets_the_amazing_life_and_death_of_an_internet_pioneer_on_flight_11/.

Rasmussen, Nicholas. *Gene Jockeys: Life Science and the Rise of Biotech Enterprise*. Baltimore: Johns Hopkins University Press, 2014.

Regalado, Antonio. "A Scientist's Contested History of CRISPR." *MIT Technology Review*, January 19, 2016.

Roach, Marilynne. "Charles Davenport." *Town Crier: Newsletter of the Historical Society of Watertown*, January 2010.

Roberts, Edward B. *Celebrating Entrepreneurship: A Half-Century of MIT's Growth and Impact*. Cambridge, MA: MIT Sloan School of Management, 2018.

Roberts, Edward B. *Entrepreneurs in High Technology: Lessons from MIT and Beyond*. Oxford: Oxford University Press, 1991.

Roberts, Sam. "Ted Stanley, Whose Son's Illness Inspired Philanthropy, Dies at 84." *New York Times*, January 8, 2016.

Rosen, Andy. "Broad Institute launches $300m initiative to fight diseases with artificial intelligence." *Boston Globe*, March 25, 2021.

Rosenberg, Ronald. "AI Alley's Longest Winter." *Boston Globe*, December 18, 1988.

Ross, Casey, and Robert Weisman. "Biogen Idec Expanding Cambridge Footprint." *Boston Globe*, July 20, 2011.

Ross, Casey, and Robert Weisman. "Novartis Doubles Plan for Cambridge." *Boston Globe*, October 27, 2010.

Roush, Wade. "Cambridge Innovation Center Turns 10; Looking Inside a Landmark for Boston-Area Entrepreneurs." Xconomy, December 3, 2009. https://xconomy.com/boston/2009/12/03/cambridge-innovation-center-turns-10-looking-inside-a-landmark-for-boston-area-entrepreneurs/.

Roush, Wade. "Google's Open House: Of Ping-Pong, the Gov, and Four Local Projects." Xconomy, May 14, 2008. https://xconomy.com/boston/2008/05/14/googles-open-house-of-ping-pong-the-gov-and-four-local-projects/.

Roush, Wade. "ITA Software: The Travel Company Everyone Uses and No One Knows Reinvents Airline Reservations, Again." Xconomy, December 17, 2008. https://xconomy.com/boston/2008/12/17/ita-software-the-travel-company-everyone-uses-and-no-one-knows-reinvents-airline-reservations-again/.

Roush, Wade. "New Business Association Looks to the Future of Kendall Square, 'The Product Cambridge Offers to the World.'" Xconomy, March 2, 2009. https://xconomy.com/boston/2009/03/02/new-business-association-looks-to-the-future-of-kendall-square-the-product-cambridge-offers-to-the-world/.

Roush, Wade. "New Microsoft Lab in Cambridge to Combine Math and Social Science; Already Besieged By Potential Research Collaborators." Xconomy, February 4, 2008. https://xconomy.com/boston/2008/02/04/new-microsoft-lab-in-cambridge-to-combine-math-and-social-science-already-besieged-by-potential-research-collaborators/2/.

Roush, Wade. "R.I.P. Orange Labs Cambridge (2002–2009): A Story of Opportunities Missed." Xconomy, October 29, 2009. https://xconomy.com/boston/2009/10/29/r-i-p-orange-labs-cambridge-2002-2009-a-story-of-opportunities-missed/.

Roush, Wade. "TechStars 'Entrepreneurship Boot Camp' Comes to Boston: An Interview with Cofounder David Cohen." Xconomy, February 17, 2009.

Roush, Wade. "What Is the Future of the Internet? Experts Predict Next 50 Years," Xconomy, July 2, 2019. https://xconomy.com/boston/2009/02/17/techstars-entrepreneurship-boot-camp-comes-to-boston-an-interview-with-co-founder-david-cohen/.

Roush, Wade. "You Don't Need Tech Companies to Reboot Your City's Economy." *Scientific American*, February 1, 2019.

Rowe, Tim. "Tragedy of the Commons: It's (Really) Time to Ban Non-Compete Agreements." Xconomy, July 15, 2009. https://xconomy.com/boston/2009/07/15/tragedy-of-the-commons-it's-really-time-to-ban-non-compete-agreements/.

Royden, Leigh, and Rosalind Williams. "MIT: Where Now?" *MIT Faculty Newsletter*, January 15, 2020.

Rude, Emelyn. "The Surprising Link Between World War II and Frozen Orange Juice." *TIME*, August 31, 2017.

Safire, William. "Location, Location, Location." *New York Times*, June 26, 2009.

Sagonowsky, Eric. "Sanofi Leases Its Soon-to-Be-Former Genzyme HQ to Bluebird Bio—at a Profit." Fierce Pharma, April 23, 2019. https://www.fiercepharma.com/pharma/sanofi-to-make-small-profit-boston-sublease-report.

Sagonowsky, Eric. "Takeda to Move U.S. Headquarters to Boston, Affecting 1,000 Employees." Fierce Pharma, September 12, 2018. https://www.fiercepharma.com/pharma/takeda-to-move-u-s-headquarters-to-boston-uprooting-1-000-employees.

Sahlman, William A. *Lotus Development Corporation*. Cambridge, MA: Harvard Business School Case Study, 1987.

Sahlman, William A. *Palladian Software*. Cambridge, MA: Harvard Business School Case Collection, February 1986.

Saltzman, Jonathan. "Biogen Conference Likely Led to 20,000 COVID-19 Cases in Boston Area, Researchers Say." *Boston Globe*, August 25, 2020.

Saltzman, Jonathan. "Brigham President Resigns from Moderna Board after Conflict of Interest Questions Raised." *Boston Globe*, July 29, 2020.

Saltzman, Jonathan. "Cambridge Biotech Started by Flagship Pioneering Raises Another $189m." *Boston Globe*, April 13, 2021.

Saltzman, Jonathan. "Despite Slumping Economy, Atlas Venture Raises $400m for Biotech Startups." *Boston Globe*, June 5, 2020.

Saltzman, Jonathan. "A Lab Bench Costs $4,600 a Month. For Startups, It's the Place to Be." STAT News, March 6, 2019. https://www.statnews.com/2019/03/06/labcentral-bench-startups-kendall/.

Saltzman, Jonathan. "Massachusetts Biopharma Firms Raised a Record $5.8 Billion in Venture Capital Last Year." *Boston Globe*, March 4, 2021.

Saltzman, Jonathan. "Moderna Seeks FDA Approval for Emergency Use of Its COVID-19 Vaccine." *Boston Globe*, November 29, 2020.

Saltzman, Jonathan. "Takeda, Already a Giant in Mass. Biopharma, Plans a Bigger Footprint." *Boston Globe*, July 4, 2019.

Saltzman, Jonathan. "VC Firm Flagship Pioneering Raises $1.1b for Biotech Startups, despite Reeling Economy." *Boston Globe*, April 2, 2020.

Saltzman, Jonathan. "Working with Area's Colleges, Broad Institute Surpasses 1 Million COVID-19 Tests." *Boston Globe*, August 31, 2020.

Sanger, David E. "A Computer Full of Surprises." *New York Times*, May 8, 1987.

Sanger, David E. "Software Shifts: As Industry Tightens, Jobs 'Hide' in Unexpected Places." *New York Times*, October 13, 1985.

Saxenian, Annalee. 1996. *Regional Advantage: Culture and Competition in Silicon Valley and Route 128*. Cambridge, MA: Harvard University Press.

Saxon, Wolfgang. "Richard S. Morse, 76, an Inventor of Orange Juice Concentrate, Dies." *New York Times*, July 4, 1988.

Schiesel, Seth. "GTE Discloses 3 Big Deals in Growth Bid." *New York Times*, May 7, 1997.

Schmeck, Harold M. Jr.. "Natural Virus-Fighting Substance Is Reported Made by Gene Splicing." *New York Times*, January 17, 1980.

Scott, Alwyn. "Boeing to Buy Autonomous and Electric Flight Firm Aurora." Reuters, October 5, 2017.

Seaburg, Alan, Thomas Dahill, and Carol Rose. *Cambridge on the Charles*. Boston: Anne Miniver Press, 2001.

Seidel, Sam. "Kendall Square, a Brief Historical Sketch." *Blog.samseidel.org* (blog), March 17, 2013. http://blog.samseidel.org/2013/03/kendall-square-brief-historical-sketch.html.

Seiffert, Don. "Alexandria Aims to Repeat Success of Tech Square with One Kendall Square." *Boston Business Journal*, August 18, 2016.

Shah, Angela. "Mark Levin Named Xconomy's 2019 Lifetime Achievement Award Winner in Boston." Xconomy, July 31, 2019. https://xconomy.com/national/2019/07/31/mark-levin-named-xconomys-2019-lifetime-achievement-award-winner-in-boston/.

Sheridan, Kate. "Among Biotech VCs, Flagship Pioneering Got the Best Single-Fund Returns." STAT News, July 22, 2020. https://www.statnews.com/2020/07/22/biotech-venture-capital-flagship-pioneering-best-fund-returns/.

"Shire to Acquire TKT." *PharmaTimes*, April 22, 2005.

"Shire to Buy Transkaryotic Therapies for $1.6 Billion." WCG FDA News. Press release, April 29, 2005.

Shulman, Seth. *The Telephone Gambit.* New York: W. W. Norton & Company, 2008.

Simha, O. Robert. "A Brief History of MIT's Land Acquisition Policies." *MIT Faculty Newsletter*, November/December 2011. http://web.mit.edu/fnl/volume/242/simha2.html.

Simha, O. Robert. "Cambridge Redevelopment Authority—History and Background." Accessed November 6, 2019. http://rwinters.com/docs/CRA.htm.

Simha, O. Robert. "A Critical Look at the Plan for MIT's East Campus." *MIT Faculty Newsletter*, January/February 2016. http://web.mit.edu/fnl/volume/283/simha.html.

Simha, O. Robert. "MIT 2030: Concerns for the Future." *MIT Faculty Newsletter*, November/December 2011. http://web.mit.edu/fnl/volume/242/simha1.html.

Simha, O. Robert. *MIT Campus Planning 1960–2000.* Cambridge, MA: MIT Press, 2003.

Smith, Harrison. "Charles Bachman, Engineer Who Devised a Better Way to Manage Data, Dies at 92." *Washington Post*, July 16, 2017.

Solomon, Ethan A. 2011. "Community Reacts to Plan for Kendall." *The Tech*, February 18, 2011.

Sperance, Cameron. "Apple, Capital One, Boeing Plan to Share a Kendall Square Tower." Bisnow, December 20, 2018.

Sperance, Cameron. "Boston Properties Gets Approval for Kendall Square Tower." Bisnow, December 6, 2018.

Sperance, Cameron. "There's No Room in Kendall Square, So Why Do Companies Keep Moving There?" Bisnow, October 16, 2018.

Spinrad, Paul. "The New Cool." *WIRED*, August 1, 1991.

Stallman, Richard. "My Lisp Experiences and the Development of GNU Emacs." GNU Operating System, October 28, 2002. https://www.gnu.org/gnu/rms-lisp.html.

Stangler, Dane. "Non-compete Agreements: The Good, the Bad, and the Ugly." Xconomy, August 14, 2014. https://xconomy.com/national/2014/08/14/non-compete-agreements-the-good-the-bad-and-the-ugly/.

Statt, Nick. "Microsoft Just Had a Stellar Three Months Thanks to Azure, Surface, and Xbox." The Verge, July 22, 2020. https://www.theverge.com/2020/7/22/21334725/microsoft-q4-2020-earnings-azure-surface-xbox-gains-growth-profits-sales.

Stockman, Farah, and Kim Barker. "How Biogen Became a Coronavirus 'Superspreader.'" *New York Times*, April 12, 2020.

Stone, Orra L. *History of Massachusetts Industries: Their Inception, Growth, and Success.* Chicago: S. J. Clarke Publishing Company, 1930.

"Subway Stations." *Cambridge Chronicle*, April 27, 1907.

Sullivan, Charles. *Draft Landmark Designation Study Report: Kendall Square Landmark Group.* Cambridge, MA: Cambridge Historical Society, 2012. https://www.cambridgema.gov/~/media/Files/historicalcommission/pdf/Landmark_reports/L_Kendall_prelimreport.pdf?la=en.

Sullivan, Charles. *Final Landmark Designation Report: Blake & Knowles Foundry, 101 Rogers Street, Cambridge.* Cambridge, MA: Cambridge Historical Commission, 2018. http://rwinters.com/council/FoundryLandmark2018Apr30c.pdf.

Taubes, Gary A. "The Rise and Fall of Thinking Machines." *Inc.*, February 6, 2020.

Taylor, Nick Paul. "Bristol Myers Inks Lease to Consolidate Cambridge R&D Sites, Join Sanofi at New Development." Fierce Biotech, August 20, 2020. https://www.fiercebiotech.com/biotech/bristol-myers-inks-lease-to-consolidate-cambridge-r-d-sites-join-sanofi-at-new-development.

Taylor, Nick Paul. "Ipsen Moves U.S. Headquarters to Kendall Square." Fierce Biotech, June 8, 2018.

Taylor, Nicole Estvanik. "A Conversation with the New Computing Dean: Alumnus Daniel Huttenlocher." *Slice of MIT*, April 27, 2020.

"Telegraphing Speech." *Boston Daily Advertiser*, October 19, 1876.

Tercyak, Thad. "Kendall Square Urban Renewal Project: Initial Years, 1963 to 1982." *Cambridge Civic Journal Forum*, July 12, 2012. http://cambridgecivic.com/?p=2102.

Tercyak, Thad. "Kendall Square Urban Renewal Project: Six Pivotal Episodes." *Cambridge Civic Journal Forum*, June 8, 2013. http://cambridgecivic.com/?p=2731.

Tercyak, Thad. "MBTA Role in Jump-Starting Development of the Cambridge Center Project Kendall Station Urban Initiatives Project, 1979–1989." *Cambridge Civic Journal Forum*, February 13, 2014. http://cambridgecivic.com/?p=3491.

Terry, Mark. "It's a Wrap! Takeda-Shire Merger Is a Done Deal, Making Takeda the Largest Mass. Biotech Employer." BioSpace, January 8, 2019.

Thiessen, Arthur E. *A History of the General Radio Company 1915–1965*. Concord, MA: General Radio Company, 1965.

"Thinking Machines Corp. (Bankruptcy)." *Wall Street Journal*, August 18, 1994.

Thomke, Stefan, and Ashok Nimgade. *Millennium Pharmaceuticals, Inc. (A)*. Boston, MA: Harvard Business Publishing, December 21, 1999.

Timmerman, Luke. "Biogen Idec CEO on Move Back to Cambridge: 'We're Working on It.'" Xconomy, June 16, 2011. https://xconomy.com/boston/2011/06/16/biogen-idec-ceo-on-move-back-to-cambridge-were-working-on-it/.

Timmerman, Luke. "The Genetics Institute Alumni: Where Are They Now?" Xconomy, March 22, 2010. https://xconomy.com/boston/2010/03/22/the-genetics-institute-alumni-where-are-they-now/.

Timmerman, Luke. "George Scangos, the Boy from Working Class Boston, on His Road back to Lead Biogen Idec." Xconomy, August 31, 2011. https://xconomy.com/boston/2011/08/31/george-scangos-the-boy-from-working-class-boston-on-his-road-back-to-lead-biogen-idec/.

Timmerman, Luke. "Gov. Patrick Travels West to Plug Massachusetts' Life Sciences Initiative at BIO." Xconomy, June 16, 2008. https://xconomy.com/boston/2008/06/16/gov-patrick-travels-west-to-tout-massachusetts-life-sciences-initiative-at-bio/.

Toffel, Michael, and Aldo Sesia. *Genzyme Center (A)*. Boston, MA: Harvard Business Publishing, September 16, 2010.

Toffel, Michael, and Aldo Sesia. *Genzyme Center (B)*. Boston, MA: Harvard Business Publishing, September 16, 2009.

Toffel, Michael, and Aldo Sesia. *Genzyme Center (C)*. Boston, MA: Harvard Business Publishing, December 16, 2009.

Trenholm, Richard. "20 Years Ago, Star Wars: The Phantom Menace's Trailer Made Web History." CNET, November 17, 2018. https://www.cnet.com/news/star-wars-the-phantom-menace-trailer-made-web-history-20-years-ago/.

Vaccaro, Adam. "Long before Pandemic, Problems Mounted for New Red and Orange Line Cars." *Boston Globe*, October 25, 2020.

Vaccaro, Adam. "New Red, Orange Line Trains Will Be at Least a Year Late." *Boston Globe*, October 5, 2020.

Vanasse Hangen Brustlin, Inc. *Kendall Square Annual Transportation Data Report*. Cambridge, MA: Cambridge Redevelopment Authority, 2020. https://static1.squarespace.com/static

/51f173a6e4b04fc573b07c0c/t/5f5a9e15e32a6b700a885b9b/1599774242079/KSURP+2019+Mobility+Data+Report+w.+appendices.pdf.

Van den Steen, Eric. *Akamai's Edge (A)*. Cambridge, MA: Harvard Business School, 2013.

Van den Steen, Eric. *Akamai's Edge (B)*. Cambridge, MA: Harvard Business School, 2013.

Van Veen, Frederick. *The General Radio Story*. Massachusetts: Frederick Van Veen, 2011.

Vellucci, Al. Interview by Rae Goodell. MIT Archives, Recombinant DNA Controversy Oral History Collection, May 9, 1977.

Vermes, Krystle. "Booming Alnylam to Take Over Vertex's Large Space in Cambridge." BioSpace, April 7, 2015.

Vest, Charles M. "Genome Research Presents Opportunity for Hub." *Boston Globe*, August 11, 2002.

Vest, Charles M. "Introductory Comments." In "A Study on the Status of Women Faculty in Science at MIT," special edition, *MIT Faculty Newsletter*, March 1999. https://fnl.mit.edu/wp-content/uploads/2020/06/fnl114xy.pdf.

Vinluan, Frank. "Alnylam Gets Quick FDA OK for Second RNAi Rare Disease Drug." Xconomy, November 20, 2019. https://xconomy.com/boston/2019/11/20/alnylam-gets-quick-fda-ok-for-second-rnai-rare-disease-drug/.

Vinluan, Frank. "Rubius Upsizes IPO and Hauls in $241M for Red Blood Cell Therapies." Xconomy, July 18, 2018. https://xconomy.com/boston/2018/07/18/rubius-upsizes-ipo-and-hauls-in-241m-for-red-blood-cell-therapies/.

Wade, Nicholas. "Scientists Complete Rough Draft of Human Genome." *New York Times*, June 26, 2000.

Wakabayashi, Daisuke, Karen Weise, Jack Nicas, and Mike Isaac. "Big Tech Continues Its Surge Ahead of the Rest of the Economy." *New York Times*, October 29, 2020.

"Walworth History." Walworth. Accessed July 31, 2019. https://walworth.com/en/historia/.

Watson, Thomas Augustus. *The Birth and Babyhood of the Telephone (Transcript of Call)*. Ashland, Ohio: Library of Alexandria, 1971.

Waugh, Alice C. "F&T Fans Recall Good Talk and Good Food." *MIT News*, June 12, 2003.

Webb, C.A., Robert Coughlin, and Tim Murray. "In Allston Interchange Debate, Don't Forget Grand Junction." *CommonWealth*, October 1, 2020.

Weintraub, Arlene. "Biogen Buys Stromedix for $75M, Regaining Fibrosis Drug." Xconomy, February 14, 2012. https://xconomy.com/boston/2012/02/14/biogen-buys-stromedix-for-75m-regaining-fibrosis-drug/.

Weisman, Robert. "At Novartis, a Winning Formula." *Boston Globe*, January 4, 2011.

Weisman, Robert. "LabCentral Will More than Double Its Space to Incubate Drug-Discovery Start-ups." *Boston Globe*, June 19, 2017.

Weisman, Robert. "New Complex Unites Vertex Employees." *Boston Globe*, February 4, 2014.

Weisman, Robert. "Report Hails Mass. Biotech Spending as Job Creator." *Boston Globe*, March 26, 2013.

Weisman, Robert. "Shire to Expand Kendall Square Research Outpost Acquired in Baxalta Buyout." *Boston Globe*, June 3, 2016.

Weisman, Robert. "Steven Holtzman Named CEO of Decibel." *Boston Globe*, July 13, 2016.

Weisman, Robert, and Jack Newsham. "Bristol-Myers Squibb to Open Research Site in Cambridge." *Boston Globe*, June 25, 2015.

"West Boston Bridge Editorial." *Columbian Centinel*, January 21, 1792.

Whitehead Institute. "Philanthropists Eli & Edythe Broad of Los Angeles Give $100m to Create Institute with MIT, Harvard, and Whitehead to Fulfill Genome's Promise for Medicine." Press release, June 19, 2003.

Wikipedia. "Applera." Last revised December 4, 2020. https://en.wikipedia.org/wiki/Applera.

Wikipedia. "Arthur D. Little." Last revised May 23, 2021. https://en.wikipedia.org/wiki/Arthur_D._Little.

Wikipedia. "Athenaeum Press." Last revised April 22, 2021. https://en.wikipedia.org/wiki/Athenaeum_Press.

Wikipedia. "Cambridge, Massachusetts." Last revised July 16, 2021. https://en.wikipedia.org/wiki/Cambridge,_Massachusetts.

Wikipedia. "Carnegie Mellon University." Last revised July 16, 2021. https://en.wikipedia.org/wiki/Carnegie_Mellon_University.

Wikipedia. "Daniel Chapman Stillson." Last revised January 4, 2019. https://en.wikipedia.org/w/index.php?title=Daniel_Chapman_Stillson&oldid=876743567.

Wikipedia. "David Farragut." Last revised July 18, 2019. https://en.wikipedia.org/w/index.php?title=David_Farragut&oldid=906810651.

Wikipedia. "Digital Equipment Corporation." Last revised June 15, 2021. https://en.wikipedia.org/wiki/Digital_Equipment_Corporation.

Wikipedia. "Draper Laboratory." Last revised June 6, 2021. https://en.wikipedia.org/wiki/Draper_Laboratory.

Wikipedia. "Edwin H. Land." Last revised May 22, 2019. https://en.wikipedia.org/w/index.php?title=Edwin_H._Land&oldid=898297583.

Wikipedia. "Embargo Act of 1807." Last revised June 20, 2021. https://en.wikipedia.org/wiki/Embargo_Act_of_1807.

Wikipedia. "Francis Dana." Last revised May 21, 2021. https://en.wikipedia.org/wiki/Francis_Dana.

Wikipedia. "Henry Lee Higginson." Last revised July 8, 2021. https://en.wikipedia.org/wiki/Human_Genome_Project.

Wikipedia. "Houghton Mifflin Harcourt." Last revised September 16, 2021. https://en.wikipedia.org/wiki/Houghton_Mifflin_Harcourt.

Wikipedia. "Human Genome Project." Last revised July 11, 2021.

Wikipedia. "Lisp Machines." Last revised July 5, 2021. https://en.wikipedia.org/wiki/Lisp_Machines.

Wikipedia. "Lotus Software." Last revised July 7, 2021. https://en.wikipedia.org/wiki/Lotus_Software.

Wikipedia. "Middlesex Canal." Last revised June 29, 2021. https://en.wikipedia.org/wiki/Middlesex_Canal.

Wikipedia. "MIT Computer Science and Artificial Intelligence Laboratory." Last revised July 13, 2021. https://en.wikipedia.org/wiki/MIT_Computer_Science_and_Artificial_Intelligence_Laboratory.

Wikipedia. "Pipe Wrench." Last revised July 4, 2021. https://en.wikipedia.org/wiki/Pipe_wrench.

Wikipedia. "Polaroid Corporation." Last revised July 11, 2021. https://en.wikipedia.org/wiki/Polaroid_Corporation.

Wikipedia. "Raytheon Company." Last revised July 5, 2021. https://en.wikipedia.org/wiki/Raytheon_Company.

Wikipedia. "Richard Greenblatt (programmer)." Last revised April 6, 2021. https://en.wikipedia.org/wiki/Richard_Greenblatt_(programmer).

Wikipedia. "Technology Square (Cambridge, Massachusetts)." Last revised February 1, 2021. https://en.wikipedia.org/wiki/Technology_Square_(Cambridge,_Massachusetts).

Wikipedia. "Thinking Machines Corporation." Last revised June 18, 2021. https://en.wikipedia.org/wiki/Thinking_Machines_Corporation.

Wikipedia. "War of 1812." Last revised July 19, 2021. https://en.wikipedia.org/wiki/War_of_1812.

Wikipedia. "Whitehead Institute." Last revised April 30, 2021. https://en.wikipedia.org/wiki/Whitehead_Institute.

Wildes, Karl L. "Electrical Engineering at Massachusetts Institute of Technology." Unpublished manuscript. 1971.

Wilke, John. "Genzyme Says It Has a Potential Hot Seller." *Boston Globe*, January 6, 1987.

Wilson, Edward O.. *Consilience: The Unity of Knowledge*. New York City: Vintage, 1999.

Wilson, Mark. "This Company Will Mail You a Walk-in Closet." *Fast Company*, November 13, 2018.

Winling, Ladale C. *Building the Ivory Tower: Universities and Metropolitan Development in the Twentieth Century*. Philadelphia: University of Pennsylvania Press, 2018.

Winn, Zach. "An Engine for Game-Changing Innovation." *MIT News*, December 13, 2019.

Yoon, J. Meejin. "Sean Collier Memorial." MIT List Visual Arts Center, Spring 2015. https://listart.mit.edu/public-art-map/sean-collier-memorial.

Zimmer, Carl. "One Meeting in Boston Seeded Tens of Thousands of Infections, Study Finds," *New York Times*, August 26, 2020.

INDEX

Page numbers followed by b indicate boxes; page numbers followed by f indicate figures.

292 Main Street, 13, 49, 310n4
314 Main Street, 12–13, 203b, 264, 298
5AM Ventures, 217
700 Main
  Allen & Endicott and, 238
  Bell and, 238
  cancer research and, 5, 241
  COVID-19 pandemic and, 246
  Davenport and, 26, 28, 34, 37, 157, 237–238
  industrial development and, 54
  Kaplan Furniture Company and, 238, 240
  LabCentral and, 5, 149, 161, 196b, 210b, 237, 241–247, 333n1, 333n15
  Land and, 5, 7, 54, 73, 197, 238–241
  McCready and, 7
  Mole's Hole and, 73
  as National Historic Chemical Landmark, 240
  Novartis and, 157, 333n1
  Polaroid and, 5, 73, 240, 243
  Shire and, 196b
  story of, 237–247
  TKT and, 197
  Walworth and, 237
  zoning and, 149
810 Main Street, 50, 336n1
911 attacks, 124, 129–130

Abbott, 165
AbbVie, 157, 196b
Abt Associates, 89
Accel, 123
Accelerator fellowships, 233
Accomplice, 216, 253
Acute hepatic porphyria (AHP), 176

Adams, Bob, 320n22
Adams, Daniel, 106–107
Adams, Jody, 119
Adams, John, 22–23
Advanced Research Projects Agency (ARPA), 72, 92, 94, 96, 122, 317n11, 323n9
Afeyan, Noubar
  corporatization and, 199
  Flagship and, 218–222, 273
  on industrial development, 118
  Moderna and, 117, 218, 259f, 334n1
Affinivax, 246
Affymetrix, 181
Agenerase, 171
Age of Living Machines, The: How Biology Will Build the Next Technology Revolution (Hockfield), 271
Aggarwal, Raj, 211
Agios, 117
AI Alley, x, 205
  convergence and, 297
  Cooper on, 320n26
  corporatization and, 202
  dissipation of, 120
  lessons from, 291
  Lisp Machines and, 95–96
  Lotus and, 7 (*see also* Lotus)
  Palladian and, 95, 97–98, 320nn25–26
  Silicon Valley and, 94–95
  Symbolics and, 95–98
  Thinking Machines and, 6–7, 91–98, 117, 119, 122, 186, 319n15, 320n29, 323n10
AIDS, 90, 171
AI Group, 71, 96
AI Horizons Network, 299

Akamai
  Apple and, 128–130
  Berners-Lee and, 124
  buildings of, 18, 82b, 126–127, 131–132, 149, 268, 270
  Conrades and, 126–132, 323n10, 324n17
  decline of, 129–130
  edge computing and, 127
  ESPN and, 127–128
  formation of, 124–125
  IBM and, 126, 130
  impact of, 13
  Infoseek and, 127–128
  Kaplan and, 126–127
  Leighton and, 5, 124–131, 180, 323n10, 324n13, 324nn19–20
  Lewin and, 125–126, 128, 324n13, 324n20
  March Madness and, 127–128
  Microsoft and, 130
  NASA and, 82b, 83
  NOCC and, 5, 127
  Sagan and, 128, 130
  Seelig and, 126–127
  success of, 129–131, 137
Alafi, Moshe, 107
Albert Einstein College of Medicine, 173
Alewife T station, 53, 89, 112, 218
Alexandria LaunchLabs, 31, 210b
Alexandria Real Estate Equities, 100, 177, 150, 325n4
Allen & Endicott, 28, 238
Allied Expert Systems, 95
Allied Health and Scientific Products, 165
Alnylam, xv
  buildings of, 3, 18, 195, 197, 330n7
  COVID-19 pandemic and, 176
  Food and Drug Administration (FDA) and, 176, 328n27
  impact of, 176–177
  Maraganore and, 175b, 176–177, 195, 250, 253, 329n4
  patisiran and, 328n27
  Sharp and, 164, 176, 180, 252, 280, 327n25
  transthyretin amyloidosis and, 3, 176, 328n27
  venture capital and, 175b
  Vertex and, 194
Alphabet, 263, 334n7
ALS, 280
Al- Suqami, Satam, 129

Altshuler, David, 181–185, 186b, 190, 194, 284
Alvarez, Luis, 57
Alzheimer's disease, 241, 246, 274, 280
Amazon, 18, 98, 122, 130, 199, 203b, 263
America Biltrite Rubber Company, 31
American Academy of Arts and Sciences, 183, 228
American Appliance Company, 53
American Chemical Society, 240
American Home Products, 112
American Railroad Journal, 27
American Red Cross, 52
American Research and Development (ARD), 61, 63, 206, 320n22
American Twine Building, 218
Amgen, 112, 163–165
Amini, Lisa, 299
Analog Devices, 155
Anderson and Stowe, 97
Andorra la Vella, 10
Andrews, Tom, 263
Android, 4, 200–201, 203b, 330n16
Angle, Colin, 123
Aniston, Jennifer, 161
Apple
  Advanced Technology Lab (ATL) and, 121–122, 323n6
  Akamai and, 128–130
  buildings of, 4, 13, 17
  challenges of, 263–264
  corporatization and, 199, 203b
  Jobs and, 73, 87, 121, 128
  Kapor and, 319n4
  Kendall Square Initiative and, 17
  Macintosh and, 121
  MIT Media Lab and, 121–122
  Newton and, 121
  QuickTime and, 128
  research labs and, 122
  rising profits of, 263
  Sculley and, 121
  success of, 18, 121–122, 129, 334n7
  Tiny Troll and, 88
Applera, 332n3
Applied Genomic Technology Capital (AGTC), 218
Applied Invention, 277
Area Four, 71
Ariad Pharmaceuticals, 195, 196b

Armony, Izhar, 208
Arnold, Frances, 4
ARPANET, 92, 96, 122, 323n9
Arrakis Therapeutics, 251
Arthur D. Little (ADL), 53, 59, 74, 95, 112, 138, 318n20, 320n22
Artificial intelligence (AI), 106
  AI Alley and, 87 (*see also* AI Alley)
  AI Horizons Network and, 299
  convergence and, 280–281, 297–300
  Gosper and, 319n19
  Levin on, 278
  machine learning and, 4, 191, 200, 232, 246, 297–299
  MIT AI Lab and, 93, 95–96, 120–121, 154, 202
  MIT-IBM Watson AI Lab and, 13, 298, 310n7, 336n5
  MIT Media Lab and, 121 (*see also* MIT Media Lab)
  Noftsker and, 319n19
  Project MAC and, 71–73, 91, 96, 120–121, 154, 199, 293, 319n19
  Stallman and, 319n19
  Stephen A. Schwarzman College of Computing and, 13–14, 285, 299
Art Technology Group (ATG), 123, 137, 215, 274
Ashdown House, 313n5
Ashland Oil & Refining Company, 53
Ashton Valve Company, 90
Asilomar Conference, 102–103, 321n14
Aspect Ventures, 232
AstraZeneca, 157
AT&T, 62, 127
Atari, 121, 203b, 323n4
Athenaeum, 51, 73, 160, 174
Athey, Susan, 200
Atlas, 215–216, 253
Attention deficit hyperactivity disorder (ADHD), 197–198
Aulet, Bill, 51, 273, 287, 289, 293
Aurora Flight Sciences, 13
Autism, 4, 187, 246
Aveo Pharmaceuticals, 171, 330n6
Avila Therapeutics, 251
Avonex, 165–166, 326n8

Bachman, Charles, 97–98
Back Bay, 21, 37, 40–41, 73, 175b, 208–209, 263, 283

Badger Building, 84, 139, 143, 174, 205
Baerkahn, Jesse, 172b, 266
Bailyn, Lotte, 226
Bain & Company, 209
Baker, Charles, 264, 287
Baltimore, David
  background of, 67
  biotech and, 101–102, 104–105, 114–117
  Whitehead Institute and, 114–117, 120, 183, 274, 323n9
Barber, Gerald, 95
Barta Building, 61
Bartel, David, 327n25
Battery Ventures, 126
Battle of Britain, 56, 58
Baty, Gordon, 205–206
Baxalta, 195, 196b, 327n19, 330n6
Baxter Healthcare, 112
Baxter International, 113, 166, 171, 330n6
Bay Bridge Bio, 221
Bay Colony Corporate Center, 206–207
Bayer, 13, 157, 173, 196b
Bayh-Dole Act, 134–135
Beacon Hill, 52
Begley, Sharon, 187, 189, 232
Behnisch, Stefan, 169
Bell, Alexander Graham, 7, 238, 251
Berg, Paul, 102–103, 105–106, 114
Berkowitz, Roger, 144b
Bermuda Race, 45
Berners-Lee, Tim, 5, 123–124
Bertozzi, Carolyn, 232
Bessemer Venture Partners, 208
Beta thalassemia, 2
Bhatia, Sangeeta
  background of, 223–225
  Hockfield and, 4–5, 223–225, 228, 232
  Hopkins and, 223–225, 228–229
  as serial entrepreneur, 159, 274
  women in business and, 223–225, 228–229, 232–236
Biden, Joe, 180, 187, 277
Bill & Melinda Gates Foundation, 246
Bill & Melinda Gates Medical Research Institute, 277
Biogen, xv, 163
  Bosley and, 151–152
  buildings of, 82b, 83, 99, 109, 111, 114, 164, 167, 179, 205–206, 284

Biogen (cont.)
　challenges for, 262, 265, 268
　CityScope and, 18
　collaboration and, 251–252, 255
　compared to big pharma, 177
　corporatization and, 193–194, 204
　COVID-19 pandemic and, 179
　culture of, 255
　Food and Drug Administration (FDA) and, 164–165
　Genzyme and, 18, 111–114, 117, 120, 136, 164, 168–169, 177, 236, 251, 254
　Gilbert and, 99–100, 106–108, 165, 321n28
　Hirulog and, 166, 175
　Idec and, 165, 169, 193
　impact of, 111, 117
　Inco and, 106–108, 112, 321n27
　LEED certification of, 164
　licensing and, 165–166, 322n33
　losses of, 166
　Maraganore and, 166, 175
　Mulligan and, 116
　ordinance on, 7, 99–100, 105, 109, 114, 167, 284, 321n22
　patents and, 108, 165, 176, 322n33, 322n37
　restructuring of, 165–167
　Scangos and, 193
　Sharp and, 99, 106–109, 179–180, 236, 252, 255, 280, 321n28
　Stromedix and, 217
　strong management of, 136
　success of, 111–113, 120, 136, 165–166
　talent war of, 169
　Tysabri and, 193
　venture capital and, 107–108
　Vincent and, 165–167
　Volpe Transportation Center and, 83, 99, 109, 167
Bio-Information Associates (BIA), 112–113, 322n6
BioInnovation Capital, 245
BioLabs, 210b, 243, 333n19
BioMed Realty, 100, 147, 171–172, 266, 330n7
BioNTech, 221
BioPharma Hub, 251
Biotech. *See also specific company*
　cancer research and, 101–105, 112, 114–116
　Celebration of Biotechnology and, 163
　Human Genome Project and, 181–182, 191

　Massachusetts Biotechnology Council and, 152, 163, 177
　ordinance on, 7, 99–100, 105, 109, 114, 167, 284, 321n22
　Sharp and, 99, 101, 105–109, 164, 176, 236, 328n26
Bischofberger, Norbert, 246–247
Blair, Henry, 112
Blair, Tony, 182
Blake and Knowles Foundry, 312n29
Bluebird Bio, 2, 18, 175b, 195, 309n3
Blue Room, 119, 323n1
Boeing, 4, 13, 17, 264
Boger, Joshua, 170–171, 274
Bolt Beranek and Newman (BBN), 95, 122, 126
Bonanos, Christopher, 73
Bonsen, Joost, xiii, 14, 19, 291
Booth, Bruce, 215–217
Borgs, Christian, 199–200, 330n15
Bosley, Katrine, 151–152, 251
Boston & Albany Railroad, 29
Boston & Worcester Railroad, 25–26
Boston Advertiser newspaper, 238
Boston Biotech Working Group, 228, 332n2
Boston Children's Hospital, 233, 246
Boston Consulting Group, 1, 137, 139, 144b
Boston Daily Advisor newspaper, 40–41
Boston Daily Globe newspaper, 47, 314n3
Boston Globe newspaper, 95, 98, 116, 152, 155, 184b, 208, 226
Boston Harbor, 21, 44, 209
Boston Herald newspaper, 23
Boston Innovation District, 7, 15, 148, 209, 261–262, 271
Boston Neck, 22
Boston Phoenix newspaper, 103
Boston Properties, 13, 83–84, 100, 144b, 149, 200, 202, 268, 270
Boston Redevelopment Authority, 75, 78
Boston Sail Loft, 119
Boston Seaport District, 111, 171, 194, 197, 211, 261
Boston Symphony Orchestra, 43
Boston University Bridge, 270, 312n17
Boston Woven Hose and Rubber, 7, 30–31, 113, 119, 167
Bourque, Janice, 152, 163, 177–178, 254, 328n30
Boyer, Herb, 100, 102, 106, 163
Brand, Stewart, 120

Breakthrough Prize in Life Sciences, 187
Brennan, Aoife, 234
Bricklin, Dan, 88
Bridges, Albert, 26, 311n7
Bridges, Alfred, 311n7
Brin, Sergey, 187
Bristol Myers Squibb, 3, 18, 157, 181, 196b, 197, 309n3
Broad, Edythe, 4, 183
Broad, Eli, 4, 183
Broad Institute
  buildings of, 4, 179–181, 185–186
  cancer research and, 4, 181, 186b, 188, 297–298
  controversy over, 179–180
  convergence and, 297
  COVID-19 pandemic and, 179, 187–189, 263, 269, 329n21
  Eric and Wendy Schmidt Center, 4, 298–299, 329n13
  formation of, 185
  founding principles of, 186b
  Genomics Platform and, 179
  Gray and, 327n20
  Human Cell Atlas and, 4
  Lander and, 5, 116, 144b, 178, 180–191, 277–278
  lessons from, 292
  McGuire and, 187–189, 278–279
  Millennium and, 181
  patents and, 189–190, 239–240, 329n25
  Sharp and, 180
  success of, 186–187
  vision for, 183, 185, 187
  Volpe Transportation Center and, 83–84, 85f
  Whitehead Institute and, 180, 182–185
Broderick, Shawn "Doody," 206
Brookline Village, 22
Brown, James, 315n19
Bucentaur barge, 44–45
Budweiser, 179
"Building Regional Advantage" (Hockfield), 262
Burroughs, 62
Bush, Vannevar, 53, 56–57, 159
Bus rapid transit (BRT), 10

Cabot, Cabot & Forbes (CC&F), 70, 284, 293
Cabot family, 50
Cambridge Athletic Club, 51

Cambridge Chamber of Commerce, 70, 143
Cambridge Chronicle newspaper, 33, 36, 44
Cambridge City Council, 43, 143, 266
Cambridge Experimental Review Board (CERB), 103
Cambridge Gateway, 84
Cambridge Historical Commission, 21, 25–26, 268, 303, 310n4
Cambridge Incubator, 138–141
Cambridge Innovation Center (CIC), 65
  block parties of, 144–145
  buildings of, 3, 137–142, 209, 210b
  CIC Health and, 137–138
  COVID-19 pandemic and, 137–138, 145, 264
  coworking space and, 137, 142
  DFJ and, 139–140
  foundry model and, 139–141
  Mamlet and, 139–144, 174
  Miner and, 201
  number of startups in, 5–6
  Olmsted and, 138–140
  remote workers and, 264
  Rowe and, 65, 137–148, 174, 243, 264, 294
  Venture Cafe and, 65, 145
  venture capital and, 137–140, 208–209
  vision for, 138–139
  zoning and, 147–150
Cambridge Redevelopment Authority (CRA), 70, 80, 83, 149–150, 269, 318n4
Cambridge Research Laboratory, 78, 122
Cambridge Residents Alliance, xiv, 267
CambridgeSide Galleria, 122
Cambridge Spirits, 17f, 18b
Cancer research
  700 Main and, 5, 241
  Ariad Pharmaceuticals and, 195, 196b
  biotech and, 101–105, 112, 114–116
  Broad Institute and, 4, 181, 186b, 188, 297–298
  Dana-Farber Cancer Institute and, 112, 186b
  drug-conjugated antibodies and, 112
  Fruehauf and, 242
  Kendall Square Association (KSA) and, 19, 269
  Koch Institute and, xv, 55, 225, 246, 271, 332n4
  LabCentral and, 5, 241, 246
  Luria and, 224
  MIT Cancer Center and, 101, 115, 176, 181
  Morse and, 240

Cancer research (cont.)
  Nixon's war on, 101
  Novartis and, 153
  Ontario Cancer Institute and, 101
  Rubius and, 221
  Sharp and, 4, 101, 176, 280
  Shire and, 196b
  Whitehead Institute and, 4, 115–116, 185b
  Xtuit Pharmaceuticals and, 162
Candy manufacturers, 49–52, 61–62, 66, 152, 155–157, 179, 295, 336n1
Capital One, 13, 17, 264
Cardinal Partners, 327n25
Carnegie, Andrew, 40–42, 56, 314n17
Carnegie Institution, 56–57
Carnegie Mellon University, 95
Carroll, John, 174–175
Carter's Ink, 51
Casablanca (film), 138
Catalano, Eduardo, 71
Cavity magnetron, 56–57, 60–61
Cawthorn, Rob, 108
CB, 211, 331n16
CBS, 94
C. Davenport & Co. Car Works, 27
CDN (content delivery network), 130
Celebration of Biotechnology, 163
Celera, 182, 219, 328n7
Cell journal, 189
Cell phones, 98, 177
Cequent, 242–243, 333n18
Ceredase, 113, 169
Cetus, 107, 164
Chain Home network, 56
Chan, Priscilla, 187
Chang, Morris, 315n13
Charles River Embankment Company, 38
Charles River Ventures (CRV), 208, 211
Charles Street Genomics Platform, 187
Charleston Chews, 49–50
Charpentier, Emmanuelle, 189–190
Chayes, Jennifer, 199–200, 330n15
Chicago Evening Post newspaper, 43
Chiron, 164
Chrome, 201
Chung, Joe, 123–124, 137, 274–275
Church, George, 159, 246
Churchill, Winston, 56
CIA, 91–92, 98

CityCar, 9
City Science, 9, 14f, 310n1
CityScope
  Biogen and, 18
  buildings of of, 9–19
  Correa and, 16–17
  Genzyme and, 18
  innovation model of, 9–19
  Kendall Square Initiative (KSI) and, 11–13, 17, 18b, 310n7
  LabCentral and, 12–13
  Larson and, 14–16
  LEED certification and, 10
  LEGO and, 14f, 15–16
  Marriott and, 11–12, 16
  MIT Media Lab and, 14–16, 19
  Volpe Transportation Center and, 9–11, 14–16
Civil War, 28–30, 34, 40, 237
Clapp, J. Emory, 52
Clarke, John, 327n25
Clark Medal, 200
Clem, Chet, 67
Clem, David, 67, 104–105, 167–169, 266, 327n13
Clinton, Bill, 182
Cohen, Daniel, 173
Cohen, David, 206
Cohen, Stanley, 102
Cold Spring Harbor, 101, 224, 250, 320n7
Collaboration
  Biogen and, 251–252, 255
  BioPhama Hub and, 251
  Broad Institute and, 184b
  COVID-19 pandemic and, 249
  The Engine and, 233, 261
  environment for, 249–256
  Genzyme and, 249, 251, 254
  impact of, 293–294
  Kendall Square Association (KSA) and, 253, 269
  mentorship and, 3, 66, 141, 181, 209, 234, 249–252, 294–295
  Millennium and, 251, 254–256
  Novartis and, 251, 255, 333n18
  Pfizer and, 251
  Sharp and, 252, 254–255, 294
  Termeer and, 249–251
  Vertex and, 251, 255

Collins, Jim, 159, 275
Columbian Centinel newspaper, 23
Comdex, 89
Commonwealth Fusion systems, 300
Commonwealth Pier, 44
Compaq Computer, 88
Compton, Karl, 57, 61
Computer Pictures, 97
Computer Science and Artificial Intelligence Laboratory (CSAIL), 5, 15, 72, 120, 123
Confectioner's Row, 49
Connection Machines, 93–94, 98
Conrades, George, 126–132, 323n10, 324n17
Consilience: The Unity of Knowledge (Wilson), 300–301
Consistent hashing, 215
Constellation Center, 266
Cook, Jodi, 232
Cooney, Charlie, 66, 113, 117
Cooper, Phil, 97, 320nn25–26
Copyright, 284–285
Corbato, Fernando "Corby," 72
Corona satellites, 239
Corporatization
  AI Alley and, 202
  Apple and, 199, 203b
  Biogen and, 193–194, 204
  COVID-19 pandemic and, 198, 202
  Genzyme and, 195–198
  Google and, 199–202, 203b
  IBM and, 199, 202, 203b
  LabCentral and, 196b
  Microsoft and, 199–202, 203b
  Novartis and, 193, 195, 196b
  Vertex and, 194–197
Correa, Charles, 16–17
COR Therepeutics, 174–175
Cosmic Background Explorer (COBE), 67
Cottage Farm, 38, 312n17
Couch, Stephanie, 275
Coughlin, Bob, 250, 270, 275, 293, 336n13
Coursera, 232
COVID-19 pandemic, x, xiv
  700 Main and, 246
  Alnylam and, 176
  BioGen and, 179
  BioNTech and, 221
  Broad Institute and, 179, 187–189, 263, 269, 329n21
  Cambridge Innovation Center (CIC) and, 137–138, 145, 264
  changing workspaces from, 263
  Charles Street Genomics Platform and, 187
  collaboration and, 249
  corporatization and, 198, 202
  estimated number of companies working on, 6
  Food and Drug Administration (FDA) and, 2–3, 176, 257, 263
  Genzyme and, 198
  Google and, 202
  Heaton on, 277
  McGuire on, 278
  MassBio and, 6, 309n9
  Moderna and, 2, 4, 188, 217, 221, 257, 263
  Murray on, 279
  need for interaction, 296
  New England Journal of Medicine and, 277
  Pfizer and, 221, 257
  project effects of, 4, 6, 13, 67, 210b, 231–233, 249, 262–270
  remote workers and, 188, 263–264, 277–278
  RNA interference and, 3, 176, 242
  surveillance testing and, 189
  Teal Bio and, 4, 162
  vaccines and, 2, 6, 176, 221, 246, 257, 263
  venture capital and, 210b, 221
  Whitehead Institute and, 4
Coworking, 3, 10, 13, 137, 142, 148, 172, 209, 210b
Crane, Ed, 70
Crick, Francis, 101
CRISPR, 7, 189–190, 278, 297
Crosby, Bing, 74
Cruise, Tom, 98
Curtis Davis Company, 30, 48
CyPhy Works, 276

Dagres, Todd, 126
Daley, George, 116, 182
Daly, Mark, 182
Dana, Francis, 22
Dana-Farber Cancer Institute, 112, 186b
DARPA, 92–94, 124
Darwin, Charles, 250
Data General, 89, 252
Davenport & Bridges, 26–28
Davenport, Alvin, 311n7

Davenport, Charles
  700 Main and, 26, 28, 34, 37, 157, 237–238
  background of, 25
  Charles River development of, 37–41, 291
  Kimball and, 25–26, 28
  patents and, 26
  railroad and, 25–29, 37, 197, 210b, 237, 312n17
  retirement of, 28
Davies, Julian, 108
Davis & Brody, 84
Davis, John, Jr., 34
Daye, Stephen, 50
Dell, 18, 148, 203b
Denali Therapeutics, 117, 221
Dennis, Jack, 72
Department of Brain and Cognitive Sciences, 5
Department of Transportation (DOT), 10, 82b, 83, 264
Deshpande Center, 310n5
Dewpoint Therapeutics, 252
Digital Equipment Corporation (DEC), 62–63, 89, 122, 184b, 252
Diovan, 152
Disney, 127
DNA
  Asilomar Conference and, 102–103, 321n14
  BioBuilder and, 246
  cancer research and, 101–103
  CERB and, 104
  Clem and, 104–105
  Crick and, 101
  CRISPR and, 7, 189–190, 278, 297
  discovery of, 272
  double helix structure of, 101
  Genetics Institute and, 111
  Necco and, 155, 156f
  recombinant DNA ordinance and, 7, 99–100, 105, 109, 114, 167, 284, 321n22
  Vellucci and, 99–100
  Watson and, 101
  Whitehead Institute and, 114
Dock Square, 24, 28, 30–31, 34–38
Dodge, Theodore, 30
Dog Patch Labs, 207
Dolphin Tank, 234–236
Doriot, Georges, 61, 63
Dot-com bubble, 129–131, 138, 140, 151, 177, 213
Doudna, Jennifer, 189–190, 329n25

Dracopoli, Nicholas, 180–181
Draper, Charles Stark, 54, 61, 239
Draper Fisher Jurveston (DFJ), 139–140
Draper Laboratory, 29, 48, 49f, 54, 74f, 78, 99, 179
Draper Prize, 200
DuBridge, Lee A., 57
Dudley Square, 22
Dunsire, Deborah, 174, 276, 291, 327n23
Du Pont, T. Coleman, 42–43
Durant, John, 101–103, 105

East Cambridge Planning Team, 268
Eastgate Apartments, 34
Eastham, Melville, 52, 57
Eastman, George, 43
Edge computing, 127
Edison, Thomas, 87
Editas Medicine, 251
Eduardo, Pablo, 250
Edwards, David, xiii–xiv
Egozy, Eran, 123
Einstein, Albert, 153
Electric vehicles, 9
Electronics Research Center (ERC), 77–83, 284, 291, 318n10
Eli Lilly, 157
Eliot, Charles, 40
Embargo Act of 1807, 24
Ember, 141
Emmens, Matt, 197–198, 329n4
Engine, The, 233, 261, 271, 279, 295–296, 300
English, Todd, 119
Enriquez, Juan, 4, 283, 288b
Entertainment Tonight (TV show), 128
Entrepreneur Walk of Fame, 16, 87
Enzytech, 150–161
Eons, 200, 295
Eric and Wendy Schmidt Center, 4, 298–299, 329n13
Esplanade, 1–2, 41–43, 45, 47, 91
ESPN, 127–128
ETH, 153
Evian, 113
Evoo, 171–172
Exelixis, 193

F&T Restaurant, 65–67
Fabrazyme, 169
Fabry disease, 169, 197

Facebook, 18, 122, 130, 199, 203b, 204, 263, 285, 334n7
Fagnan, Jeff, 215, 253
Fairchild Camera and Instrument, 62
Fano, Robert, 72
Farber, Dana, 182
Farber, Sidney, 112, 186b
Fariborz Maseeh Hall, 40
Farragut, David Glasgow, 237–238
Farris, Lee, xiv
Fast Search & Transfer, 200
Feinstein Kean Healthcare, 328n30
Feld, Brad, 123, 206, 291
Feldman, Maryann, 104–105, 109
Feynman, Richard, 94
Fidelity, 11, 71, 310n3
Fifty Years of Technology exhibit, 44
Fink, Gerald, 116
Firouzbakht, Mahmood, 172b
Fischer, Bobby, 96
Fishman, Mark, 153–155
Five Cambridge Center, 84, 90, 97, 99, 109, 119–122, 201, 203b
Flagship Pioneering, 2, 117, 217–222, 259f, 273, 332n2
Flash crowd, 124
Flint, Jon, 126
Florida, Richard, 207, 287, 289–291, 301
Flybridge Capital Partners, 331n22
Foghorn Therapeutics, 221
Fonstad, Jennifer, 139
Food and Drug Administration (FDA)
　Agenerase and, 171
　Alnylam and, 176, 328n27
　Aveo and, 330n6
　Avonex and, 166
　BioGen and, 164–165
　Bourque and, 177
　Ceredase and, 113, 169
　COVID-19 pandemic and, 2–3, 176, 257, 263
　Fabrazyme and, 169
　Genzyme and, 164, 169
　Gleevec and, 153
　Hirulog and, 166
　interferon and, 165
　Millennium and, 174
　Moderna and, 257, 263
　Novartis and, 153, 255
　Velcade and, 174

Formerly Known as Atlas (FKA), 215–216
Forrester, Jay, 62
"Forty missing companies," 229, 236
Forum Pharmaceuticals, 276
Foundry model, 139–141
Fountain Hill, 28
Fox, Isaac, 65
Fox, Marvin, 65–67
Fox Cross Company, 49
F-Prime Capital, 231, 233
Frankston, Bob, 88
Friedman, Jeffery, 173
Friend Connect, 201
Frost, Robert, 163
Fruehauf, Johannes
　background of, 242
　BioLabs and, 243
　Cequent and, 242–243
　graduation lab of, 245
　LabCentral and, 150, 210b, 242–245, 264, 276, 292
　Li and, 242
　Mission BioCapital and, 245, 276
　MLSC and, 243–244
　Parker and, 243, 245
　Vithera and, 242–243
Fulcrum Therapeutics, 245
Functional genomics, 181–182
Fuseyard, 216
Future Founders Initiative, 223, 225, 228, 236, 251
Future of Industrial Research, The (conference), 240–241

Gandhi, Sameer, 123
Garage, 73
Gates, Bill, 87, 246, 277
Gaucher disease, 113, 169
Gawande, Atul, 138
Genentech, 164
　Boyer and, 100, 102, 106, 163
　Langer and, 159
　Levin and, 173
　success of, 112, 165
　Swanson and, 87, 106–107, 163
General Doriot, 206, 320n22
General Electric (GE), 62, 97
General Motors, 9, 130
General Radio, 52–53, 57, 315n26o

Genetics Institute (GI), 94, 111–112, 114, 136, 164, 218, 251, 254
Genex, 100
"Genome Research Presents Opportunity for Hub" (*Boston Globe*), 184b
GenRad, 52
Genzyme
  Biogen and, 18, 111–114, 117, 120, 136, 164, 168–169, 177, 236, 251, 254
  Blair and, 113
  buildings of, 111, 113–114, 119, 122, 169–172, 177, 327n13
  Ceredase and, 113
  challenges for, 249, 251, 254, 266
  CityScope and, 18
  Clem and, 167–169, 266, 327n13
  Cooney and, 66, 113, 117
  corporatization and, 195–198
  COVID-19 pandemic and, 198
  Fabrazyme and, 169
  Food and Drug Administration (FDA) and, 164, 169
  Langer and, 161
  LEED certification and, 169–170
  Lodish and, 113, 223, 236, 322n6
  as Sanofi Genzyme, 2–3, 18, 157, 195–198, 249, 309n3
  Snyder and, 113
  strong management of, 136
  success of, 112, 120, 167
  talent war of, 169
  Termeer and, 3, 112–114, 144b, 163, 169–170, 197, 249–251, 294
  Vinter and, 201–204, 330n20
Genzyme Center, 168f, 170, 172b, 185, 197–198, 249, 266, 330n7
Gerstner, Lou, 130
Gilbert, Wally
  background of, 99, 101
  Biogen and, 99–100, 106–108, 165, 321n28
  commercialism and, 106
  licensing and, 165
  patents and, 176
  proexperiment stance of, 104
Gilead Sciences, 246–247
Gillis, Steve, 173, 327n22
Gilman, Arthur, 51
Ginn & Company, 73
Glass House, 330n7

GlaxoSmithKline, 165, 171, 198
Gleevec, 153
GlympseBio, 232
GNS Healthcare, 297
GNU project, 72
Goin, Suzanne, 119
Gold Hill, 95
Golub, Todd, 182, 186b, 187
Goodell, Rae, 321n17
Google, 274
  Alphabet and, 263, 334n7
  Android and, 4, 200–201, 203b, 330n16
  buildings of, 3–4, 13, 28, 35f, 149
  challenges and, 263
  corporatization and, 199–202, 203b
  COVID-19 pandemic and, 202
  massive growth of, 201
  research labs and, 122
  rising profits of, 263
  Schmidt and, 4, 298–299, 329n13
  success of, 3–4, 18, 130, 202
  venture capital and, 141, 207, 211
  Wertheimer and, 330n20
  YouTube and, 4, 202
  zoning and, 150
Google Ads, 202
Google Books, 201
Google Chrome, 4, 201–202
Google Computer Engine, 202
Google Flights, 202
Google Play, 4, 202
Google Search, 4
Google Ventures (GV), 141, 202, 211, 234
Go programming language, 4, 202
Gosper, Bill, 319n19
Gouw, Theresia, 232
Graffito SP, 172b, 266
Grand Junction Railroad, 29, 49f, 70, 270, 312n17
Grand Metropolitan, 108
Granges, Donald Des, 315n13
Graphical user interface (GUI), 121
Graves' Neck, 21–22
Gray, Paul, 327n20
Great Depression, 54, 65–66
Great Marsh, 22–23
Great Recession, 202, 219
Greenblatt, Richard, 96, 319n19
Green Line, 90, 275

GreenWheel, 9
Greif, Irene, 122, 199, 330n12
Greiner, Helen, 276
Groove Networks, 200
GTE, 126
Guitar Hero (game), 123

H2X, 58, 61
Habitat 67 (Safdie), 84
Hackers, 72, 91–92, 95–96, 319n19
Hackers (Levy), 92, 96
Haley, Jim, 322n37
Hall, Stephen, 103, 107, 189–190
Hamilton, David, 139–140
Handler, Sheryl, 94
Harleston Parker Medal, 170
Harmonix, 123
Harrison Street Real Estate, 333n1
Harthorne, John, 209
Hartley, Brian, 107
Harvard, John, 138
Harvard Bridge, 38, 39f, 42, 45
Harvard Business School, 40, 88, 130, 169, 227, 280, 293, 314n29
Harvard Coop, 119
Harvard Graduate School of Design, 139
Harvard Kennedy School, 287
Harvard Medical School, 42, 102, 112, 116, 153, 186b
Harvard Square, 21–23, 33, 40, 73, 95, 112, 123, 138, 280
Hayes, Daniel "Dan," 321n20
Heatherington, Anne, 276, 297–298
Heaton, Penny, 277
Henri A. Termeer Legacy Program, 249
Henri A. Termeer Square, 3, 176, 242
Hepatitis, 166, 326n6
Heroes of CRISPR, The (Lander), 189
Hewlett, Bill, 87
Higginson, Henry, 43–44
High, Katherine, 232
Highland Capital Partners, 208
High Voltage Engineering Corporation, 61, 63
Hill, Colin, 297
Hillis, Danny
  background of, 91–92
  CIA and, 91–92
  Logo Lab and, 91
  master thesis of, 92

parallel computing and, 72
Thinking Machines and, 6, 93–95, 98, 277, 320n29
Hinds, Chuck, 268
Hiriko Fold, 9
Hirulog, 166, 175
History of Massachusetts Industries (Stone), 47
HIV/AIDS, 90, 171
Hockfield, Susan, xv
  The Age of Living Machines and, 271
  Bhatia and, 4–5, 223–225, 228, 232
  "Building Regional Advantage" and, 262
  Hopkins and, 223–225, 228
  Kendall Square Association and, 144b
  on Kendall Square's appearance, 15–16, 261
  Koch and, 332n4
  Lodish and, 232–233
  Menino and, 261–262, 271, 334n1
  as president of MIT, 15, 223–224
  vision of future technology by, 271–272
  women founding companies and, 231
Hofschneider, Peter Hans, 106–107
Holmes, Oliver Wendell, 47
Hopkins, Nancy, 5
  background of, 224
  Bhatia and, 223–225, 228–229
  biotech and, 101
  Hockfield and, 223–225, 228
  marginalization of, 226
  women's issues and, 223–232, 235–236
Hotspot, 124
Houghton, Mifflin & Co., 30, 50
Housing Act, 70, 79, 318n7
Howe, Elias, 30
HubSpot, 123, 202, 215
Hughes, 62
Hughes, Tom, 154–155, 157, 255–256
Human Cell Atlas, 4, 187
Human Genome Project, 181–182, 191
Hunt, Aisling, 144
Hunt, Walter, 30
Huttenlocher, Daniel, 299

IBM
  air defense and, 62
  Akamai and, 126, 130
  Armstrong and, 247
  as Big Blue, 3, 13, 122, 199, 247, 298
  buildings of, 2–4, 13, 72–73, 95, 121

IBM (cont.)
   Conrades and, 126, 130
   convergence and, 298–299
   core memory and, 62
   corporatization and, 199, 202, 203b
   Greif and, 122
   Kapor and, 87–88, 91
   Lotus and, 87, 90–91, 199, 203b
   MIT-IBM Watson AI Lab and, 13, 298, 310n7, 336n5
   MIT Media Lab and, 121–122
   personal computers and, 88–89
   success of, 62–63
   Vandebroek and, 280–281
IBM Cambridge Scientific Center, 72
IBM Research, 3, 122, 199, 247, 280–281, 298–299, 323n8, 330n12
IBM Resilient, 3, 18
IBM Watson Health, 2–3, 298
Ibsen, 327n19
Icon Corp., 205
Idea Barn, 138
Idec Pharmaceuticals, 165, 169, 193
Immunex, 173
Immunogen, 111–112
Inari Agriculture, 221
Indigo Agriculture, 117
Industrial development
   700 Main and, 54
   candy manufacturing and, 49
   faculty workday for, 105–106
   flight to suburbs and, 69
   Kendall Square Initiative (KSI) and, 49, 52
   post-World War II, 40, 53–55, 58–60
   Stone on, 47–54
Infinity Pharmaceuticals, 180
Inflammatory bowel disease (IBD), 183
Infoseek, 127–128
Interferon, 108, 165–166, 322n37, 326n8
International Nickel Company (Inco), 106–108, 112, 321n27
Internet Archive, 94
Invention That Changed the World, The (Buderi), 60
Invisible Frontiers (Hall), 103, 107
Ipsen, 157
IPVision, 138
iRobot, 123, 276
ITA, 330n20

Itakura, Keiichi, 106
ITA Software, 202

Jaenisch, Rudolf, 116
James Beard awards, 323n1
James O. Welch Company, 60
Jefferson, Thomas, 24
J. J. Walworth & Company, 237
J. L. Hammett Building, 310n4
Jobs, Steve, 73, 87, 121, 128
John F. Cuneo Press, 73
John Harvard's Brewery & Ale House, 138
Johns Hopkins, 102
Johnson & Johnson, 157
Johnson, Carolyn, 52, 232
Johnson, Lyndon B., 77, 81
Joint Services Electronics Program, 60
Junior Mints, 49–50, 295, 336n1
Jurassic Park (film), 98

K2 Plan, 148
Kahle, Brewster, 94
Kaleido Biosciences, 221
Kane, Bill, 147, 266–267
Kania, Ed, 219
Kaplan, Isaac, 238
Kaplan, Randall, 126–127
Kaplan Furniture Company, 238, 240
Kapor, Mitch, 87–91, 319n4, 319n9
Karlen, Jon, 215
Kay, Alan, 121
Kayak, 202
Kean, Marcia, 328n30
Kelley, Albert, 78
Kelley, Paul, 205
Kendall & Davis, 34
Kendall & Roberts, 34
Kendall, Edward, 31–36, 38, 44
Kendall, George, 34
Kendall, James, 34
Kendall Center, 3, 330n17
Kendall Square. *See also specific company*
   CityScope and, 9–19
   corporatization of, 193–204
   Davenport and, 25–31
   as ecosystem, ix-xv
   eleven decisions that shaped, 283–285
   first economic vision of, 21–24
   industrial development and, 47–54

K2 Plan and, 148
layout of, 1–7
lessons from, 287–294
regional advantage and, 261–271
zoning and, 82b, 83–84, 147–150, 168–169, 293–294, 325n3
Kendall Square Association (KSA), xiv, 3
  cancer research and, 19, 269
  challenges for, 266, 269–270
  charter of, 143
  collaboration and, 253, 269
  Hockfield and, 144b
  labor demographics and, 5
  McCready and, 278
  McGuire and, 188, 278–279
  main priority of, 33
  Rowe and, 143
  transportation and, 269
  Webb and, 234, 270, 281
Kendall Square Initiative (KSI)
  292 Main Street and, 13, 49, 310n4
  314 Main Street and, 12–13
  approval of, 11–12
  challenges for, 264–266
  CityScope and, 11–13, 17, 18b, 310n7
  Eastgate and, 34
  industrial development and, 49, 52
  zoning and, 150
Kendall Square Manufacturing Association, 143
"Kendall Square—The Most Innovative Square Mile on Earth" presentation, 144b
Kendall Square Urban Redevelopment Area, 149
Kendall Square Urban Renewal Project, 79–80, 82b, 99, 318n4, 325n3
Kennedy, Caroline, 67
Kennedy, John F., 55, 77–78, 81, 318n1, 318n10
Kennedy, Ted, 78, 154
Kennedy Biscuit Company, 49, 315n14
Kidder, Tracy, 1
Killian, James, 70, 79
Kim, Peter, 116
Kimball, Ebenezer, 25–26
Kimball, Edward, 28
Kimball, Ranch, 293
King, Jonathan, 104
Kinsella, Kevin, 170
Kinto Care, 274
Kleespies, Gavin, 23, 30
Kleiner, Perkins, Caulfield & Byers, 88, 97

KnicKrehm, Glenn, 266
Knight, Steve, 231
Knight, Tom, 72
Knight Science Journalism, ix
Knowledge Navigator, 121
Koch Center for Integrated Cancer Research, 55
Koch Institute for Integrative Cancer Research, xv, 332n4
  Bhatia and, 225, 228
  Hockfield and, 271
  Koehler and, 246–247
  Langer and, 159–160
  lessons from, 292
  Sharp and, 4, 180
Kodak, 43, 74, 240
Koehler, Angela, 246–247
Koller, Daphne, 232
Koop, Bryan, 270
Kresge Auditorium, 43
Kronos Bio, 246–247
Kucherlapati, Raju, 173

LabCentral
  2019 Impact Report of, 241–242
  buildings of, 5, 12–13, 149, 161, 196b, 210b, 237, 241–247, 292, 295, 333n1, 333n15
  cancer research and, 5, 241, 246
  CityScope and, 12–13
  corporatization and, 196b
  Fruehauf and, 150, 210b, 242–245, 264, 276, 292
  Kronos Bio and, 246–247
  Learning Lab of, 246
  MLSC and, 243–244
  Pfizer and, 245
  Rowe and, 243, 264
  success of, 264, 294
  venture capital and, 210b
  Vivtex and, 161
  zoning and, 149–150
Laboratory for Computer Science (LCS), 96, 120, 123–124
Lafayette Square, 23, 28
Lampson, Butler, 200
Land, Edwin
  700 Main and, 5, 7, 54, 73, 197, 238–241
  company doctrine of, 241
  impact of, 238–239
  McElheny and, 239–240

Land, Edwin (cont.)
  Mole's Hole and, 239–240
  patents and, 239
  Polaroid and, 5, 72–73, 239–240, 333n10
  Rowland Institute, 93, 119, 240
Land Discovery (Grieg), 45
Lander, Eric
  Altshuler and, 181–182, 185, 186b
  Biden and, 180, 277
  Breakthrough Prize in Life Sciences and, 187
  Broad Institute and, 5, 116, 144b, 178, 180–191, 277–278
  criticism of, 189–190
  functional genomics and, 181–182
  Heroes of CRISPR, The, and, 189
  IBD research and, 183
  MacArthur "genius" grant and, 180
  Millennium and, 178, 180–181
  Obama and, 187
  Office of Science and Technology Policy and, 180, 277
  personal achievements of, 180, 187
  President's Council of Advisors on Science and Technology and, 187
  Westinghouse science competition and, 5, 180
  Whitehead Institute and, 4, 80, 84, 94, 119, 173, 178–185, 186b
  as whiz kid, 5
Landry, Kevin, 107, 112
Land-Wheelwright Laboratories, 73
Langer, Bob
  companies of, 161–162
  email response time of, 251
  Genzyme and, 161
  as inventor, 4
  Koch Institute and, 4, 159–160
  list of companies of, 161–162
  methodology of, 160
  Moderna and, 4, 160–162, 259f
  patents and, 159
  as serial entrepreneur, 4, 160–161, 251, 259f
Larson, Kent, xiii, 14–16
Larson, Michela, 119
Laser Interferometer Gravitational-Wave Observatory (LIGO), 67
Lawrence, E. O., 57
Lawrence Scientific School, 40
Lechmere Station, 90–91

LEED certification, 10, 113, 164, 170
Legal Sea Foods, 35f, 84, 97, 119, 144b, 209
Leighton, Frank Thomson "Tom"
  Akamai and, 5, 124–131, 180, 323n10, 324n13, 324n20
  Westinghouse science competition and, 180
Lemeleson-MIT Program, 275
Lenat, Douglas, 94
Lettvin, Jerome, 67, 317n5
LeukoSite, 174
Lever Brothers, 48, 49f, 66, 70–71, 74, 101, 315n12, 318n20
Levin, Mark
  on artificial intelligence, 278
  Genentech and, 173
  Millennium and, 1, 164–165, 173–174, 175b, 178, 256, 278, 327n21
Levy, Steve, 92, 96, 334n11
Lewin, Daniel "Danny"
  911 attacks and, 124, 129
  Akamai and, 125–126, 128, 324n13, 324n20
  MIT 50K Entrepreneurship Competition and, 125–126
Lewis, Herman, 102
Li, Chiang, 242
Library of Congress, 55
Licenses
  BioGen and, 165–166, 322n33
  Boston's Children's Hospital and, 246
  Flagship and, 220
  Hiriko Fold and, 9
  landlords and, 136
  lessons from, 291–292
  Lisp and, 96
  Millennium and, 328n26
  Nelsen and, 133–136, 292
  SmithKline Beecham and, 326n6
  Patent, Copyright, and Licensing Office and, 284–285
  Startup 101 and, 232
  Technology Licensing Office (TLO) and, 133–134
Licklider, J. C. R., 72
Lifebuoy, 48
Lin, Maya, 10, 18
Lincoln Lab, 62, 78, 134, 184b
Linde, Doug, 144b
Link, 149
Lisp Machines, 95–96

Little, Brown and Company, 50–51, 315n19
Little, Charles, 315n19
Lives of the Cell, The: Notes of a Biology Watcher (Thomas), 115
Living Proof, 161
Localytics, 211
Lockwood, Jeff, 157
Lodish, Harvey, 113–116, 159, 223, 232–236, 322n6
Logo Lab, 91
Lola.com, 6
Longfellow, Henry Wadsworth, 47
Longfellow Bridge, 23, 38, 43, 45, 47, 74, 91, 119, 123, 199, 259f
Loomis, Alfred, 56–57
Loran, 58
Los Alamos, 55, 94
Lotus, 2, 7, 117
   buildings of, 88–91, 98, 119, 122, 150, 203b, 259f
   convergence and, 297
   early success of, 87–89
   employee benefits of, 90
   Greif and, 122, 199
   Kapor and, 87–91, 319n4, 319n9
   lessons from, 291
   Manzi and, 90–91
   Microsoft and, 89
   Sachs and, 87–88, 98, 319n4
   sells to IBM, 87, 90–91, 199, 203b
   Sturtevant and, 200
   success of, 87, 291
   vacant land around, 98
Lovering, Alexandra, 149
Lowe, Nichola, 104–105, 109
Lower Port, 22, 24, 28–29
Lundbeck, 276, 291
Luria, Salvador "Salva," 101, 114, 224, 254, 281–282
Lux, 48
Lycos, 295
Lyme Properties, 167–172, 197, 330n7
Lynch, Barbara, 119
Lynch, Chris, 215

MacArthur "genius" grant, 116, 180
MacBird, Bonnie, 121
Mach, Bernard, 106
Mac Hack, 96

Machine learning, 4, 191, 200, 232, 246, 297–299
Macintosh computer, 121
Maki, Fumihiko, 9
Mamlet, Geoffrey, 139–144, 174
Manhattan Project, 74
Maniatis, Tom, 101, 111
Manufacturer's National Bank, 51–52
Manzi, Jim, 90–91
Maraganore, John
   Alnylam and, 175b, 176–177, 195, 250, 253, 329n4
   BioGen and, 166, 175
   Hirulog and, 166, 175
   Millennium and, 175–176, 327n21, 328n26
   Termeer and, 250–251
March Madness, 127–128
Marcuse, Herbert, 88
Maris, Bill, 202
Marquardt, Charles, 17f, 18b
Marriott, 133, 200, 209
   BioGen and, 262
   Cambridge Command Center (CIC) and, 143, 201
   CityScope and, 11–12, 16
   layout of Kendall Square, 1, 3, 11–12, 16, 24, 28, 80
   Lyme and, 168
   NASA buildings near, 77, 80, 84–87
   Safdie and, 1, 16, 85f, 319n15
   urban renewal and, 119
   zoning and, 325n3
Marsh, Steve, xiv
Marshall, Laurence, 53
Martin Trust Center for MIT Entrepreneurship, 273, 287
Maseeh, Fariborz, 40, 313n5
Massachusetts Bay Colony, 34
Massachusetts Biotechnology Council (MassBio)
   Bourque and, 152, 163, 177, 254, 328n30
   Coughlin and, 250, 270, 275, 293, 336n13
   COVID-19 pandemic and, 6, 309n9
   Hockfield and, 262, 271
   Route 128, and, 253
   State of the Possible Conference and, 262
Massachusetts General Hospital, 153, 182, 186b, 224, 262
Massachusetts Historical Society, 23

Massachusetts Institute of Technology (MIT). *See also specific building*
  Altshuler and, 181–185, 186b, 190
  Compton and, 57, 61
  founding of, 37–46
  Hockfield and, 223, 261, 271
  Killian and, 70
  patents and, 135, 159
  Pritchett and, 40–41, 43
  Vest and, 151, 154, 163, 183, 184b, 226, 228
  Wiesner and, 94, 120
Massachusetts Life Science Initiative, 293
Massachusetts Life Sciences Center (MLSC), 7, 243–244, 278, 293–294
Massachusetts State House, 51
MassChallenge, 209, 261, 292, 331n11
Massie, Thomas, 51
Mather, John, 67
Max Planck Institute of Infection Biology, 189
Maycock, Susan, 24, 47, 63, 80
Mayfield Fund, 173
MBTA Kendall/MIT subway station, 11, 16, 65, 269, 318n4
McCall, Samuel, 45
McCarthy, John, 72
McCready, Travis, 7, 278
McElheny, Victor, ix, 101, 239–240
McGovern Institute for Brain Research, 5, 17, 29
McGuire, Lee, 187–189, 278–279
McGuire, Terry, 126, 231
McKinsey & Company, 90, 137
McMillan, Ed, 57
MDRNA, 242
Mead Hall, 84
Media Lab, The: Inventing the Future at MIT (Brand), 120
Memorial Drive Trust, 320n22
Menino, Thomas, 209, 261–262, 271, 334n1
Mentoring, 3, 66, 141, 181, 209, 234, 249–252, 294–295
Merck, 108, 116, 152, 157, 170, 196b, 198
Meselson, Mathew, 104
Mesirov, Jill, 94, 186
Metcalfe, Bob, 288
Micro Finance Systems, 88
Microsoft
  Akamai and, 130
  buildings of, 119, 219, 330n15, 331n4
  corporatization and, 199–202, 203b
  impact of, 18, 199, 263, 295
  Lotus and, 89
  research labs and, 122
  rising profits of, 263
  venture capital and, 206–207, 219
  zoning and, 147, 150
Microsoft New England Research and Development (NERD), 119, 199–200
Microwave Committee, 57
Microwaves, 56–61
Mid-Continent Railway Museum, 25
Middlesex Canal, 23
Mifflin, George, 50
Millennium Pharmaceuticals
  board of, 327nn22
  Broad Institute and, 181
  buildings of, 164, 173–174
  collaboration and, 251, 254–256
  Davenport and, 237
  Dunsire and, 174, 276, 291, 327n23
  Food and Drug Administration (FDA) and, 174
  impact of, 174–175
  Lander and, 178, 180–181
  Levin and, 1, 164–165, 173–174, 175b, 178, 256, 278, 327n21
  Maraganore and, 175–176, 327n21, 328n26
  Rowe and, 141, 174
  Takeda and, 1, 174, 195, 196b, 198, 327n23
  talent war of, 169
  Velcade and, 174
  Vertex and, 164
Miller's River, 22, 23
Millikan, Max, 60
Millikan, Robert, 60
Milner, Yuri, 187
Miner, Rich, 201–202
Minority Report (film), 123
Minsky, Marvin, 71–72, 91, 94, 120–121, 319n19, 323n4
Minute Maid, 74
Mission BioCapital, 245, 276
Mission: Impossible (film), 98
MIT 100K, 125, 234, 292, 324n13
MIT 50K Entrepreneurship Competition, 125–126
MIT Artificial Intelligence Lab (AI Lab), 93, 95–96, 120–121, 154, 202

MIT Center for Cancer Research, 101, 115, 176, 181, 185
MIT Center for Genome Research, 173, 180–181, 184b
MIT Center for International Studies, 60
MIT Corporation, 45, 70, 78, 281–282
MIT Entrepreneurship Center, 75
MIT Faculty Newspaper, 226, 232, 235
MIT-IBM Watson AI Lab, 13, 298, 310n7, 336n5
MIT Innovation HQ, 12, 310n5
MIT Investment Management Company (MITIMCo), xiv, 15–16, 150, 264, 310n7
MIT Jameel Clinic for Machine Learning in Health (J-Clinic), 298
MIT Media Lab, xiii, 43
  Apple and, 121–122
  Brand on, 120
  Celebration of Biotechnology and, 163
  City Science and, 9, 14f, 310n1
  CityScope and, 9, 14–16, 19
  Ember and, 141
  formation of, 120
  IBM and, 121–122
  impact of, 120–121
  Larson and, 14
  Minsky and, 323n4
  Negroponte and, 55, 120–122, 303
  Pei and, 55, 117, 120
  Whitehead Institute and, 120
MIT Museum, 4, 12–13, 17, 52, 67, 101, 264
MIT News, 186
MIT Press Bookstore, 49, 310n4
MIT Regional Entrepreneurship Acceleration Program (MIT REAP), 271, 287, 289
MIT Sloan School of Management, 314n29
  Aulet and, 273, 289
  Bailyn and, 226
  buildings of, 43, 48, 53
  Cooper and, 320n25
  Hadzima and, 138
  Kapor and, 88
  Morse and, 75
  Murray and, 6, 228, 279
  Nijhawan and, 125
  Noftsker and, 97
  Roberts and, 125, 205, 218, 285
  Rowes and, 137
  Seelig and, 126
  Siegal and, 97

Mitsubishi, 122, 199
Moderna
  Afeyan and, 117, 218f, 259f, 334n1
  buildings of, 5, 18, 71, 195, 257, 258f–259f
  COVID-19 pandemic and, 2, 4, 188, 217, 221, 257, 263
  culture of, 255–256
  Food and Drug Administration (FDA) and, 257, 263
  Langer and, 4, 160–162, 259f
  network of, 257, 258f–259f
  success of, 195, 221, 255, 257–258, 259f
  venture capital and, 217, 221, 243
Mole's Hole, 73, 239–240
Monsanto, 108
Moore, Ryan, 215
Morris and Sophie Chang Building, 315n13
Morse, Ken, 163
Morse, Meroë, 240
Morse, Richard Stetson, 74–75
Morss, Henry, 45
Motive Labs, 216
Mount Auburn Street, 73
Muddy Charles Pub, xiii
Mullen, Jim, 166, 193
Multiple sclerosis, 165–166, 187, 193
Murray, Fiona, 6, 13, 228, 279, 287, 289–290, 292, 294
Murray, Kenneth, 106
Murray, Tim, 270
Murray Printing Company, 51
Museum of Modern Art, 81
Museum of Science, 29
Musk, Elon, 140
Myers, Christopher, 119

Napoleon, 24
NASA
  Akamai and, 82b, 83
  buildings of, 77
  Cosmic Background Explorer (COBE) and, 67
  Electronics Research Center (ERC) and, 77, 80–81, 82b, 284, 291, 318n10
  Johnson and, 77, 81
  Kelley and, 78
  Kennedy and, 77–78, 318n1, 318n10
  myths about, 77
  Nixon and, 81–83, 318n10

NASA (cont.)
  Office of Advanced Research and Technology, 78
  Rowland and, 75, 78–83
  Rubins and, 116
  satellites and, 67, 77, 318n9
  urban renewal and, 77–86
  Volpe National Transportation Center and, 10
  Webb and, 77–78
Nathans, Daniel, 102
National Academies of Sciences, 4, 225
National Academy of Engineering, 4, 225
National Academy of Inventors, 225
National Academy of Medicine, 4, 225
National Cancer Institute, 101
National Defense Research Committee (NDRC), 57, 316n5
National Historic Chemical Landmark, 240
National Institutes of Health, 103–104, 293
National Practice Leader, 278
National Register of Historic Places, 155
National Research Corporation (NRC), 74–75
National Science Foundation, 102
National September 11 Memorial, 86
Navitor Pharmaceuticals, 255
Necco Wafers, 49–50, 52, 62, 152–157, 315n16
Negroponte, Nicholas, 55, 120–122, 303
Neighborhood Conservation District, 268
Nelsen, Lita, 133–136, 292
Nelson, Teresa, 228
Network Operations Command Center (NOCC), 5, 127
New Atlantic Ventures, 140
NewcoGen, 218–219
New England Journal of Medicine, 277
New Yorker magazine, 60
New York Times newspaper, 97–98, 100, 103, 108, 187, 226, 262
Nibber. *See* Novartis
Nigam, Akhil, 209
Nijhawan, Preetish, 125
Nissan, 122
Nixon, Richard, 81–83, 101, 318n10
Noftsker, Russell, 96–97, 319n19, 320n22
Novartis
  700 Main and, 157, 333n1
  Bosley and, 151–152
  Bourque and, 152, 177
  buildings of, 5, 50, 100, 152, 155, 315n16
  cancer research and, 153
  Cequent and, 333n18
  collaboration and, 251, 255, 333n18
  corporatization and, 193, 195, 196b
  Diovan and, 152
  Dunsire and, 174
  Fishman and, 153–155
  Food and Drug Administration (FDA) and, 153, 255
  formation of, 326n5
  Gleevec and, 153
  Hughes and, 154–155, 157, 255–256
  impact of, 151–152, 163–165, 169, 171, 178, 193, 285
  inclisiran and, 176, 328n27
  as Novartis Institutes for BioMedical Research (NIBR) (Nibber), 50, 151–158, 196b, 326n7
  Stowe and, 151
  success of, 157, 163
  Vasella and, 152–154
  Venture Fund, 212f, 233
  Vest and, 151

Oakley Country Club, 37
Oak Ridge National Laboratory, 74
Obama, Barack, 187
Oblong Industries, 123
O'Donnell, Joseph, 181
Office of Science and Technology Policy, 180
Olmsted, Andy, 138–140
Olsen, Ken, 62, 63
One Broadway, 15, 43, 139, 141, 174–175, 196b, 206, 328n30
O'Neil, Tip, 67
One Laptop Per Child (OLPC), 120, 138
OneLiberty Ventures, 219
Ong, Betty, 129
Ontario Cancer Institute, 101
Orbimed, 221
Orbitz, 202
Ordinances
  biosafety, 7, 99–100, 105, 109, 114, 167, 284, 321n22
  zoning, 147–150
Ori Living, 9
Osborn Triangle, 27, 150, 333n1
Oyster Bank Bay, 21

Packard, David, 87
Page, David, 116
Paley, William, 94
Palladian, 95, 97–98, 320nn25–26
Palm Pilot, 177
Papadopoulos, Greg, 94
Papert, Seymour, 91, 120–121, 323n4
Parker, Peter, 243, 245
Parkinson's disease, 4, 280
Partners HealthCare, 233
Patent, Copyright, and Licensing Office, 284–285
Patents
  Bhatia and, 232
  Biogen and, 108, 165, 322n33, 322n37
  Broad Institute and, 189–190, 239–240, 329n25
  CRISPR, 189–190, 329n25
  Davenport drawbar and, 26
  Emmens and, 197
  Flagship and, 217, 220
  Genetics Institute and, 112
  Howe and, 30
  institutional patent agreements (IPAs), 135
  Land and, 239
  Langer and, 159
  licenses and, 165 (*see also* Licenses)
  Massachusetts Institute of Technology (MIT) and, 135, 159
  Nelsen and, 135
  Polaroid and, 239–240
  Rogers and, 240
  royalties and, 30, 88, 134–135, 166, 176, 238, 323n14, 328n27
  Shire and, 197
  Stillson and, 237–238
  Whitehead Institute and, 323n14
Patient capital, 295
Patisiran, 328n27
Patrick, Deval, 195, 201, 293
Patricof Associates, 320n22
Pearl Harbor, 55
Pearl Street Hotel, 25
Pei, I. M., 9, 55, 117, 120
PeopleStreet, 140
Perkin Elmer, 182, 218–219
PerSeptive Biosystems, 218–219
Personal digital assistants (PDAs), 122, 177
Personal Software Inc. (PSI), 88

Peter Walker and Partners, 86
Pfizer, 161
  buildings of, 152, 157, 196b, 210b, 244–245, 333n1
  Cawthorn and, 108
  collaboration and, 251
  COVID-19 pandemic and, 221, 257
  LabCentral and, 245
  success of, 222t
Phillip A. Sharp Building, 99, 179
Picture a Scientist (PBS Nova), 224
Pilgrim Church, 34
Pillar, 233
Pine Island Capital Partners, 95–98
Pi Tower, 13, 17, 264, 298
Plexuss, 271
Plug and Play, 325n8
Plump, Andy, 198
Polaris, 126, 176, 207, 211, 217, 231, 233–234, 331n14
Polaroid
  Adams and, 240
  Bonanos on, 73
  buildings of, 5, 73, 195, 199, 239–240, 243, 293, 295, 317n14
  convergence and, 297
  decline of, 272, 317n15
  Land and, 5, 72–73, 239–240, 333n10
  Morse and, 250
  patents and, 239–240
  success of, 291, 297
Polaroid Retirees Association, 240
President's Council of Advisors on Science and Technology, 187
Preston, John, 133–134
Prime Computer, 89, 252
Pritchett, Henry, 40–41, 43
Privacy Sandbox, 202
Project MAC, 71–73, 91, 96, 120–121, 154, 199, 293, 319n19
Ptashne, Mark, 101, 104, 111
PTC Therapeutics, 232
Pulitzer Prize, ix, 1
Purcell, Edward, 57
PWP Landscape Architecture, 86

Quaise, 300
Quanterix, 117
Quincy Market, 24

Rabi, I. I., 57
Radar, 10, 56–62, 69, 284
Radarange, 61
Radiation Laboratory (Rad Lab)
  Basic Research Division and, 60
  buildings of, 55–56, 72
  Bush and, 56–57
  Draper and, 61
  end of work at, 69
  formation of, 284
  microwaves and, 56–61
  Millikan and, 60
  radar systems and, 56–62, 69, 284
  Raytheon and, 56, 59–60
  recruitment for, 55–58
  Research Laboratory of Electronics (RLE) and, 60–62
  Ridenour and, 59, 62
  Robinson and, 61
  Stratton and, 69–70
  Valley and, 61
  Wiesner and, 55, 60–61, 69
Rae, Katie, xiii, 206, 279, 295–296, 300–301
Ragon, Phillip "Terry," 330n15
Ramsey, Norman, 57
Randall, George W., 25
Raytheon, 159, 316n31
  Bush and, 56
  growth of, 53–54
  post–World War II era and, 69–70, 73–74
  Radiation Laboratory and, 56, 59–60
  Smith and, 53
Reagan, Ronald, 96
Red Line, 90, 209, 269, 275, 279
Redstar Ventures, 274
Regional Advantage: Culture and Competition in Silicon Valley and Route 128 (Saxenian), 252, 262, 290
Reif, Rafael, 233
Reimers, Niels, 133–134
Repertoire Immune Systems, 221
Research Dataware, 140
Research Laboratory of Electronics (RLE), 59b, 60–62, 72
Revlon, 114
Rhames, Ving, 98
Ridenour, Louis, 59, 62
Rig Assembly, 216
Riggs, Arthur, 106

Rise of the Creative Class, The (Florida), 207, 287, 301
Riverbank Court Hotel, 40
Riverside Press, 30, 50, 315n18
RNA interference (RNAi), 3, 176, 242, 297, 328n26
Roberts, Edward, 125, 205, 218, 285
Roberts, George, 34
Robinson, Denis, 61, 63
Roblin, Richard, 102
Rogers Block, 70
Roosevelt, Franklin D., 45, 56–57
Rose, Dick, 120, 138, 143
Rosen, Ben, 88, 90
Roskill, S. W., 58
Rowe, Amy, 137
Rowe, Tim
  background of, 138
  Cambridge Chamber of Commerce and, 143
  Cambridge Innovation Center (CIC) and, 65, 137–148, 174, 243, 264, 294
  Compton and, 57
  LabCentral and, 243, 264
  Mamlet and, 139–144, 174
  Millennium and, 141, 174
  Olmsted and, 138–140
  zoning and, 147
Rowland, Robert, 75, 78–83, 318n4
Rowland Institute, 93, 119, 240
Royal Air Force, 57–58
Royalties, 30, 88, 134–135, 166, 176, 238, 323n14, 328n27
Rubins, Kathleen, 116
Rubius Therapeutics, 221, 245
Ruiz, Israel, 12, 16, 310n7
Ryan, Nancy, 267

Sachs, Jonathan, 87–88, 98, 319n4
Safdie, Moshe, 1, 16, 84–86, 91, 319n15
Sagan, Paul, 128, 130
Sahlman, William, 88, 280
Sandia, 94
Sanofi Genzyme, 2–3, 18, 157, 195–198, 249, 309n3
Sapient, 123
Saxenian, AnnaLee, 252–254, 262, 290
Sayare, Mitch, 112
Sayeret Matkal, 124
Scangos, George, 193

Schaefer, Ray, 106–107, 112
Schaller, Heinz, 106
Schering-Plough, 108
Schimmel, Paul, 327n25
Schlamminger, Karl, 86
Schmergel, Gabriel, 112
Schmidt, Eric, 4, 298–299, 329n13
Schmidt, Wendy, 4, 298–299, 329n13
School of Science, 226–229, 332n7
Schreiber, Stuart, 186b
Schroeter, Sven, 139–140
Schulman, Amy, 231
Science for the People, 104
ScienceWatch, 187
Scientific advisory board (SAB), 108, 231, 322n32
Scientific American journal, 27, 189
Scientific Data Systems, 320n22
Sculley, John, 121
Seelig, Jonathan, 126–127
Semi-Automatic Ground Environment (SAGE), 62
SensAble Technologies, 51, 315n23
Servier, 157
Sevin Rosen Partners, 88
Shark Tank (TV show), 234
Sharp, Anne, 106
Sharp, Phillip, xv
   Alnylam and, 164, 176, 180, 252, 280, 327n25
   Biogen and, 99, 106–109, 179–180, 236, 252, 255, 280, 321n28
   biotech and, 99, 101, 105–109, 164, 176, 236, 328n26
   Broad Institute and, 180
   building of, 179
   cancer research and, 4, 101, 176, 280
   collaboration and, 252, 254–255, 294
   Dewpoint and, 252
   Koch Institute and, 4, 180
   RNAi and, 328n26
   as serial entrepreneur, 159
Shire, 195–198, 327n19, 330n6
Shreve, Lamb, and Harmon, 315n13
Siegal, Abraham, 97
Siemens Nixdorf, 201–204, 330n20
Sigilon Thearpeutics, 221
Silicon Valley
   AI Alley and, 94–95
   comparisons to, 6, 9, 94–95, 124, 184b, 207, 252–253, 261, 274, 294

Draper Fisher Jurveston (DFJ) and, 139
Facebook and, 204
Kapor and, 88
Lewin on, 126
Plug and Play, 325n8
Rae and, 295
Simha, O. Robert, xiv, 17, 69–71, 75, 79, 86, 268, 284, 291, 327n20
Singer, Isaac, 30
Singer, Maxine, 104
Singh Mahendrajeet "Jeet," 123, 137
Sklar, Pamela, 182
Slice of MIT (online publication), 299
Smith, Charles, 53
Snyder, Sherry, 113
Social media, 6, 124, 130, 189, 285
SoftGrid, 200
Softricity, 200
Soldiers Field, 41
Solomon, Cynthia, 323n4
Soul of a New Machine, The (Kidder), 1
Spark Capital, 126
Spark Therapeutics, 232
Spencer, Percy, 60–61
Springer, Timothy, 159
Squirrel Brand Company, 49
SS Bunker Hill, 44
Stalder, Carrie, 65
Stallman, Richard, 72, 319n19
Standard Oil, 48
Stanley Center for Psychiatric Research, 186–187
Stanton, Fred, 94
Starr, Kevin, 175b
Startup 101, 232
Star Wars Episode I: The Phantom Menace (film), 128
STAT, 187, 189, 221–222, 231–232
Stata Center, 5, 17, 55, 123
Stella Artois, 113
Stephen A. Schwarzman College of Computing, 13–14, 285, 299
Stillson, Daniel, 237–238
Stone, Charles, 41–42, 44
Stone, Edward Durell, 81
Stone, Orra I., 47–54
Story of Polaroid, The (Bonanos), 73
Stowe, Barbara Gunderson, 151
Strategic Defense Initiative, 96, 98

Stratton, Julius, 69–70
Street, J. Curry, 57
Stromedix, 217
Sturtevant, Reed, 91, 200
Suffolk Engraving & Electrotyping Company, 310n4
Sugar Babies, 50
Sugar Daddies, 49
Sullivan, Charles, 21–24, 29–30, 38, 48, 52, 54, 303, 311n10, 312n29
Summers, Larry, 183, 184b
Sun Microsystems, 94
Surface Oncology, 245
Swanson, Robert, 87, 106, 107, 163
Sweeney, Madeline "Amy," 129
Symbolics, 95–98, 320n22
Synologic, 234

TA Associates, 107, 112
Takeda, 157, 330n11
  corporation and, 195–200
  Data Sciences Institute and, 297
  Heatherington and, 276
  J-Clinic and, 298
  Millennium and, 1, 174, 195, 196b, 198, 327n23
  Shire and, 195–198, 327n19, 330n6
  success of, 3
Teal Bio, 4, 159, 162, 214t
Tech Model Railroad Club (TMRC), 72
Technicon, 114
Technology Corporation, 41, 283
Techstars, 123, 206–207, 211, 291–292, 295
Telephones, 5, 7, 98, 120, 177, 237–238, 246
Tepper, Bob, 175b
Termeer, Adriana, 249
Termeer, Belinda, 249
Termeer, Henri A.
  death of, 249
  Genzyme and, 3, 112–114, 144b, 163, 169–170, 197, 249–251, 294
  Maraganore and, 250–251
  mentoring of, 3, 249–251
  sculpture of, 249–250
Termeer Foundation Fellows, 249–251
Tertill, 276
Tessera Therapeutics, 221
Tessier-Lavigne, Marc, 221

Texas Instruments, 315n13
Thinking Machines
  AI Alley and, 6–7, 91–98, 117, 119, 122, 186, 277, 319n15, 320n29, 323n10
  bankruptcy of, 98
  buildings of, 93–94
  Connection Machines, 93–94, 98
  Hillis and, 6, 93–95, 98, 277, 320n29
  Minsky and, 91, 94
  success of, 92–98
  venture capital and, 92–94, 98
  Wiesner and, 94
Third Rock Ventures, 1, 175b, 256, 278
Thomas, Lewis, 115
ThomsonReuters, 187
Three Cambridge Center, 28, 201, 331n21
TIME magazine, 74
Time Warner, 128
Tiny Troll, 88
Tishman, Robert, 65
Tobin, James, 166
Toomey, Tim, 268–269
Tootsie Roll Industries, 50
Toyota, 199
Transkaryotic Therapies (TKT), 196b, 197–198
Transthyretin amyloidosis, 3, 176, 328n27
TravelFit, 137
Tron (film), 121
Trump, John, 61
Tufts Medical Center, 155
Turing Award, 72, 121
Tuschl, Thomas, 327n25, 328n26
Twitter, 15, 189, 199, 203b
Tysabri, 193

U-2 spy plane, 239
U-boats, 58
Uhlenbeck, George, 55
Underkoffer, John, 123
Unicorn, 128–129
University Residential Communities, 327n20
Unum Therapeutics, 245
Urban renewal
  Boston Redevelopment Authority and, 75, 78
  Cambridge Redevelopment Authority (CRA) and, 70, 80, 83, 149–150, 269, 318n4
  Catalano and, 71

Crane and, 70
Housing Act and, 70, 79
impact of, 284
Kendall Square Urban Renewal Project, 79–80, 82b, 99, 318n4, 325n3
NASA and, 77–86
Rogers Block and, 70
Simha and, 69–71, 75
US News & World Report magazine, 271

Vaccines, 2, 6, 176, 221, 246, 257, 263
Valcade, 174
Valley, George, 61
Vandebroek, Sophie, 280–281
Van de Graaff, Robert J., 61
Vasella, Daniel, 152–154
Vellucci, Al, 99–100, 103, 111, 321n17
Venrock and Welsh, Carson, Anderson & Stowe, 97
Venter, J. Craig, 182, 219
Venture Cafe, 65, 145
Venture capital
  accelerator fellowships and, 233
  AI Alley and, 88 (*see also* AI Alley)
  Alnylam and, 175b
  BioGen and, 107–108
  bootcamps for, 232–233
  Cambridge Innovation Center (CIC) and, 137–140
  collaboration and, 249–256
  COVID-19 pandemic and, 210b, 221
  Dolphin Tank competition and, 234–236
  The Engine and, 233, 261, 271, 279, 295–296, 300
  Genetics Institute and, 111
  Google and, 141, 207, 211
  GV and, 141, 202, 211, 234
  LabCentral and, 210b
  Microsoft and, 206–207, 219
  MIT REAP and, 271, 287, 289
  Moderna and, 217, 221, 243
  Novartis and, 212f, 233
  pledge for, 233–234
  regional advantage and, 261–271
  startup squeeze and, 205–222
  Thinking Machines and, 92–94, 98
  women and, 223–236
  Zuckerberg and, 285

Venture Nights, xiii
Vertex
  Agenerase and, 171
  Altshuler and, 190
  beginnings of, 111, 173
  Boger and, 170–171, 274
  buildings of, 111–112, 167, 170–171, 177, 185, 209, 261, 329n4, 330n7
  collaboration and, 251, 255
  corporatization and, 194–197
  Emmens and, 197
  Millennium and, 164
  relocation of, 194–195
  success of, 120, 164, 171
  talent war of, 169
Vest, Charles, 151, 154, 163, 183, 184b, 226, 228
Viaux, Frederick, 38
Vietnam Veterans Memorial, 10
Vincent, Ji, 165–167
Vinter, Steve, 201–204, 330n20
VisiCalc, 88
Vithera Pharmaceuticals, 242–243
Vivtex, 161
Voice of America, 55
Volpe, John, 83
Volpe, Mike, 6
Volpe Transportation Center, 172
  Biogen and, 83, 99, 109, 167
  buildings of, 24, 91, 99
  CityScope and, 9–11, 14–16
  Department of Transportation and, 10, 82b, 264–265
  Loughrey Walkway and, 83
  Marsh and, 16
  NASA and, 10, 83–84, 85f
  overhaul of, 264–266, 269
  Stone proposal for, 81f
  Twitter and, 15
  zoning and, 150

Wald, George, 103–104, 114
Wald, Ruth Hubbard, 103–104
Walker, Francis A., 44
Walt, David, 159
Wang, 89, 252
Wang, Danny, 117
War of 1812, 24
Warren Brothers, 53

Washington Post newspaper, 232
Waterloo, 24
Watermark West, 171–172
Watson, James, 101–102, 114, 224, 254, 320n7
Watson, Thomas, 238
Webb, C. A., xiv, 18–19, 235, 256, 269–270, 272, 281
Webb, James, 77–78
Weber, Christophe, 198
Weinberg, Robert, 116
Weiskopf, Edwin, 113
Weiss, Rainer, 66–67
Weissman, Sherman, 102
Weissmann, Charles, 106–108
Wellcome Sanger Institute, 187
Wertheimer, Jeremy, 330n20
West Boston Bridge, 23–24, 28–29, 33–34, 38, 46, 251
Westinghouse Science Competition, 5, 180
West Kendall Street, 3
Westphal, Christoph, 176
Wheeler, Andy, 141
Wheeler, Cornelia "Connie," 321n20
Wheelwright, George, 73
Whirlwind, 62
White, Kevin, 250
Whitehead, Edwin C. "Jack," 114–115, 117
Whitehead, John, 115
Whitehead, Peter, 115
Whitehead, Susan, 115, 281–282
Whitehead Genome Center, 182–183
Whitehead Institute
  affiliation agreement of, 115
  Baltimore and, 114–117, 120, 183, 274, 323n9
  biotech and, 114–117, 154, 163, 173, 180, 323n14, 327n25
  Broad Institute and, 180, 182–185
  buildings of, 4, 94, 115, 120, 270
  cancer research and, 4, 115–116, 185b
  COVID-19 pandemic and, 4
  Daley and, 182
  DNA and, 114
  Dracopoli and, 180–181
  Lander and, 4, 80, 84, 94, 119, 163, 173, 178–185, 186b
  Lodish and, 115
  MIT Center for Genome Research and, 173, 180
  MIT Media Lab and, 119–120

  patents and, 323n14
  success of, 116–117, 180
  Weinberg and, 116
Whitesides, George, 112, 159
"Why Eric Lander Morphed from Science God to Punching Bag" (Begley), 189
Wiesner, Jerry, 55, 60–61, 69, 94, 120
Wiesner Building, 9
Wilson, E. O., ix–x, 7, 296–297, 300–301
Windham-Bannister, Susan, 243–244, 293–294
Wired magazine, 4, 18, 318n20
Wojcicki, Anne, 187
Wolfram, Stephen, 94
WordsWorth, 123
World Trade Center, 124, 129
World War II era, 241, 283–284, 316n10
  Battle of Britain and, 56
  collaboration and, 252
  computers and, 78, 89, 199, 207, 272
  corporatization and, 199
  industrial development and, 40, 53–55, 58–60
  Los Alamos and, 55, 94
  National Defense Research Committee (NDRC) and, 57, 316n5
  Pearl Harbor and, 55
  Radiation Laboratory and, 55–63
  Raytheon and, 69–70, 73–74
  Sandia and, 94
  urban renewal and, 69–70, 73–74
World Wide Web Consortium (W3C), 5, 123–124, 330n12
Worthington Pump Works, 312n29
Wyeth, 112

Xconomy, 312n29, 315n23
Xerox, 121, 280–281
X Institute, 183
X-ray equipment, 52
Xtuit Pharmaceuticals, 162, 276

Young, Richard, 116
YouTube, 4, 202

Zacharias, Jerrold, 57, 66
Zamore, Phillip, 327n25
Zapata Computing, 300
Zero Stage Capital, 205, 208, 218
Zhang, Feng, 189, 329n25

Zinder, Norton, 102
Zoning
  700 Main and, 149
  CIC and, 147–150
  Clem and, 168–169
  Google and, 150
  Kendall Square Initiative (KSI) and, 150
  LabCentral and, 149–150
  lessons from, 293–294
  Marriott and, 325n3
  Microsoft and, 147, 150
  NASA and, 82b, 83–84
  Volpe Transportation Center and, 150
Zuber, Maria, 233, 332n11
Zuckerberg, Mark, 187, 204, 207, 285
Zuckerman, Mort, 84–85
Zymera, 225